D1195094

Statistics for Biology and Health

Series Editors
K. Dietz, M. Gail, K. Krickeberg, J. Samet, A. Tsiatis

Statistics for Biology and Health

Eric Vittinghoff   Stephen C. Shiboski
David V. Glidden   Charles E. McCulloch

Regression Methods in Biostatistics

Linear, Logistic, Survival, and Repeated Measures Models

With 54 Illustrations

 Springer

Eric Vittinghoff
Department of Epidemiology and Biostatistics
University of California
San Francisco, CA 94143
USA
eric@biostat.ucsf.edu

David V. Glidden
Department of Epidemiology and Biostatistics
University of California
San Francisco, CA 94143
USA
dave@biostat.ucsf.edu

Stephen C. Shiboski
Department of Epidemiology and Biostatistics
University of California
San Francisco, CA 94143
USA
steve@biostat.ucsf.edu

Charles E. McCulloch
Department of Epidemiology and Biostatistics
University of California
San Francisco, CA 94143
USA
chuck@biostat.ucsf.edu

Series Editors
M. Gail
National Cancer Institute
Rockville, MD 20892
USA

K. Krickeberg
Le Chatelet
F-63270 Manglieu
France

J. Samet
Department of Epidemiology
School of Public Health
Johns Hopkins University
615 Wolfe Street
Baltimore, MD 21205-2103
USA

A. Tsiatis
Department of Statistics
North Carolina State University
Raleigh, NC 27695
USA

W. Wong
Department of Statistics
Stanford University
Stanford, CA 94305-4065
USA

Library of Congress Cataloging-in-Publication Data
Regression methods in biostatistics : linear, logistic, survival, and repeated measures models / Eric Vittinghoff ... [et al.].
    p. cm. — (Statistics for biology and health)
    Includes bibliographical references and index.
    ISBN 0-387-20275-7 (alk. paper)
        1. Medicine—Research—Statistical methods.   2. Regression analysis.   3. Biometry.
    I. Vittinghoff, Eric.   II. Series.
    R853.S7R44   2004
    610′.72′7—dc22   2004056545

ISBN 0-387-20275-7   Printed on acid-free paper.

Printed in the United States of America.   (EB)

9 8 7 6 5 4 3 2 1   SPIN 10946190

springeronline.com

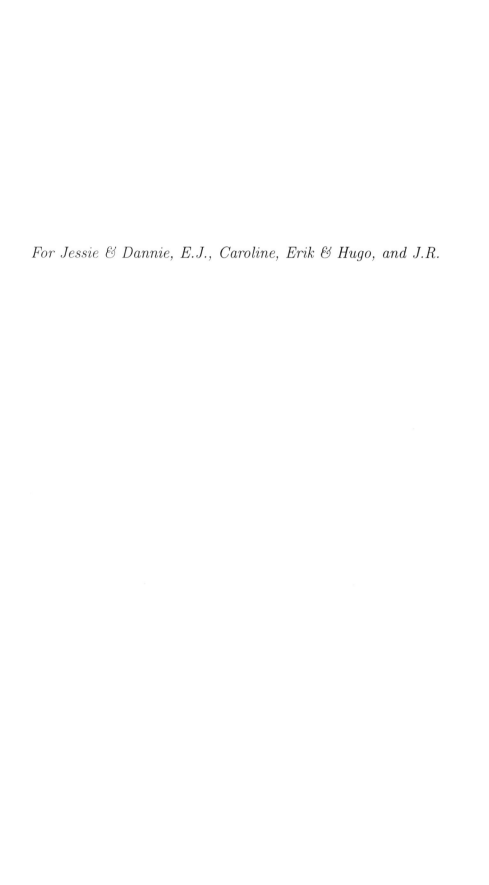

For Jessie & Dannie, E.J., Caroline, Erik & Hugo, and J.R.

Preface

The primary biostatistical tools in modern medical research are single-outcome, multiple-predictor methods: multiple linear regression for continuous outcomes, logistic regression for binary outcomes, and the Cox proportional hazards model for time-to-event outcomes. More recently, generalized linear models and regression methods for repeated outcomes have come into widespread use in the medical research literature. Applying these methods and interpreting the results requires some introduction. However, introductory statistics courses have no time to spend on such topics and hence they are often relegated to a third or fourth course in a sequence. Books tend to have either very brief coverage or to be treatments of a single topic and more theoretical than the typical researcher wants or needs.

Our goal in writing this book was to provide an accessible introduction to multipredictor methods, emphasizing their proper use and interpretation. We feel strongly that this can only be accomplished by illustrating the techniques using a variety of real datasets. We have incorporated as little theory as feasible. Further, we have tried to keep the book relatively short and to the point. Our hope in doing so is that the important issues and similarities between the methods, rather than their differences, will come through. We hope this book will be attractive to medical researchers needing familiarity with these methods and to students studying statistics who would like to see them applied to real data. The methods we describe are, of course, the same as those used in a variety of fields, so non-medical readers will find this book useful if they can extrapolate from the predominantly medical examples.

A prerequisite for the book is a good first course in statistics or biostatistics or an understanding of the basic tools: paired and independent samples t-tests, simple linear regression and one-way ANOVA, contingency tables and χ^2 (chi-square) analyses, Kaplan–Meier curves, and the logrank test.

We also think it is important for researchers to know how to interpret the output of a modern statistical package. Accordingly, we illustrate a number of the analyses with output from the Stata statistics package. There are a number of other packages that can perform these analyses, but we have chosen this one because of its accessibility and widespread use in biostatistics and epidemiology.

This book grew out of our teaching a two-quarter sequence to post-graduate physicians training for a research career. We thank them for their feedback and patience. Partial support for this came from a K30 grant from the National Institutes of Health awarded to Stephen Hulley, for which we are grateful.

We begin the book with a chapter introducing our viewpoint and style of presentation and the big picture as to the use of multipredictor methods. Chapter 2 presents descriptive numerical and graphical techniques for multi-predictor settings and emphasizes choice of technique based on the nature of the variables. Chapter 3 briefly reviews the statistical methods we consider prerequisites for the book.

We then make the transition in Chapter 4 to multipredictor regression methods, beginning with the linear regression model. This chapter also covers confounding, mediation, interaction, and model checking in the most detail. Chapter 5 deals with predictor selection, an issue common to all the multi-predictor models covered. In Chapter 6 we turn to binary outcomes and the logistic model, noting the similarities to the linear model. Ties to simpler, contingency table methods are also noted. Chapter 7 covers survival outcomes, giving clear indications as to why such techniques are necessary, but again emphasizing similarities in model building and interpretation with the previous chapters. Chapter 8 looks at the accommodation of correlated data in both linear and logistic models. Chapter 9 extends Chapter 6, giving an overview of generalized linear models. Finally, Chapter 10 is a brief introduction to the analysis of complex surveys.

The text closes with a summary, Chapter 11, attempting to put each of the previous chapters in context. Too often it is hard to see the "forest" for the "trees" of each of the individual methods. Our goal in this final chapter is to provide guidance as to how to choose among the methods presented in the book and also to realize when they will not suffice and other techniques need to be considered.

San Francisco, CA
October, 2004

Eric Vittinghoff
David V. Glidden
Stephen C. Shiboski
Charles E. McCulloch

Contents

1

Introduction

The book describes a family of statistical techniques that we call *multipredictor* regression modeling. This family is useful in situations where there are multiple measured factors (also called predictors, covariates, or independent variables) to be related to a single outcome (also called the response or dependent variable). The applications of these techniques are diverse, including those where we are interested in prediction, isolating the effect of a single predictor, or understanding multiple predictors. We begin with an example.

1.1 Example: Treatment of Back Pain

Korff *et al.* (1994) studied the success of various approaches to treatment for back pain. Some physicians treat back pain more aggressively, with prescription pain medication and extended bed rest, while others recommend an earlier resumption of activity and manage pain with over-the-counter medications. The investigators classified the aggressiveness of a sample of 44 physicians in treating back pain as low, medium, or high, and then followed 1,071 of their back pain patients for two years. In the analysis, the classification of treatment aggressiveness was related to patient outcomes, including cost, activity limitation, pain intensity, and time to resumption of full activity,

The primary focus of the study was on a single categorical predictor, the aggressiveness of treatment. Thus for a continuous outcome like cost we might think of an analysis of variance, while for a categorical outcome we might consider a contingency table analysis and a χ^2-test. However, these simple analyses would be incorrect at the very least because they would fail to recognize that multiple patients were *clustered* within physician practice and that there were *repeated outcome measures* on patients.

Looking beyond the clustering and repeated measures (which are covered in Chap. 8), what if physicians with more aggressive approaches to back pain also tended to have older patients? If older patients recover more slowly (regardless of treatment), then even if differences in treatment aggressiveness have no effect, the age imbalance would nonetheless make for poorer outcomes in the patients of physicians in the high-aggressiveness category. Hence, it would be misleading to judge the effect of treatment aggressiveness without correcting for the imbalances between the physician groups in patient age and, potentially, other prognostic factors – that is, to judge without *controlling for confounding*. This can be accomplished using a model which relates study outcomes to age and other prognostic factors as well as the aggressiveness of treatment. In a sense, multipredictor regression analysis allows us to examine the effect of treatment aggressiveness while *holding the other factors constant*.

1.2 The Family of Multipredictor Regression Methods

Multipredictor regression modeling is a family of methods for relating multiple predictors to an outcome, with each member of the family suitable for a different type of outcome. The cost outcome, for example, is a numerical measure and for our purposes can be taken as *continuous*. This outcome could be analyzed using the linear regression model, though we also show in Chapter 9 why a generalized linear model might be a better choice.

Perhaps the simplest outcome in the back pain study is the yes/no indicator of moderate-to-severe activity limitation; a subject's activities are limited by back pain or not. Such a categorical variable is termed *binary* because it can only take on two values. This type of outcome is analyzed using the logistic regression model.

In contrast, pain intensity was measured on a scale of ten equally spaced values. The variable is numerical and could be treated as continuous, although there were many tied values. Alternatively it could be analyzed as a categorical variable, with the different values treated as ordered categories, using extensions of the logistic model.

Another potential outcome might be time to resumption of full activity. This variable is also continuous, but what if a patient had not yet resumed full activity at the end of the follow-up period of two years? Then the time to resumption of full activity would only be known to exceed two years. When outcomes are known only to be greater than a given value (like two years), the variable is said to be *right-censored* – a common feature of time-to-event data. This type of outcome can be analyzed using the Cox proportional hazards model.

Furthermore, in the back pain example, study outcomes were measured on groups, or clusters, of patients with the same physician, and on multiple occasions for each patient. To analyze such *hierarchical* or *longitudinal* outcomes, we need to use extensions of the basic family of regression modeling techniques suitable for repeated measures data. Related extensions are also required to analyze data from complex surveys.

The various regression modeling approaches, while differing in important statistical details, also share important similarities. Numeric, binary, and categorical predictors are accommodated by all members of the family, and are handled in a similar way: on some scale, the systematic part of the outcome is modeled as a linear function of the predictor values and corresponding *regression coefficients*. The different techniques all yield estimates of these coefficients that summarize the results of the analysis and have important statistical properties in common. This leads to unified methods for selecting predictors and modeling their effects, as well as for making inferences to the population represented in the sample. Finally, all the models can be applied to the same broad classes of practical questions involving multiple predictors.

1.3 Motivation for Multipredictor Regression

Multipredictor regression can be a powerful tool for addressing three important practical questions. These include *prediction, isolating the effect of a single predictor,* and *understanding multiple predictors.*

1.3.1 Prediction

How can we identify which patients with back pain will have moderate-to-severe limitation of activity? Multipredictor regression is a powerful and general tool for using multiple measured predictors to make useful predictions for future observations. In this example, the outcome is binary and thus a multipredictor logistic regression model could be used to estimate the predicted probability of limitation for any possible combination of the observed predictors. These estimates could then be used to classify patients as likely to experience limitation or not. Similarly, if our interest was future costs, a continuous variable, we could use a linear regression model to predict the costs associated with new observations characterized by various values of the predictors.

1.3.2 Isolating the Effect of a Single Predictor

In settings where multiple, related predictors contribute to study outcomes, it will be important to consider multiple predictors even when a single predictor is of interest. In the von Korff study the primary predictor of interest was how aggressively a physician treated back pain. But incorporation of other predictors was necessary for the clearest interpretation of the effects of the aggressiveness of treatment.

1.3.3 Understanding Multiple Predictors

Multipredictor regression can also be used when our aim is to identify multiple independent predictors of a study outcome – independent in the sense that they appear to have an effect over and above other measured variables. Especially in this context, we may need to consider other complexities of how predictors jointly influence the outcome. For example, the effect of injuries on activity limitation may in part operate through their effect on pain; in this view, pain *mediates* the effect of injury and should not be adjusted for, at least initially. Alternatively, suppose that among patients with mild or moderate pain, younger age predicts more rapid recovery, but among those with severe pain, age makes little difference. The effects of both age and pain severity will both potentially be misrepresented if this *interaction* is not taken into account. Fortunately, all the multipredictor regression methods discussed in this book easily handle interactions, as well as mediation and confounding, using essentially identical techniques. Though certainly not foolproof, multipredictor models are well suited to examining the complexities of how multiple predictors are associated with an outcome of interest.

1.4 Guide to the Book

This text attempts to provide practical guidance for regression analysis. We interweave real data examples from the biomedical literature in the hope of capturing the reader's interest and making the statistics as easy to grasp as possible. Theoretical details are kept to a minimum, since it is usually not necessary to understand the theory to use these methods appropriately. We avoid formulas and keep mathematical notation to a minimum, instead emphasizing selection of appropriate methods and careful interpretation of the results.

This book grew out a two-quarter sequence in multipredictor methods for physicians beginning a career in clinical research, with a focus on techniques appropriate to their research projects. For these students, mathematical explication is an ineffective way to teach these methods. Hence our reliance on real-world examples and heuristic explanations.

Our students take the course in the second quarter of their research training. A beginning course in biostatistics is assumed and some understanding of epidemiologic concepts is clearly helpful. However, Chapter 3 presents a review of topics from a first biostatistics course, and we explain epidemiologic concepts in some detail throughout the book.

Although theoretical details are minimized, we do discuss techniques of practical utility that some would consider advanced. We treat extensions of basic multipredictor methods for repeated measures and hierarchical data, for data arising from complex surveys, and for the broader class of *generalized linear models*, of which logistic regression is the most familiar example. We address model checking as well as model selection in considerable detail.

The orientation of this book is to *parametric* methods, in which the systematic part of the model is a simple function of the predictors, and substantial assumptions are made about the distribution of the outcome. In our view parametric methods are usually flexible and robust enough, and we show how model adequacy can be checked. The Cox proportional hazards model covered in Chapter 7 is a *semi-parametric* method which makes few assumptions about an important component of the systematic part of the model, but retains most of the efficiency and many of the advantages of fully parametric models. *Generalized additive models*, briefly reviewed in Chapter 6, go an additional step in this direction. However, fully *nonparametric* regression methods in our view entail losses in efficiency and ease of interpretation which make them less useful to researchers. We do recommend a popular bivariate nonparametric regression method, LOWESS, but only for exploratory data analysis.

Our approach is also to encourage exploratory data analysis as well as thoughtful interpretation of results. We discourage focusing solely on P-values, which have an important place in statistics but also important limitations. In particular, P-values measure the strength of the evidence for an effect, but not its size. In our view, data analysis profits from considering the estimated effects, using confidence intervals to quantify their precision.

We recommend that readers begin with Chapter 2, on exploratory methods. Since Chapter 3 is largely a review, students may want to focus only on unfamiliar material. Chapter 4, on multipredictor regression methods for continuous outcomes, introduces most of the important themes of the book, which are then revisited in later chapters, and so is essential reading. Similarly, Chapter 5 covers predictor selection, which is common to the entire family of regression techniques. Chapters 6 and 7 cover regression methods specialized for binary and time-to-event outcomes, while Chapters 8–10 cover extensions of these methods for repeated measures, counts and other special types of outcomes, and complex surveys. Readers may want to study these chapters as the need arises. Finally, Chapter 11 reprises the themes considered in the earlier chapters and is recommended for all readers.

For interested readers, Stata code and selected data sets used in examples and problems, plus errata, are posted on the website for this book:

http://www.biostat.ucsf.edu/vgsm

2

Exploratory and Descriptive Methods

Before beginning any sort of statistical analysis, it is imperative to take a preliminary look at the data with three main goals in mind: first, to check for errors and anomalies; second, to understand the distribution of each of the variables on its own; and third, to begin to understand the nature and strength of relationships among variables. Errors should, of course, be corrected, since even a small percentage of erroneous data values can drastically influence the results. Understanding the distribution of the variables, especially the outcomes, is crucial to choosing the appropriate multipredictor regression method. Finally, understanding the nature and strength of relationships is the first step in building a more formal statistical model from which to draw conclusions.

2.1 Data Checking

Procedures for data checking should be implemented before data entry begins, to head off future headaches. Many data entry programs have the capability to screen for egregious errors, including values that are out the expected range or of the wrong "type." If this is not possible, then we recommend regular checking for data problems as the database is constructed.

Here are two examples we have encountered recently. First, some values of a variable defined as a proportion were inadvertently entered as percentages (i.e., 100 times larger than they should have been). Although they made up less than 3% of the values, the analysis was completely invalidated. Fortunately, this simple error was easily corrected once discovered. A second example involved patients with a heart anomaly. Those whose diagnostic score was poor enough (i.e., exceeded a numerical threshold) were to be classified according to type of anomaly. Data checks revealed missing classifications for patients whose diagnostic score exceeded the threshold, as well as classifications for patients whose score did not, complicating planned analyses. Had the data been screened as they were collected, this problem with study procedures could have been avoided.

2.2 Types Of Data

The proper description of data depends on the nature of the measurement. The key distinction for statistical analysis is between numerical and categorical variables. The number of diagnostic tests ordered is a numerical variable, while the gender of a person is categorical. Systolic blood pressure is numerical, whereas the type of surgery is categorical.

A secondary but sometimes important distinction within numerical variables is whether the variable can take on a whole continuum or just a discrete set of values. So systolic blood pressure would be continuous, while number of diagnostic tests ordered would be discrete. Cost of a hospitalization would be continuous, whereas number of mice able to successfully navigate a maze would be discrete. More generally,

> Definition: A numerical variable taking on a continuum of values is called *continuous* and one that only takes on a discrete set of values is called *discrete*.

A secondary distinction sometimes made with regard to categorical variables is whether the categories are ordered or unordered. So, for example, categories of annual household income (<$20,000, $20,000–$40,000, $40,000–$100,000, >$100,000) would be ordered, while marital status (single, married, divorced, widowed) would be unordered. More exactly,

> Definition: A categorical variable is *ordinal* if the categories can be logically ordered from smallest to largest in a sense meaningful for the question at hand (we need to rule out silly orders like alphabetical); otherwise it is unordered or *nominal*.

Some overlap between types is possible. For example, we may break a numerical variable (such as exact annual income in dollars and cents) into ranges or categories. Conversely, we may treat a categorical variable as a numerical score, for example, by assigning values one to five to the ordinal responses Poor, Fair, Good, Very Good, and Excellent. In the following sections, we present each of the descriptive and exploratory methods according to the types of variables involved.

2.3 One-Variable Descriptions

We begin by describing techniques useful for examining a single variable at a time. These are useful for uncovering mistakes or extreme values in the data and for assessing distributional shape.

2.3.1 Numerical Variables

We can describe the distribution of numerical variables using either numerical or graphical techniques.

Example: Systolic Blood Pressure

The Western Collaborative Group Study (WCGS) was a large epidemiological study designed to investigate the association between the "type A" behavior pattern and coronary heart disease (Rosenman *et al.*, 1964). We will revisit this study later in the book, focusing on the primary outcome, but for now we want to explore the distribution of systolic blood pressure (SBP).

Numerical Description

As a first step we obtain basic descriptive statistics for SBP. Table 2.1 gives detailed summary statistics for the systolic blood pressure variable, sbp. Several

Table 2.1. Numerical Description of Systolic Blood Pressure

```
. summarize sbp, detail

                              systolic BP
-----------------------------------------------------------------
          Percentiles      Smallest
 1%           104               98
 5%           110              100
10%           112              100        Obs              3154
25%           120              100        Sum of Wgt.      3154

50%           126                         Mean          128.6328
                             Largest      Std. Dev.     15.11773
75%           136              210
90%           148              210        Variance      228.5458
95%           156              212        Skewness      1.204397
99%           176              230        Kurtosis      5.792465
```

features of the output are worth consideration. The largest and smallest values should be scanned for outlying or incorrect values, and the mean (or median) and standard deviation should be assessed as general measures of the location and spread of the data. Secondary features are the skewness and kurtosis, though these are usually more easily assessed by the graphical means described in the next section. Another assessment of skewness is a large difference between the mean and median. In *right-skewed* data the mean is quite a bit larger than the median, while in *left-skewed* data the mean is much smaller than the median. Of note: in this data set, the largest observation is more than six standard deviations above the mean!

Graphical Description

Graphs are often the quickest and most effective way to get a sense of the data. For numerical data, three basic graphs are most useful: the histogram, boxplot, and normal quantile-quantile (or Q-Q) plot. Each is useful for different purposes. The histogram easily conveys information about the location, spread, and shape of the frequency distribution of the data. The boxplot is a schematic identifying key features of the distribution. Finally, the normal quantile-quantile (Q-Q) plot facilitates comparison of the shape of the distribution of the data to a normal (or bell-shaped) distribution.

The histogram displays the frequency of data points falling into various ranges as a bar chart. Fig. 2.1 shows a histogram of the SBP data from WCGS. Generated using an earlier version of Stata, the default histogram uses five intervals and labels axes with the minimum and maximum values only. In this

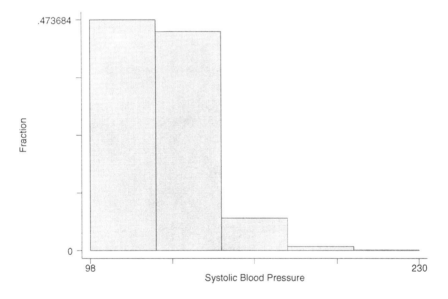

Fig. 2.1. Histogram of the Systolic Blood Pressure Data

figure, we can see that most of the measurements are in the range of about 100 to 150, with a few extreme values around 200. The percentage of observations in the first interval is about 47.4%.

However, this is not a particularly well-constructed histogram. With over 3,000 data points, we can use more intervals to increase the definition of the histogram and avoid grouping the data so coarsely. Using only five intervals, the first two including almost all the data, makes for a loss of information, since we only know the value of the data in those large "bins" to the limits of the interval (in the case of the first bin, between 98 and 125), and learn nothing about how the data are distributed within those intervals. Also, our preference is to provide more interpretable axis labeling. Fig. 2.2 shows a modified histogram generated using the current version of Stata that provides much better definition as to the shape of the frequency distribution of SBP.

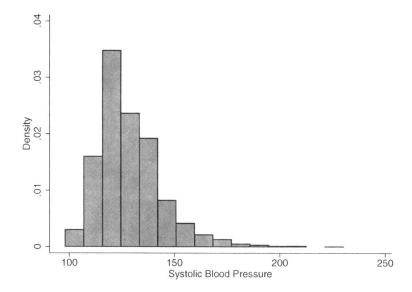

Fig. 2.2. Histogram of the Systolic Blood Pressure Data Using 15 Intervals

The key with a histogram is to use a sufficient number of intervals to define the shape of the distribution clearly and not lose much information, without using so many as to leave gaps, give the histogram a ragged shape, and defeat the goal of summarization. With 3,000 data points, we can afford quite a few bins. A *rough* rule of thumb is to choose the number of bins to be about $1 + 3.3 \log_{10}(n)$, (Sturges, 1926) where n is the sample size (so this would suggest 12 or 13 bins for the WCGS data). More than 20 or so are rarely needed. Fig. 2.2 uses 15 bins and provides a clear definition of the shape as well as a fair bit of detail.

A boxplot represents a compromise between a histogram and a numerical summary. The boxplot in Fig. 2.3 graphically displays information from the summary in Table 2.1, specifically the minimum, maximum, and 25th, 50th (median), and 75th percentiles. This retains many of the advantages of a graphical display while still providing fairly precise numerical summaries. The "box" displays the 25th and 75th percentiles (the lower and upper edges of the box) and the median (the line across the middle of the box). Extending from the box are the "whiskers" (this colorful terminology is due to the

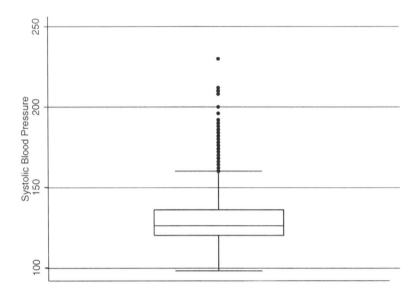

Fig. 2.3. Boxplot of the Systolic Blood Pressure Data

legendary statistician John Tukey, who liked to coin new terms). The bottom whisker extends to the minimum data value, 98, but the maximum is above the upper whisker. This is because Stata uses an algorithm to try to determine if observations are "outliers," that is, values a large distance away from the main portion of the data. Data points considered outliers (they can be in either the upper or lower range of the data) are plotted with symbols and the whisker only extends to the most extreme observation not considered an outlier.

Boxplots convey a wealth of information about the distribution of the variable:

- location, as measured by the median
- spread, as measured by the height of the box (this is called the interquartile range or IQR)
- range of the observations
- presence of outliers
- some information about shape.

This last point bears further explanation. If the median is located toward the bottom of the box, then the data are *right-skewed* toward larger values. That is, the distance between the median and the 75th percentile is greater than that between the median and the 25th percentile. Likewise, right-skewness will be indicated if the upper whisker is longer than the lower whisker or if there are more outliers in the upper range. Both the boxplot and the histogram show evidence for right-skewness in the SBP data. If the direction of the inequality is reversed (more outliers on the lower end, longer lower whisker, median toward the top of the box), then the distribution is *left-skewed.*

Our final graphical technique, the normal Q-Q plot, is useful for comparing the frequency distribution of the data to a normal distribution. Since it is easy to distinguish lines that are straight from ones that are not, a normal Q-Q plot is constructed so that the data points fall along an approximately straight line when the data are from a normal distribution, and deviate *systematically* from a straight line when the data are from other distributions. Fig. 2.4 shows the Q-Q plot for the SBP data. The line of the data points shows a distinct

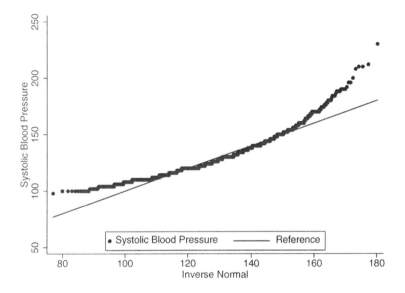

Fig. 2.4. Normal Q-Q Plot of the Systolic Blood Pressure Data

curvature, indicating the data are from a non-normal distribution.

The shape and direction of the curvature can be used to diagnose the deviation from normality. Upward curvature, as in Fig. 2.4, is indicative of right-skewness, while downward curvature is indicative of left-skewness. The other two common patterns are S-shaped. An S-shape as in Fig. 2.5 indicates a *heavy-tailed* distribution, while an S-shape like that in Fig. 2.6 is indicative of a *light-tailed* distribution.

Heavy- and light-tailed are always in reference to a hypothetical normal distribution with the same spread. A heavy-tailed distribution has more observations in the middle of the distribution and way out in the tails, and fewer a modest way from the middle (simply having more in the tails would just mean a larger spread). Light-tailed means the reverse: fewer in the middle and

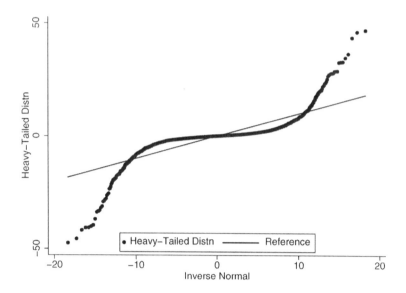

Fig. 2.5. Normal Q-Q Plot of Data From a Heavy-Tailed Distribution

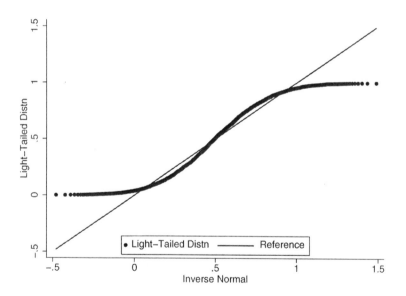

Fig. 2.6. Normal Q-Q plot of Data From a Light-Tailed Distribution

far out tails and more in the mid-range. Heavy-tailed distributions are generally more worrisome than light-tailed since they are more likely to include outliers.

Transformations of Data

A number of the techniques we describe in this book require the assumption of approximate normality or, at least, work better when the data are not highly skewed or heavy-tailed, and do not include extreme outliers. A common method for dealing with these problems is to transform such variables. For example, instead of the measured values of SBP, we might instead use the logarithm of SBP. We first consider why this works and then some of the advantages and disadvantages of transformations.

Transformations affect the distribution of values of a variable because they emphasize differences in a certain range of the data, while de-emphasizing differences in others. Consider a table of transformed values, as displayed in Table 2.2. On the original scale the difference between .01 and .1 is .09, but

Table 2.2. Effect of a $\log_{10}$ Transformation

Value	Difference	$\log_{10}$ value	Difference
0.01	0.09	-2	1
0.1	0.9	-1	1
1	9	0	1
10	90	1	1
100	900	2	1
1000	–	3	–

on the $\log_{10}$ scale, the difference is 1. In contrast, the difference between 100 and 1,000 on the original scale is 900, but this difference is also 1 on the $\log_{10}$ scale. So a log transformation de-emphasizes differences at the upper end of the scale and emphasizes those at the lower end. This holds for the natural log as well as $\log_{10}$ transformation. The effect can readily be seen in Fig. 2.7, which displays histograms of SBP on the original scale and after natural log transformation. The log-transformed data is distinctly less right-skewed, even though some skewness is still evident. Essentially, we are viewing the data on a different scale of measurement.

There are a couple of other reasons to consider transforming variables, as we will see in later sections and chapters: transformations can simplify the relationships between variables (e.g., by making a curvilinear relationship linear), can remove interactions, and can equalize variances across subgroups that previously had unequal variances.

A primary objection to the use of transformations is that they make the data less interpretable. After all, who thinks about medical costs in log dol-

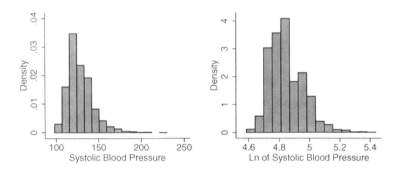

Fig. 2.7. Histograms of Systolic Blood Pressure and Its Natural Logarithm

lars? In situations where there is good reason to stay with the original scale of measurement (e.g., dollars) we may prefer alternatives to transformation including generalized linear models and weighted analyses. Or we may appeal to the robustness of normality-based techniques: many perform extremely well even when used with data exhibiting fairly serious violations of the assumptions.

In other situations, with a bit of work, it is straightforward to express the results on the original scale when the analysis has been conducted on a transformed scale. For example, Sect. 4.7.5 gives the details for log transformations in linear regression.

A compromise when the goal is, for example, to test for differences between two arms in a clinical trial is to plan ahead to present basic descriptive statistics in the original scale, but perform tests on a transformed scale more appropriate for statistical analysis. After all, a difference on the transformed scale is still a difference between the two arms.

Finally we remind the reader that different scales of measurement just take a bit of getting used to: consider pH.

2.3.2 Categorical Variables

Categorical variables require a different approach, since they are less amenable to graphical analyses and because common statistical summaries, such as mean and standard deviation, are inapplicable. Instead we use tabular descriptions. Table 2.3 gives the frequencies, percents, and cumulative percents for each of the behavior pattern categories for the WCGS data. Note that cumulative percentages are really only useful with ordinal categorical data (why?).

When tables are generated by the computer, there is usually little latitude in the details. However, when tables are constructed by hand, thought should be given to their layout; Ehrenberg (1981) is recommended reading. Three

Table 2.3. Frequencies of Behavior Patterns

```
behavioral |
pattern (4 |
    level) |    Freq.      Percent        Cum.
-----------+----------------------------------
        A1 |     264         8.37         8.37
        A2 |    1325        42.01        50.38
        B3 |    1216        38.55        88.93
        B4 |     349        11.07       100.00
-----------+----------------------------------
     Total |    3154       100.00
```

easy-to-follow suggestions from that article are to arrange the categories in a meaningful way (e.g., not alphabetically), report numbers to two effective digits, and to leave a gap every three or four rows to make it easier to read across the table. Table 2.4 illustrates these concepts. With the table arranged

Table 2.4. Characteristics of Top Medical Schools

School	Rank	NIH research ($10 millions)	Tuition ($thousands)	Average MCAT
Harvard	1	68	30	11.1
Johns Hopkins	2	31	29	11.2
Duke	3	16	31	11.6
Penn	4(tie)	33	32	11.7
Washington U.	4(tie)	25	33	12.0
Columbia	6	24	33	11.7
UCSF	7	24	20	11.4
Yale	8	22	30	11.1
Stanford	9(tie)	19	30	11.1
Michigan	9(tie)	20	29	11.0

Source: US News and World Report (http://www.usnews.com, 12/6/01)

in order of the rankings, it is easy to see values that do not follow the pattern predicted by rank, for example, out-of-state tuition.

2.4 Two-Variable Descriptions

Most of the rest of this book is about the relationships among variables. An example from the WCGS is whether behavior pattern is related to systolic blood pressure. In investigating the relationships between variables, it is often useful to distinguish the role that the variables play in an analysis.

2.4.1 Outcome Versus Predictor Variables

A key distinction is whether a variable is being predicted by the remaining variables, or whether it is being used to make the prediction. The variable singled out to be predicted from the remaining variables we will call the *outcome variable*; alternate and interchangeable names are *response variable* or *dependent variable*. The variables used to make the prediction will be called *predictor variables*. Alternate and equivalent terms are *covariates* and *independent variables*. We slightly prefer the outcome/predictor combination, since the term *response* conveys a cause-and-effect interpretation, which may be inappropriate, and *dependent/independent* is confusing with regard to the notion of statistical independence. ("Independent variables do not have to be independent" is a true statement!)

In the WCGS example, we might hypothesize that change in behavior pattern (which is potentially modifiable) might cause change in SBP. This would lead us to consider SBP as the outcome and behavior pattern as the predictor.

2.4.2 Continuous Outcome Variable

As before, it is useful to consider the nature of the outcome and predictor variables in order to choose the appropriate descriptive technique. We begin with continuous outcome variables, first with a continuous predictor and then with a categorical predictor.

Continuous Predictor

When both the predictor and outcome variables are continuous, the typical numerical description is a correlation coefficient and its graphical counterpart is a scatterplot. Again considering the WCGS study, we will investigate the relationship between SBP and weight.

Table 2.5 shows the Stata command and output for the correlation coefficient, while Fig. 2.8 shows a scatterplot. Both the graph and the numerical summary confirm the same thing: there is a weak association between the

Table 2.5. Correlation Coefficient for Systolic Blood Pressure and Weight

```
. correlate sbp weight (obs=3154)

             |      sbp   weight
-------------+------------------
         sbp |   1.0000
      weight |   0.2532   1.0000
```

two variables, as measured by the correlation of 0.25. The graph conveys important additional information. In particular, there are quite a few outliers,

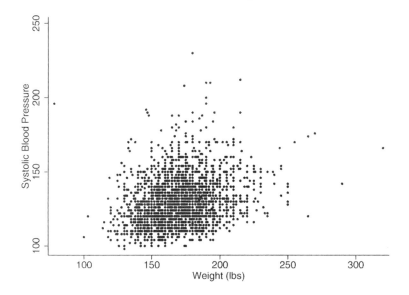

Fig. 2.8. Scatterplot of Systolic Blood Pressure Versus Weight

including an especially anomalous data point with high blood pressure and the lowest weight in the data set.

The Pearson correlation coefficient r, more fully described in Sect. 3.2, is a scale-free measure of association that does not depend on the units in which either SBP or weight is measured. The correlation coefficient varies between –1 and 1, and correlations of absolute value 0.7 or larger are considered strong associations in many contexts. In fields where data are typically noisy, including our SBP example, much smaller correlations may be considered meaningful.

It is important to keep in mind that the Pearson correlation coefficient only measures the strength of the *linear* relationship between two variables. To determine whether the correlation coefficient is a reasonable numerical summary of the association, a graphical tool that helps to assess linearity in the scatterplot is a *scatterplot smoother*. Fig. 2.9 shows a scatterplot smooth superimposed on the graph of SBP versus weight. The figure was generated by the Stata command `lowess sbp weight, bw(0.25)` (with a few embellishments to make it look nicer). This uses the LOWESS technique to draw a smooth (but not necessarily straight) line representing the average value of the variable on the y-axis as a function of the variable on the x-axis. LOWESS is short for LOcally WEighted Scatterplot Smoother. The `bw(0.25)` option specifies that for estimation of the height of the curve at each point, 25% of the data nearest that point should be used. This is all just a fancy way of drawing a flexible curve through a cloud of points. Fig. 2.9 shows that the relationship between SBP and weight is very close to linear. The small upward

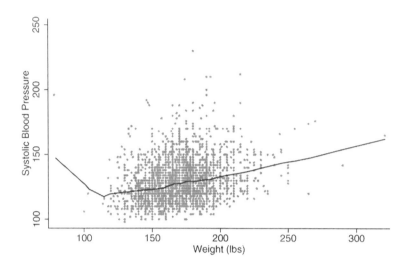

Fig. 2.9. LOWESS Smooth of Systolic Blood Pressure Versus Weight

bend at the far left of the graph is mostly due to the outlying observation at the lowest weight and is a warning as to the instability of LOWESS (or any scatterplot smoother) at the edges of the data.

Choice of bandwidth is somewhat subjective. Small bandwidths like 0.05 often give very bumpy curves, which are hard to interpret. At the other extreme, bandwidths too close to one force the curve to be practically a straight line, obviating the advantage of using a scatterplot smoother. See Problem 2.6.

Categorical Predictor

With a continuous outcome and a categorical predictor, the usual strategy is to apply the same numerical or graphical methods used for one-variable descriptions of a continuous outcome, but to do so separately within each category of the predictor. As an example, we describe the distribution of SBP in WCGS, within levels of behavior pattern. Table 2.6 shows the most direct way of doing this in Stata. Alternatively, the `table` command can be used to make a more compact display, with command options controlling which statistics are listed. The results are shown in Table 2.7.

Side-by-side boxplots, as shown in Fig. 2.10, are an excellent graphical tool for examining the distribution of SBP in each of the behavior pattern categories and making comparisons among them. The four boxplots are quite similar. They each have about the same median, interquartile range, and a

Table 2.6. Summary Data for Systolic Blood Pressure by Behavior Pattern

```
. bysort behpat: summarize sbp


------------------------------------------------------------------------
-> behpat = A1

    Variable |     Obs        Mean    Std. Dev.       Min         Max
-------------+----------------------------------------------------------
         sbp |     264    129.2462    15.29221        100         200


------------------------------------------------------------------------
-> behpat = A2

    Variable |     Obs        Mean    Std. Dev.       Min         Max
-------------+----------------------------------------------------------
         sbp |    1325    129.8891    15.77085        100         212


------------------------------------------------------------------------
-> behpat = B3

    Variable |     Obs        Mean    Std. Dev.       Min         Max
-------------+----------------------------------------------------------
         sbp |    1216    127.5551    14.78795         98         230


------------------------------------------------------------------------
-> behpat = B4

    Variable |     Obs        Mean    Std. Dev.       Min         Max
-------------+----------------------------------------------------------
         sbp |     349    127.1547    13.10125        102         178
```

Table 2.7. Descriptive Statistics for Systolic Blood Pressure by Behavior Pattern

```
. table behpat, contents(mean sbp sd sbp min sbp max sbp)


------------------------------------------------------------------
Behaviora |
l Pattern |   mean(sbp)      sd(sbp)     min(sbp)     max(sbp)
----------+-------------------------------------------------------
       A1 |    129.2462     15.29221          100          200
       A2 |    129.8891     15.77085          100          212
       B3 |    127.5551     14.78795           98          230
       B4 |    127.1547     13.10125          102          178
------------------------------------------------------------------
```

slight right-skewness. At least on the basis of this figure, there appears to be little relationship between SBP and behavior pattern.

2.4.3 Categorical Outcome Variable

With a categorical outcome variable, the typical method is to tabulate the outcome within levels of the predictor variable. To do so first requires breaking any continuous predictors into categories. Suppose, for example, we wished to treat behavior pattern as the outcome variable and weight as the predictor. We might first divide weight into four categories: ≤140 pounds, >140–170, >170–200, and >200. As with histograms, we need enough categories to avoid

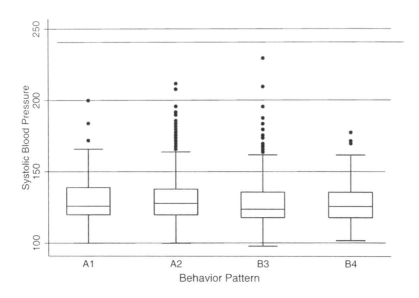

Fig. 2.10. Boxplots of Systolic Blood Pressure by Behavior Pattern

loss of information, without defining categories that include too few observations. Familiar clinical categories are often useful (e.g., glucose <110, 110–125, >125). In this table we have requested percentages for each column to facili-

Table 2.8. Behavior Pattern by Weight Category

```
. tabulate behpat wghtcat, column
```

behavioral pattern (4 level)	< 140	wghtcat 140-170	170-200	> 200	Total
A1	20	125	98	21	264
	8.62	8.13	8.37	9.86	8.37
A2	100	612	514	99	1325
	43.10	39.79	43.89	46.48	42.01
B3	90	610	443	73	1216
	38.79	39.66	37.83	34.27	38.55
B4	22	191	116	20	349
	9.48	12.42	9.91	9.39	11.07
Total	232	1538	1171	213	3154
	100.00	100.00	100.00	100.00	100.00

tate the comparison of the percentages in each behavior pattern between the weight categories. Row percentages or percentages out of the total of 3,154 could also have been requested.

In choosing cutoff points for categorical variables it is entirely fair to look at the distribution of that variable to try to obtain, for example, roughly equal sample sizes in each of the categories. Splitting the data into 3, 4, 5, or 10 groups of equal size is a common approach. However, fishing for cutpoints that prove a point is an easy way to arrive at misleading conclusions.

A different strategy with a categorical outcome and a continuous predictor is to "turn the problem around" and treat the continuous variable as the outcome, using the methods of the previous section. If the only goal is to determine whether the two variables are associated, this may suffice. But when the categorical variable is clearly the outcome, this may lead to awkward models and hard-to-interpret conclusions.

2.5 Multivariable Descriptions

Description of more than two or three variables simultaneously quickly becomes difficult. One approach is to look at pairwise associations, e.g., for categorical variables, looking at a series of two-way tables, taking each pair of variables in turn. If a number of the variables are continuous, a correlation matrix (giving all the pairwise correlations) or a scatterplot matrix (giving all the pairwise plots) can be generated. Table 2.9 and Fig. 2.11 show these for the variables SBP, age, weight, and height. The correlation matrix shows

Table 2.9. Correlation Matrix for Systolic Blood Pressure, Age, Weight, and Height

```
. correlate sbp age weight height (obs=3154)

             |      sbp      age   weight   height
-------------+------------------------------------
        sbp |   1.0000
        age |   0.1657   1.0000
     weight |   0.2532  -0.0344   1.0000
     height |   0.0184  -0.0954   0.5329   1.0000
```

that SBP is very weakly correlated with age and weight and essentially uncorrelated with height.

The scatterplot matrix supports the correlation calculation. If one of the variables is clearly the outcome variable it is useful to list this variable first in the command. That way the first row of the matrix shows the outcome variable on the y-axis, plotted against each of the predictor variables on the x-axis. The matrix of scatterplots for these four variables additionally displays the modest positive correlation between weight and height, indicating the people come in all sizes and shapes!

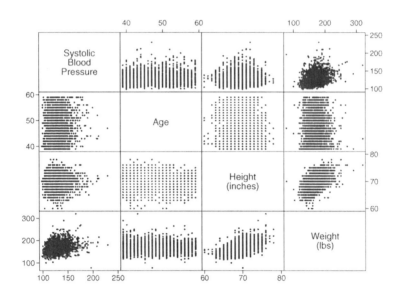

Fig. 2.11. Scatterplot Matrix of Systolic Blood Pressure, Age, Weight, and Height

Multi-way tables that go beyond pairwise relationships can be generated with multiple categorical variables. For example, Table 2.10 shows whether or not the subject had a coronary event (**chd69**), by behavior pattern within weight category. Options in the Stata command are used to obtain the row

Table 2.10. CHD Events and Behavior Pattern by Weight Category

```
. table chd69 behpat wghtcat, row col
```

```
-----------------------------------------------------------------------------
          |           wghtcat and behavioral pattern (4 level)
          | ------------ < 140 ------------    ----------- 140-170 -----------
CHD event |    A1    A2    B3    B4  Total       A1    A2    B3    B4  Total
----------+------------------------------------------------------------------
       no |    18    93    84    22    217      115   559   582   184  1,440
      yes |     2     7     6           15       10    53    28     7     98
          |
    Total |    20   100    90    22    232      125   612   610   191  1,538
-----------------------------------------------------------------------------
```

```
-----------------------------------------------------------------------------
          |           wghtcat and behavioral pattern (4 level)
          | ----------- 170-200 ------------    ------------- > 200 -----------
CHD event |    A1    A2    B3    B4  Total       A1    A2    B3    B4  Total
----------+------------------------------------------------------------------
       no |    81   438   422   108  1,049       20    87    67    17    191
      yes |    17    76    21     8    122        1    12     6     3     22
          |
    Total |    98   514   443   116  1,171       21    99    73    20    213
-----------------------------------------------------------------------------
```

and column totals. With some study, it is possible to extract information from this three-way table, but it is more difficult than with a one- or two-way table. An advantage of a three-way table is the ability to assess *interaction*, the topic of Sect. 4.6. That is, is the relationship between CHD and behavior pattern the same for each weight category?

Analogous graphical displays are also possible. For example, we could look at the relationship between SBP and weight separately by behavior pattern,

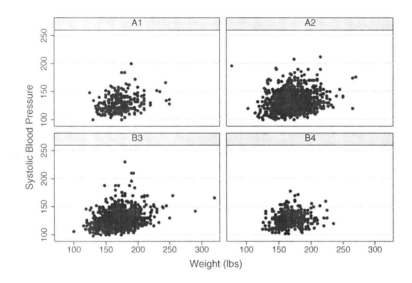

Fig. 2.12. Scatterplot of SBP Versus Weight by Behavior Pattern

as displayed in Fig. 2.12. This indicates that the relationship seems to be the same for each behavior pattern, indicating a lack of interaction.

2.6 Problems

Problem 2.1. Classify each of the following variables as numerical or categorical. Then further classify the numerical variables as continuous or discrete, and the categorical variables as ordinal or nominal.

1. gender
2. race
3. age (in years)
4. age in categories (0–20, 21–35, 36–45, 45–60, 60–85, 85+)
5. zipcode
6. toxicity (mild, moderate, life-threatening, dead)
7. number of hospitalizations in the past year
8. change in HIV-RNA
9. weeks on treatment
10. treatment (placebo vs. estrogen)

Problem 2.2. Generate pseudo-random data from a normal distribution using a computer program or statistics package. In Stata this can be done using the `generate` command and the function `invnorm(uniform())`. Now generate a normal Q-Q plot for these data. Do this for several samples of size 10, 50, and 200. How well do the Q-Q plots approximate straight lines? This is valuable practice for judging how well an actual data set can be expected to approximate a straight line.

Problem 2.3. Generate pseudo-random samples of size 50 from a normal distribution (see Problem 2.2 for how to do this in Stata). Construct histograms of the data using 5, 7, and 15 bins. What do you notice? Do the shapes look like a normal distribution?

Problem 2.4. Warfarin is a drug used to prevent blood clots, for example in patients with irregular heartbeat and after heart surgery. However, too much warfarin can cause unusual bleeding or bruising, so calibration of the dose is important. A study contrasting calibration times (in hours) in two ethnic groups had the following results. For the sample of 18 Caucasians, the times were 2, 4, 6, 7, 8, 9, 10, 10, 12, 14, 16, 19, 21, 24, 26, 30, 35, 44, and 70; for the 18 Asian–Americans, the times were 2, 2, 3, 3, 4, 5, 5, 6, 6, 6, 7, 7, 8, 9, 10, 12, 19, and 32.

1. Display the data numerically to compare the two ethnic groups.
2. Display the data graphically to compare the two ethnic groups.
3. Describe the distribution of the data within ethnic group.
4. Log transform the data and repeat the graphical display. How do the displays with and without log transformation compare?
5. Can you think of other variables you might want to adjust for to help understand the ethnic differences better?

Problem 2.5. The timing of various stages in the contraction of the heart, determined by electro-cardiogram (EKG), can be used to diagnose heart problems. A commonly measured time interval in the contraction of the ventricles is the so-called QRS wave. A study was conducted to see if longer QRS times were related to the ability to induce rapid heart rhythms (called inducible ventricular tachycardia or IVT), which have been associated with adverse outcomes. In a study of 53 subjects, the 18 with IVT had QRS times (in milliseconds) of 70, 75, 86, 90, 96, 102, 110, 114, 116, 117, 120, 130, 136, 142, 145, 152, 170, and 182. The 35 patients without IVT had QRS times of 40, 50, 65, 70, 76, 78, 80, 82, 85, 88, 88, 89, 90, 94, 95, 96, 98, 98, 100, 102, 105, 107, 109, 110, 114, 115, 120, 125, 130, 135, 138, 150, 165, 170, and 180.

1. Display the data numerically to help understand whether QRS time is related to IVT.
2. Display the data graphically to help understand whether QRS time is related to IVT.
3. QRS time is commonly considered as abnormal if the value is greater than 120 msec. Generate a numerical display to help understand if abnormal QRS is related to IVT.
4. What are the advantages and disadvantages of treating QRS as binary (above 120 msec) instead of continuous?

Problem 2.6. Using the WCGS data set, generate a LOWESS (or equivalent) scatterplot smooth of SBP versus weight, comparable to Fig. 2.9. Next try the plot with bandwidths of 0.05, 0.15, and 0.50. How do they compare? Which is most useful for judging the linearity or lack of linearity of the relationship? The WCGS data are available at http://www.biostat.ucsf.edu/vgsm.

3

Basic Statistical Methods

Statistical analyses involving multiple predictors are generalizations of simpler techniques developed for investigating associations between outcomes and single predictors. Although many of these should be familiar from basic statistics courses, we review some of the key ideas and methods here as background for the methods covered in the rest of the book and to introduce some basic notation.

Sects. 3.1–3.3 review basic methods for continuous outcomes, including the t-test and one-way analysis of variance, the correlation coefficient and the linear regression model for a single predictor. Sect. 3.4 focuses on contingency table methods for investigating associations between binary outcomes and categorical predictors, including a discussion of basic measures of association. Sect. 3.5 introduces descriptive methods for survival time outcomes, including Kaplan–Meier survival curves and the logrank test. In Sect. 3.6 we introduce the use of the bootstrap as a method to obtain confidence intervals for estimates in situations where traditional methods are inappropriate. Finally, Sect. 3.7 discusses the importance of properly interpreting negative findings from statistical analyses, focusing on the use of point estimates and confidence intervals rather than P-values.

3.1 t-Test and Analysis of Variance

The t-test and one-way analysis of variance (ANOVA) are basic tools for assessing the statistical significance of differences between the average values of a continuous outcome across two or more samples. Both the t-test and one-way ANOVA can be seen as methods for assessing the association of a categorical predictor – binary in the case of the t-test, with more than two levels in the case of one-way ANOVA – with a continuous outcome. Both are based in statistical theory for normally distributed outcomes, but work well for many other types of data; and both turn out to be special cases of linear regression models.

3.1.1 t-Test

The basic t-test is used in comparing two independent samples. The t-statistic on which the test is based is the difference between the two sample averages, divided by the standard error of that difference. The t-test is designed to work in small samples, whereas Z-tests are not. Table 3.1 shows the result of a t-test comparing average fasting glucose levels among women without diabetes, according to exercise. This is the first of many examples in Chapters 3 and 4 using data from the Heart and Estrogen/Progestin Study (HERS), a clinical trial of hormone therapy for prevention of recurrent heart attacks and death among 2,763 post-menopausal women with existing coronary heart disease (CHD) (Hulley *et al.*, 1998). Average glucose is 97.4 mg/dL among the 1,191 women who do not exercise as compared to 95.7 mg/dL among the 841 women who do. The difference of 1.7 mg/dL is statistically significant ($P = 0.0001$) in the two-sided test shown in the column headed `Ha: diff != 0` (`!=` is Stata notation for "not equal to.") The P-value gives the probability – under the null hypothesis that mean glucose levels are the same in the two populations being compared – of observing a t-statistic more extreme, or larger in absolute value, than the observed value.

Table 3.1. t-Test of Difference in Average Glucose by Exercise

```
. ttest glucose if diabetes == 0, by(exercise)

Two-sample t test with equal variances
```

Variable	Obs	Mean	Std. Err.	Std. Dev.	[95% Conf. Interval]	
no	1191	97.36104	.2868131	9.898169	96.79833	97.92376
yes	841	95.66825	.3258672	9.450148	95.02864	96.30786
combined	2032	96.66043	.2162628	9.74863	96.23631	97.08455
diff		1.692789	.4375862		.8346243	2.550954

```
Degrees of freedom: 2030

                  Ho: mean(no) - mean(yes) = diff = 0

    Ha: diff < 0                Ha: diff != 0                Ha: diff > 0
      t =   3.8685                t =   3.8685                t =   3.8685
    P < t =  0.9999          P > |t| =  0.0001          P > t =  0.0001
```

3.1.2 One- and Two-Sided Hypothesis Tests

In clinical research, unlike some other areas of science, two-sided hypothesis tests are almost always used. In the two-sided t-test, we are testing the null hypothesis (`Ho`) of equal population means against the alternative hypothesis

(Ha) that the one mean is either smaller or larger than the other. The two-sided test is appropriate, for example, when a new treatment might turn out to be beneficial *or* to have adverse effects.

In contrast, only one of these alternatives is considered in a one-sided test. As a result, the smaller of the one-sided *P*-values is half the magnitude of the two-sided *P*-value. The resulting advantage of the one-sided test is that at a given significance level, less evidence in favor of the alternative hypothesis is required to reject the null. For example, using a one-sided test in a sample of 100 observations, we would declare statistical significance at the 5% level if the *t*-statistic exceeds 1.66; using a two-sided test it would need to exceed 1.98 (in absolute value). A direct benefit is that a somewhat smaller sample size is required when a study is designed to be analyzed using a one-sided test.

Use of a one-sided test is sometimes motivated by prior information that makes only one of the alternatives of interest. An example might be in testing an existing treatment known to be safe for evidence of benefit on a new end-point. One-sided tests are also used in *non-inferiority* trials comparing a new to a standard treatment; in this setting the alternative hypothesis is that the new treatment performs almost as well or better than the standard treatment, as against the null hypothesis of clearly performing worse.

However, in part because they make it possible to reject the null hypothesis on weaker evidence, one-sided tests are not commonly used in clinical research. Even in non-inferiority trials where one-sided tests are clearly appropriate, a standard text on the conduct of clinical trials (Friedman *et al.*, 1998) recommends that the tests be carried out at a significance level of 2.5%. Thus to claim non-inferiority, the same strength of evidence would required as in a two-sided test. Furthermore, Fleiss (1988) argues that the other alternative *ought* generally to be of interest, and that in treatment trials adverse effects can rarely be ruled out with sufficient certainty to justify a one-sided test.

The Stata ttest command gives *P*-values for both one-sided tests as well as the two-sided test. In Table 3.1, the one-sided *P*-value on the right (Ha: diff > 0) gives the probability (again, under the null hypothesis) of observing a *t*-statistic larger than the observed value, while the one on the left (Ha: diff < 0) gives the probability of observing one that is smaller. In this example, there is strong evidence ($P = 0.0001$) that the mean glucose level is higher in the population of women who do not exercise, as compared to those who do, and essentially no evidence ($P = 1.0$) that it is smaller.

3.1.3 Paired *t*-Test

The paired *t*-test is for use in settings where individuals or observations are linked across the two samples. Examples include measurements taken at two time points on the same individuals, or on other naturally linked pairs, as in a clinical trial where one eye is treated and the other serves as a control. In this case, the two samples are not independent and failure to take account of the pairwise relationships wastes information and is potentially erroneous.

The paired t-test procedure first computes the pairwise differences for each individual or linked pair. In the first example, this is the change in the outcome from the first time point to the second, and in the second, the difference between the outcomes for the treated and control eyes. Then a t-test is used to assess whether the population mean of these paired differences differs from zero. An increase in power results because between-individual variability is eliminated in the first step. The paired t-test is also implemented using the `ttest` command in Stata. The more complicated case where we want to examine the influence of some other factor on within-individual changes is covered in Sect. 8.3.

3.1.4 One-Way Analysis of Variance (ANOVA)

Suppose that we need to compare sample averages across the arms of a clinical trial with multiple treatments, or more generally across more than two independent samples. For this purpose, one-way analysis of variance (ANOVA) and the F-test take the place of the t-test. The F-test, presented in more detail in Sect. 4.3, assesses the null hypothesis that the mean value of the outcome is the same across all the populations sampled from, against the alternative that the means differ in at least two of the populations. For example, the one-way ANOVA shown in Table 3.2, the F-test for `Between groups` ($P = 0.0371$), suggests that mean systolic blood pressure (SBP) differs by ethnicity in the population represented in the HERS cohort.

Table 3.2. One-Way ANOVA Assessing Differences in SBP by Ethnicity

```
. oneway sbp ethnicity, tabulate
```

ethnicity	Summary of systolic blood pressure Mean	Std. Dev.	Freq.
White	134.78376	18.831686	2451
Afr Amer	138.23394	19.992518	218
Other	135.18085	21.259767	94
Total	135.06949	19.027807	2763

Source	Analysis of Variance SS	df	MS	F	Prob > F
Between groups	2384.26992	2	1192.13496	3.30	0.0371
Within groups	997618.388	2760	361.455938		
Total	1000002.66	2762	362.057443		

3.1.5 Pairwise Comparisons in ANOVA

The statistically significant F-test in the one-way ANOVA indicates the overall importance of ethnicity for predicting SBP. In addition, Stata implements the Bonferroni, Scheffé, and Sidak procedures for assessing the statistical significance of all possible pairwise differences between groups, without inflation of the overall or experiment-wise type-I error rate (EER), which can arise from testing multiple null hypotheses. These and other methods for controlling the EER are discussed in Sect. 4.3.4. All three methods implemented in the `oneway` command show that the difference in average SBP between the African American and white groups is statistically significant after correction for multiple comparisons, but that the other pairwise differences are not.

3.1.6 Multi-Way ANOVA and ANCOVA

Multi-way ANOVA is an extension of the one-way procedure to deal simultaneously with more than one categorical predictor, while analysis of covariance (ANCOVA) is commonly defined as an extension of ANOVA that includes continuous as well as categorical predictors. The t- and F-tests retain their central importance in these procedures. However, one-way ANOVA and the t-test implicitly estimate the different population means by the sample averages; in contrast, the population means in multi-way ANOVA and ANCOVA are usually *modeled*. Thus these procedures are most easily understood as multipredictor linear regression models, which are covered in Chapter 4.

3.1.7 Robustness to Violations of Assumptions

The t- and F-tests are fairly robust to violations of the normality assumption, especially in larger samples. By robust we mean that the type-I error rate, or probability of rejecting the null hypothesis when it holds, is not seriously affected. They are primarily sensitive to outliers, which tend to decrease efficiency and make it harder to detect real differences between groups. Thus the effect is conservative, in the sense of making it more likely that we will accept the null hypothesis when some real difference exists.

Large samples reduce sensitivity of the t-test to the assumption that the outcome is normally distributed because the distribution of the difference between the sample averages, which directly underlies the test, converges to a normal distribution even when the outcome itself has some other distribution. Analogous large-sample behavior holds for the regression coefficients estimated in multipredictor linear models as well as the other regression models that are the primary topic of this book.

When sample sizes are unequal, the t-test is less robust to violations of the assumption of equal variance across samples. Violations of this assumption can seriously affect the type-I error rate, and not always in a conservative direction. In contrast, the overall F-test in ANOVA loses efficiency, but the type-I error rate is generally not increased. However, subsequent pairwise comparisons using t-tests remain vulnerable.

In the two-sample case, this problem is easily addressed using a version of the t-test for unequal variances. This is based on a modified estimate of the standard error of the difference in sample averages. In the analysis shown in Table 3.1, the standard deviation of glucose is 9.9 mg/dL among women who do not exercise, as compared to 9.5 mg/dL among the women who do. In this case the re-analysis allowing for unequal variances, shown in Table 3.3, gives qualitatively the same result ($P = 0.0001$). We recommend systematic use of this version of the t-test, since the increase in robustness comes at very little cost in efficiency. Analogous extensions of ANOVA in which the variance is allowed to vary by group are also possible, though not implemented in the Stata oneway or anova commands.

Table 3.3. t-Test Allowing for Unequal Variances

```
. ttest glucose if diabetes == 0, by(exercise) unequal;

Two-sample t test with unequal variances

------------------------------------------------------------------------------
Variable |    Obs       Mean    Std. Err.   Std. Dev.   [95% Conf. Interval]
---------+--------------------------------------------------------------------
     no  |   1191    97.36104   .2868131    9.898169    96.79833    97.92376
    yes  |    841    95.66825   .3258672    9.450148    95.02864    96.30786
---------+--------------------------------------------------------------------
combined |   2032    96.66043   .2162628    9.74863     96.23631    97.08455
---------+--------------------------------------------------------------------
    diff |           1.692789   .4341096                 .8413954    2.544183
------------------------------------------------------------------------------
Satterthwaite's degrees of freedom:   1858.33

              Ho: mean(no) - mean(yes) = diff = 0

    Ha: diff < 0                Ha: diff != 0               Ha: diff > 0
       t =    3.8995               t =    3.8995               t =    3.8995
    P < t =   1.0000            P > |t| =   0.0001          P > t =   0.0000
```

One commonly recommended solution for violations of the normality assumption is to use nonparametric Wilcoxon rank-sum or Kruskal–Wallis tests rather than the t-test or one-way ANOVA. Two other nonparametric methods are discussed below in Sect. 3.2 on the correlation coefficient. While they avoid specific parametric distributional (i.e., normality) assumptions, these methods are not assumption-free. For example, the Wilcoxon and Kruskal–Wallis tests are based on the assumption that the outcome distributions being compared differ in *location* (mean and/or median) but not in *scale* (variance) or *shape*, as might be captured by a histogram. Furthermore, these two tests do not provide an interpretable measure of the strength of the association. More generally, nonparametric methods sometimes result in loss of efficiency, and do not easily accommodate multiple predictors, unlike the regression methods which are the focus of this book. As an alternative to nonparametric approaches, transformations discussed in Sect. 4.7 can be used to "normalize" a continuous outcome, thus correcting a violation of the normality assumption. More important, we show that this assumption can be relaxed in larger samples.

3.2 Correlation Coefficient

The Pearson correlation coefficient r is a scale-free measure of linear association between two variables x and y, and is defined as follows:

$$r(x, y) = \frac{\text{Cov}(x, y)}{\text{SD}(x)\text{SD}(y)}$$

$$= \frac{\sum_{i=1}^{n}(x_i - \bar{x})(y_i - \bar{y})/(n-1)}{\sqrt{\sum_{i=1}^{n}(x_i - \bar{x})^2/(n-1)}\sqrt{\sum_{i=1}^{n}(y_i - \bar{y})^2/(n-1)}}. \qquad (3.1)$$

In (3.1), $\text{Cov}(x, y)$ is the sample covariance of x and y, $\bar{x}$ and $\bar{y}$ are their sample means, $\text{SD}(x)$ and $\text{SD}(y)$ their standard deviations, and n is the sample size. The covariance reflects the degree to which observations on the two variables differ from their respective means in the same degree and direction. Dividing by the standard deviations of x and y in (3.1) makes $r(x, y)$ scale-free in the sense that it always takes on values between –1 and 1 and does not vary with the units of measurement used for either variable (Problem 3.2).

The correlation coefficient is a measure of *linear* association, in a sense that will become clearer in Sect. 3.3 on the simple linear model. Values of r near zero denote the absence of linear association, while values near 1 mean that x and y increase almost in lockstep, their paired values in a scatterplot falling close to a straight line with increasing slope. Correlations between –1 and zero mean that y tends to *decrease* as x increases. Note that powerful *nonlinear* associations between x and y – for example, if y is proportional to x^2 – are often consistent with correlations near zero; in the example, this can happen if $\bar{x} \approx 0$.

Spearman Rank Correlation Coefficient

Like the t-test (and the coefficients of the linear regression model described below), the correlation coefficient is sensitive to outliers. In this case, a robust alternative is the Spearman correlation coefficient, which is equivalent to the Pearson coefficient applied to the *ranks* of x and y. This measure of correlation also takes on values between –1 and 1. By rank we mean position in the ordered sequence of the values of a variable; if x takes on values 1.2, 0.5, 18.3, and 2.7, then the ranks of these values are 2, 1, 4, and 3, respectively. Thus the rank of the outlier 18.3 is only 1 unit larger than the rank of the next largest value 2.7, the same distance that separates the ranks of any two sequential values of x, thus depriving the outlier of undue influence in estimating the correlation between x and y. Ties are handled by computing the average rank of the tied values. Ranks are used in a range of nonparametric methods, in no small part because of their robustness when the data include outliers. Their disadvantage is that any information contained in the measured values of the outcome beyond the ranks is lost.

Kendall's τ

Another rank-based alternative to Pearson's correlation coefficient is Kendall's τ, defined as the difference in the number of *concordant* and *discordant* pairs of data points, as a proportion of the number of evaluable pairs. In the absence of ties, the pair of data points (x_i, y_i) and (x_j, y_j) for observations i and j is concordant if $x_i > x_j$ and $y_i > y_j$, or if $x_i < x_j$ and $y_i < y_j$, and discordant otherwise. It is easy to see that we need only know the ranks of the x and y values, not their actual values, to evaluate the conditions for concordance. If the numbers of concordant and discordant pairs are about equal, then $\tau \approx 0$; essentially this means that the fact that $x_i > x_j$ gives little information about whether $y_i > y_j$. But as the proportion of concordant pairs grows, τ approaches 1, reflecting the fact that the ordering of the x pairs is highly associated with the ordering of the y pairs. Conversely, if most pairs are discordant, then τ approaches –1; again, the orderings of the x and y pairs are highly associated. Kendall's τ is sometimes used as a measure of correlation for time-to-event outcomes (Chap. 7).

3.3 Simple Linear Regression Model

Here we present the simple linear regression model with a continuous outcome and a single continuous predictor variable.

3.3.1 Systematic Part of the Model

The main purpose of this model is to determine how the average value of the continuous outcome y varies with the value of a single predictor x. The average values of the outcome are assumed to lie on a "regression line" or "line of means." Fig. 3.1 shows values of baseline systolic blood pressure (SBP) by age in the HERS trial of hormone therapy. To make the idea of a line of means more concrete, the square symbols in the plot show the average SBP within each decile of age. Naturally, there is some noise in these local means, although much less than in the raw data. Moreover, the continuous regression line, assumed to be linear, captures the increasing trend rather well. Its slope represents the systematic dependence of the outcome on the predictor, and is thus usually the focus of interest.

The formula for the regression line is simple and has interpretable parameters:

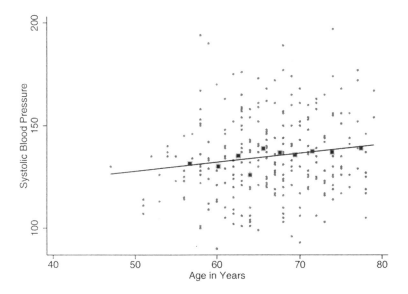

Fig. 3.1. Linear Regression Model for SBP and Age

$$E[y|x] = \text{average value of SBP for a given age}$$
$$= \beta_0 + \beta_1 \texttt{age}$$
$$= 105.7 + 0.44\texttt{age}. \tag{3.2}$$

In (3.2), $E[y|x]$ is shorthand for the <u>E</u>xpected or average value of the outcome y at a given value of the predictor x. β_1 gives the slope of the regression line, and is interpretable as the change in average SBP for a one-year increase in age. The estimate of β_1 from the sample shown in the plot suggests that among women with heart disease, average SBP increases 0.44 mmHg for each one-year increase in age. This estimate is the best fitting value in a sense explained below in Sect. 3.3.4.

It is also easy to see that the estimate of the intercept parameter $\beta_0 = 105.7$ gives the average value of the outcome when age is zero. While not meaningless in this case, these data obviously provide no direct information about SBP at age zero. This illustrates the more general point that while regression models are often approximately true within the range of the observed data, extrapolation is usually risky. "Centering" the predictor by subtracting off a value within the range of the data can resolve this problem. One reasonable choice in this example would be the sample average age of 67; then the centered age variable would have value zero for women at age 67, and the new intercept, 135.2 mmHg, estimates average SBP among women this age. The slope estimate is unaffected by centering the age variable.

3.3.2 Random Part of the Model

It is also clear from Fig. 3.1 that at any given age, SBP varies considerably. Possible sources of this variability include measurement error, diurnal patterns, and a potentially broad range of unmeasured determinants of SBP, including the immediate circumstances when the measurement was made. These factors are combined in an error term ε, so that for observation i

$$\text{SBP}_i = \text{mean SBP for subjects of } \text{age}_i + \text{error}_i$$
$$= \beta_0 + \beta_1 \text{age}_i + \varepsilon_i. \tag{3.3}$$

The statistical assumptions of the linear regression model concern the distribution of ε. Specifically, we assume that $\varepsilon_i \sim$ i.i.d $\mathcal{N}(0, \sigma_\varepsilon^2)$, meaning that ε is independently and identically distributed and has a

- normal distribution
- mean zero at every value of age
- constant variance σ_ε^2 at every value of age
- values that are statistically independent.

In Sect. 4.7 we will see that the first assumption may sometimes be relaxed. The second assumption is important to checking whether the relationship between a numerical predictor and the outcome is linear, as assumed in (3.2), (3.3), and Fig. 3.1; violations can be examined and repaired using methods also introduced in Sect. 4.7. The third assumption, of constant variance, is sometimes called *homoscedasticity*; data which violate this assumption are called *heteroscedastic*, and can be dealt with using methods also discussed in Sect. 4.7 as well as Chapter 9. Chapters 8 and 10 introduce methods for data where the fourth assumption, of independence, does not hold. Some examples include samples with repeated measures on individuals, cluster samples where patients are selected from within a sample of physician practices, and complex survey samples such as the National Health and Nutrition Examination Survey (NHANES).

3.3.3 Assumptions About the Predictor

In contrast to the outcome, no distributional assumptions are made about the predictor in the linear regression model. In the case of the linear model with a single continuous predictor, we do not assume that the predictor has a normal distribution, although we will see in Sect. 4.7 that outlying values of the predictor can cause trouble in some circumstances. In addition, binary, categorical, and discrete numeric variables including counts are easily accommodated as predictors in these models.

Although we do not need to make assumptions about the distribution of the predictor, these models do perform better when it is relatively variable. For example, it would be more difficult to estimate the age trend in average

SBP if the sample were limited to women aged 65–70. For binary and categorical predictors, the analogous limitation is that the subgroups defined by the predictor should not be too small. The impact of the variability of the predictor, or lack of it, is reflected in the standard error of the regression coefficient, as shown below in Sect. 3.3.7.

Finally, when we want to assess the relationship of the outcome with the true values of the predictor, we effectively assume that the predictors are measured without error. This is often not very realistic, and the effects of violations are the subject of ongoing statistical research. Random measurement errors unrelated to the outcome result in attenuation of estimated slope coefficients toward zero, sometimes called *regression dilution bias* (Frost and Thompson, 2000). Despite some loss of efficiency, reasonable estimation is often possible in the presence of mild-to-moderate error in the measurement of the predictors. Moreover, for prediction of new outcomes, values of the predictor measured with error may suffice.

3.3.4 Ordinary Least Squares Estimation

The model (3.3) refers to the population of women with heart disease from which the sample shown in Fig. 3.1 was drawn. The regression line in the figure is an estimate of the population regression line that was found using *ordinary least squares* (OLS). Of all the lines that could be drawn though the scatterplot of the data to represent the trend in SBP with increasing age, the OLS estimate minimizes the sum of the squared vertical differences between the data points and the line.

Since the regression line is uniquely determined by β_0 and β_1, the intercept and slope parameters, fitting the regression model amounts to finding estimates $\hat{\beta}_0$ and $\hat{\beta}_1$ which meet the OLS criterion. In addition to being easy to compute, these OLS estimates have desirable statistical properties. If model assumptions hold, $\hat{\beta}_0$ and $\hat{\beta}_1$ are unbiased estimates of the population parameters.

> Definition: An estimate is *unbiased* if, over many repeated samples drawn from the population, the average value of the estimates based on the different samples would equal the population value of the parameter being estimated.

OLS estimates are also minimally variable and well behaved in large samples when the distributional assumptions concerning ε are not precisely met. However, a drawback of the OLS estimation criterion is sensitivity to outliers, which arises from squaring the vertical differences (Problem 3.1). Sect. 4.7 will show how to diagnose and deal with influential points.

Table 3.4 shows Stata results for an OLS regression of SBP on age. The estimate of β_1, the slope coefficient (`Coef.`) for `age`, is 0.44 mmHg per year, and the intercept estimate $\hat{\beta}_0$ is 105.7 mmHg (`_cons`).

Table 3.4. OLS Regression of SBP on Age

```
. reg SBP age

      Source |       SS       df       MS              Number of obs =     276
-------------+------------------------------           F(  1,    274) =    5.58
       Model |  2179.70702      1   2179.70702         Prob > F       =  0.0188
    Residual |  106991.347    274   390.47937          R-squared      =  0.0200
-------------+------------------------------           Adj R-squared  =  0.0164
       Total |  109171.054    275   396.985652         Root MSE       =  19.761


------------------------------------------------------------------------------
         sbp |      Coef.   Std. Err.      t    P>|t|     [95% Conf. Interval]
-------------+----------------------------------------------------------------
         age |   .4405286    .186455     2.36   0.019     .0734621    .8075952
       _cons |    105.713   12.40238     8.52   0.000      81.2969     130.129
------------------------------------------------------------------------------
```

3.3.5 Fitted Values and Residuals

The OLS estimates $\hat{\beta}_0$ and $\hat{\beta}_1$ in turn determine the *fitted value* $\hat{y}$ corresponding to every data point:

$$\hat{y}_i = \hat{\beta}_0 + \hat{\beta}_1 x_i. \tag{3.4}$$

It should be plain that the fitted value $\hat{y}_i$ lies on the estimated regression line at the point where $x = x_i$. For a woman at the average age of 67, the fitted value is

$$105.713 + 0.4405286 \times 67 = 135.2 \text{ mmHg}. \tag{3.5}$$

The *residuals* are defined as the difference between observed and fitted values of the outcome:

$$r_i = y_i - \hat{y}_i. \tag{3.6}$$

The residuals are the sample analogue of ε, the error term introduced earlier in Sect. 3.3, and as such are particularly important in fitting the model, in estimating the variability of the parameter estimates, and in checking model assumptions and fit (Sect. 4.7).

3.3.6 Sums of Squares

Various *sums of squares* are central to understanding OLS estimation and to reading the Stata regression model output in Table 3.4. First is the *total sum of squares* (TSS):

$$\text{TSS} = \sum_{i=1}^{n}(y_i - \bar{y})^2, \tag{3.7}$$

where $\bar{y}$ is the sample average of the outcome y. TSS captures the total variability of the outcome about its mean. In Table 3.4, TSS = 109,171 and appears in the row and column labeled Total and SS (for Sum of Squares), respectively.

In an OLS model, TSS is split into two components. The first is the *model sum of squares* (MSS), or the part of the variability of the outcome about its mean that can be accounted for by the model:

$$\text{MSS} = \sum_{i=1}^{n}(\hat{y}_i - \bar{y})^2. \tag{3.8}$$

The second component of outcome variability, the part that cannot be accounted for by the model, is the *residual sum of squares* (RSS):

$$\text{RSS} = \sum_{i=1}^{n}(y_i - \hat{y}_i)^2. \tag{3.9}$$

By definition, RSS is minimized by the fitted regression line. In Table 3.4 MSS and RSS appear in the rows labeled `Model` and `Residual` of the `SS` column. The identity TSS = MSS + RSS is a central property of OLS, but more difficult to prove than it may seem.

3.3.7 Standard Errors of the Regression Coefficients

MSS and RSS also play an important role in estimating the standard errors of $\hat{\beta}_0$ and $\hat{\beta}_1$ and in testing the null hypothesis of central interest, H_0: $\beta_1 = 0$. These standard errors depend on the variance of ε – that is, the variance of the outcome about the regression line – which is estimated in our single predictor model by

$$\hat{\text{Var}}(\varepsilon) = s_{y|x}^2 = \text{RSS}/(n-2). \tag{3.10}$$

In Table 3.4 $s_{y|x}^2$ equals 390.5, and appears in the column and row labeled `MS` (for Mean Square) and `Residual`, respectively.

The variance of $\hat{\beta}_1$ is estimated by

$$\hat{\text{Var}}(\hat{\beta}_1) = \frac{s_{y|x}^2}{(n-1)s_x^2}, \tag{3.11}$$

where s_x^2 is the sample variance of the predictor x. The square root of the variance of an estimate is referred to as its *standard error* (SE). In Table 3.4, the standard error of the estimated slope coefficient for `age`, found in the column labeled `Std.Err.`, is approximately 0.187.

From the numerator and denominator of (3.11), it is clear that the variance of the slope estimate *increases* with the residual outcome variance not explained by the model, but *decreases* with larger sample size and with the variance of the predictor (as we pointed out earlier in Sect. 3.3.3). In our example of SBP and age, estimation of the trend in age is helped by the relatively large age range in the sample. It should also be intuitively clear that the precision of the slope estimate is increased in samples where the data are tightly clustered about the regression line – in other words, if the residual variance of the outcome is small. Fig. 3.1 shows that this is not the case with our example; SBP varies widely about the regression line at every value of age.

3.3.8 Hypothesis Tests and Confidence Intervals

When the outcome is normally distributed, the parameter estimates $\hat{\beta}_0$ and $\hat{\beta}_1$ have a normal distribution, and the ratio of the slope estimate to its standard error has a t-distribution with $n - 2$ degrees of freedom. This leads directly to a test of the null hypothesis of no slope: that is, $H_0: \beta_1 = 0$, or in substantive terms, no systematic relationship between predictor and outcome. In Table 3.4, the t-statistic and corresponding P-value for age are shown in the columns labeled t and P>|t|. In the example, we are able to reject the null hypothesis that SBP does not change with age at the usual 5% level of significance ($P = 0.019$).

The t-distribution also leads to 95% confidence intervals for the population parameter β_1, shown in Table 3.4 in the columns labeled [95% Conf. Interval]. The confidence interval does not include 0, in accord with the result of the t-test of the null hypothesis. Under the assumptions of the model, a confidence interval computed this way would, on average, include the population value of the parameter in 95 of 100 random samples. In a more intuitive interpretation, we could exclude with 95% confidence age trends in SBP smaller than 0.07 mmHg/year or larger than 0.81 mmHg/year.

Relationship Between Hypothesis Tests and Confidence Intervals

Hypothesis tests and confidence intervals provide overlapping information about the parameter or association being assessed. Common ground is that when a two-sided test is statistically significant at $P < 0.05$, then the corresponding 95% confidence interval will exclude the null parameter value. However, the P-value, especially if it is small, does give a more direct sense of the strength of the evidence against the null hypothesis. Likewise, only the confidence interval provides information about the range of parameter values that are consistent with the data. In Sect. 3.7 below, we argue that confidence intervals are particularly important in the interpretation of negative findings – that is, cases where the null hypothesis is not rejected. Both the P-value and the confidence interval are important for understanding statistical results in depth, and getting beyond the simple question of whether or not an association is statistically significant. This overlapping relationship between hypothesis tests and confidence intervals holds in many settings in addition to linear regression.

Hypothesis Tests and Confidence Intervals in Large Samples

The hypothesis tests and confidence intervals in this section follow from basic statistical theory for data with normally distributed outcomes. However, linear regression models are commonly used with outcomes that are at best approximately normal, even after transformation. Fortunately, in large samples the t-tests and confidence intervals for $\hat{\beta}_0$ and $\hat{\beta}_1$ are valid even when the underlying outcome is not normal. How large a sample is required depends on how far and in what way the outcome departs from normality. If the outcome is uniformly distributed, meaning that every value in its range is equally likely, then the t-tests and confidence intervals may be valid with as few as 30–50 observations. However, with long-tailed outcomes, samples of at least 100 and sometimes much larger may be required for hypothesis tests and confidence intervals to be valid.

3.3.9 Slope, Correlation Coefficient, and R^2

The slope coefficient estimate $\hat{\beta}_1$ in a simple linear model is systematically related to the sample Pearson correlation coefficient r, reviewed in Sect. 3.2:

$$r = \hat{\beta}_1 \mathrm{SD}(x)/\mathrm{SD}(y), \tag{3.12}$$

where $\mathrm{SD}(x)$ and $\mathrm{SD}(y)$ are the standard deviations of the predictor and outcome, respectively. Thus we can get r from $\hat{\beta}_1$ by factoring out the scales on which x and y are measured (Problem 3.3), scales which are reflected in the standard deviations. Furthermore, the t-test of H_0: $\beta_1 = 0$ is equivalent to a test of H_0: $r = 0$.

However, the correlation coefficient is not simply interchangeable with the slope coefficient in a simple linear model. In particular, the slope coefficient distinguishes the roles of the predictor x and outcome y, with differing assumptions applying to each, and would change if those roles were reversed, but $r(x, y) = r(y, x)$. Note that reversing the roles of predictor and outcome becomes even more problematic with multipredictor models. In addition, the slope coefficient β_1 depends on the units in which both predictor and outcome are measured, so that if either or both were measured in different units, β_1 would change. For example, our estimate of the age trend in SBP would be 4.4 mmHg per decade if age were measured in ten-year units. While both versions are interpretable, this dependence on the scale of both predictor and outcome can make it difficult to assess the strength of the association. In addition the dependence on scale would make it hard to judge whether age is a stronger predictor of SBP than other variables. From this point of view, the scale-free correlation coefficient r is easier to interpret.

The correlation coefficient r and thus the slope coefficient β_1 are also systematically related to the *coefficient of determination* R^2

$$R^2 = r^2 = \frac{\mathrm{MSS}}{\mathrm{TSS}}. \tag{3.13}$$

R^2 is interpretable as the proportion of the total variability of the outcome (TSS) that is accounted for by the model (MSS). As such, it is useful for comparing models (Sect. 5.3). In Table 3.4 the value of R-squared is only 0.0200, which you can easily verify is equal to MSS/TSS $= 2,179/109,171$. This shows that age only explains a very small proportion of the variability of SBP, even though it is a statistically significant predictor in a sample of moderate size.

3.4 Contingency Table Methods for Binary Outcomes

In Chapter 2 we reviewed exploratory techniques for categorical outcome variables. We expand that review here to include contingency table methods for assessing associations between binary outcomes and categorical predictors.

3.4.1 Measures of Risk and Association for Binary Outcomes

In the Western Collaborative Group Study (WCGS) (Rosenman *et al.*, 1964) of coronary heart disease (CHD) introduced in Chapter 2, an association of interest to the original investigators was the relationship between CHD risk and the presence/absence of corneal arcus senilis among participants upon entry into the study. Because each participant could be unambigously classified as having developed CHD or not during the ten-year course of the study, the indicator variable that takes on the value one or zero according to whether or not participants developed the disease is a legitimate binary outcome for the analysis. Corneal arcus is a whitish annular deposit around the iris that occurs in a small percentage of older adults, and is thought to be related to serum cholesterol level. Table 3.5 presents the results of a basic two-by-two table analysis for this example. The results were obtained using the cs command in Stata, which provides a number of useful quantities in addition to a simple crosstabulation of the binary CHD outcome chd69 with the binary indicator of the presence of arcus.

The Risk estimates (0.108 and 0.069) summarize outcome risk for individuals with and without arcus and are simply the observed proportions of individuals with CHD in these groups at the baseline visit of the study. The output also includes several standard epidemiological measures of association between outcome risk and the predictor variable, along with corresponding 95% confidence intervals. These are numerical comparisons of the risk estimates between the two groups defined by the predictor.

The Risk difference or *excess risk* is defined as the difference between the estimated risk in the groups defined by the predictor. For the table, we can verify that the risk difference is

$$0.1084 - 0.0692 = 0.039$$

Table 3.5. Two-by-Two Contingency Table for CHD and Arcus

```
. cs CHD69 arcus, or

                   | arcus senilis          |
                   |  Exposed   Unexposed   |      Total
-------------------+------------------------+-----------
          Cases  |      102          153   |       255
       Noncases  |      839         2058   |      2897
-------------------+------------------------+-----------
          Total  |      941         2211   |      3152
                   |                        |
           Risk  |  .1083953     .0691995   |   .080901
                   |                        |
                   |  Point estimate        |   [95% Conf. Interval]
                   |------------------------+-----------------------
Risk difference  |        .0391959         |   .0166915    .0617003
     Risk ratio  |        1.566419         |  1.233865    1.988603
  Attr. frac. ex.  |      .3616011         |  .1895387    .4971343
  Attr. frac. pop  |      .1446404         |
     Odds ratio  |         1.63528         |  1.257732    2.126197  (Cornfield)
                   +----------------------------------------------
                        chi2(1) =   13.64  Pr>chi2 = 0.0002
```

The **Risk ratio** or *relative risk* is the ratio of these risks – for the example in the table,

$$0.1084/0.0692 = 1.57.$$

The **Odds ratio** is the ratio between the corresponding odds in the two groups. The odds of an outcome occurring are computed as the probability of occurrence divided by the complementary probability that the event does not occur. Since the denominators of these two probabilities are identical, the odds can be also be calculated as the ratio of the number of outcomes to non-outcomes. Frequently used in games of chance, "even odds" obtains when these two probabilities are equal.

In Table 3.5, the odds of CHD occurrence in the two arcus groups are $0.1084/(1 - 0.1084) = 102/839$ and $0.0692/(1 - 0.0692) = 153/2058$, respectively. The ratio of these two numbers yields the estimated odds ratio (1.635) comparing the odds of CHD occurrence among participants with arcus to the odds of those without this condition. Although the odds ratio is somewhat less intuitive as a risk measure than the risk difference and relative risk, we will see that it has properties that make it useful in a wide range of study designs, and (in Chapter 6) that it is fundamental in the definition and interpretation of the *logistic regression* model.

Finally, note that Table 3.5 provides two auxiliary summary measures of *attributable risk* (i.e., **Attr. frac. ex.** and **Attr. frac. pop**), which estimate the fraction of outcomes which can be attributed to the predictor in the subgroup with the predictor (sometimes referred to as "exposed" individuals) and in the overall population, respectively. Although these measures can easily be estimated from the data in the table, their validity and interpretability depends on a number of factors, including study design and the causal connections between measured and unmeasured predictors and the outcome. See Rothman and Greenland (1998) for further discussion of these measures.

In the last example we saw that the observed outcome proportions for groups defined by different values of a predictor are the fundamental components of the three summary measures of association: the excess risk, relative risk, and odds ratio. To discuss these further, it will be useful to have symbolic definitions. Following the notation introduced in Sect. 3.3 for a continuous outcome measure, we will denote the binary outcome variable CHD by y, and let the values 1 and 0 represent individuals with and without the outcome, respectively. We will symbolize the outcome probability for an individual associated with a particular value x of a single predictor as

$$P(x) = \Pr(y = 1|x)$$

and estimate this using the proportion of individuals with the outcome $y = 1$ among all those in the sample with the value x of the predictor. For example, $P(0)$ and $P(1)$ symbolize the outcome probability or risk associated with two levels of the binary predictor arcus in Table 3.5 (where we follow the usual convention that individuals possessing the characteristic have the values $x = 1$, and individuals without the characteristic have $x = 0$). The following equation defines all three summary risk measures introduced above using this notation:

$$
\begin{aligned}
ER &= P(1) - P(0) \\
RR &= P(1)/P(0) \\
OR &= \frac{P(1)/\left[1 - P(1)\right]}{P(0)/\left[1 - P(0)\right]},
\end{aligned}
\tag{3.14}
$$

where ER, RR, and OR denote the excess risk, relative risk, and odds ratio, respectively.

Like the correlation coefficient, these measures provide a convenient single number summary of the direction and magnitude of the association. The major distinction between them is that the ER is a measure of the difference in risk between the two groups (with no difference indicated by a value of zero), while both the RR and OR compare the risks in relative terms (with no difference indicate by a value of one). Note that because the component risks range between zero and one, the ER can take on values between -1 and 1. By contrast, the RR and OR range between 0 and ∞.

Relative measures are appealing because they are dimensionless, and convey a clear impression of how outcome risk is increased/decreased by exposure. The RR in particular is favored by epidemiologists because of its interpretability as a ratio of risks. However, relative measures are less desirable when the goal is to convey the "importance" of a particular risk in absolute terms: In the example, the estimated RR for the risk of CHD is approximately 1.6 times higher for men with arcus. The ER tells us that this corresponds to a 4% difference in absolute risk. Note that if the risk of the outcome were ten times lower in both groups we would have the same estimated RR, but the corresponding ER would also be ten times smaller (or 0.4%).

A further feature of the RR worth remembering is that its maximum value is constrained by the level of risk in the comparison group. For example, if $\Pr(0) = 0.5$, $RR \leq 2$ must hold. The OR has the advantages of a relative measure, and in addition is not constrained by the level of the risk in the reference group. However, being based on the odds of the outcome rather than the probability, the OR lacks the intuitive interpretation of RR. The only exception is when the outcome risk is quite small. For such rare outcomes, the OR closely approximates the RR and can be interpreted similarly. (This property can be seen from the above definition by noting that if outcome risk is close to zero, then $[1-\Pr(0)]$ and $[1-\Pr(1)]$ will both be approximately one.) Unfortunately, the odds ratio is often inappropriately reported as a relative risk even when this condition isn't met (Holcomb *et al.*, 2001). Because the value of the OR is always more extreme than the value of the RR (except when both equal one), this can be misleading. For these reasons, we recommend that the measure of association reported in research findings be that actually used in the analysis.

A final important property of all three measures of association introduced above is that their interpretation depends on the underlying study design. In the WCGS example the outcome risks represent the *incidence proportion* of CHD over the entire duration of the study (approximately ten years). The measures of association in the table should be interpreted accordingly. By contrast, the sexually transmitted infection example mentioned at the beginning of this chapter referred to a cross-sectional sample. Outcome risk in this setting is measured by the *prevalence* of the outcome among the groups defined by the predictor. In this case, the terms "prevalence odds," "prevalence ratio," and "excess prevalence" provide unambiguous alternative labels for OR, RR, and ER, respectively.

The relative merits of the ER, RR, and OR are discussed at length in most epidemiology textbooks (e.g., Rothman and Greenland, 1998). For our purposes, they are equally valid and the choice is dependent on the nature and goals of the research investigation. In fact, for prospective and cross-sectional study designs, we'll see that we can freely convert between measures. (Case-control designs are a special case which will be covered in Sect. 6.3.) However, from the standpoint of regression modeling, we'll see in Chapter 6 that the OR has clear advantages.

3.4.2 Tests of Association in Contingency Tables

Addressing the research question posed in the example presented in Table 3.5 involves more than simply summarizing the degree of the observed association between CHD and arcus. We would also like to account for uncertainty in our estimates before concluding that the association reflects more than just a chance finding in this particular sample of individuals. The 95% confidence intervals associated with the measures of association in the table help in this regard. For example, the fact that the confidence interval for the odds ratio excludes the value 1.0 allows us to conclude that the true value for this measure is greater than one, and indicates a statistically significant positive association between the presence of arcus and CHD occurrence. This corresponds to testing the null hypothesis that the true odds ratio is not equal to one, with the alternative hypothesis being that this odds ratio is different than one. The fact that the value of one is excluded from the confidence interval corresponds to rejection of this hypothesis at the 5% significance level. Of course, establishing the possible causal connection between these two variables is a more complex issue.

The χ^2 (chi-squared) test of association is an alternative way to make inferences about an observed association. Note that the result of this test (presented in Table 3.5) agrees with the conclusions drawn for the 95% confidence intervals for the various measures of association. The statistic addresses the null hypothesis of no association, and is computed using the squared differences between the observed proportions of individuals in each cell of the two-way table and the corresponding proportions that would be expected if the null hypothesis were true. Large values of the statistic indicate departure from this hypothesis, and the associated P-value is computed using the χ^2 distribution with degrees of freedom specified. The χ^2 statistic for a two-by-two table is less appealing as a measure of association than the alternative measures discussed above. However, in cases where predictors have more than two levels (as discussed below) and a single summary measure of association can't be calculated, the χ^2 statistic is useful as a global indicator of whether or not an association may be present.

The validity of the χ^2 test is dependent on available sample size; like many commonly used statistical tests, the validity of the reference χ^2 distribution for the test statistic is approximate, with the approximation improving with increasing number of observations. Consequently, for small sample sizes, approximate P-values and associated inferences may be unreliable. An alternative in these cases is to base inferences on *exact* methods. Table 3.6 presents an example from a cross-sectional study of sexual transmission of human immunodeficiency virus (HIV) in monogamous female partners of males infected from contaminated blood products (O'Brien et al., 1994). The outcome of this study was HIV status of the female partner at recruitment. Males were known to have been infected first (via medical records) and exposure of females was limited to contact with male partners. The available sample size ($n = 31$) was limited by the availability of couples meeting the strict eligibility criteria.

Table 3.6 addresses the hypothesis that more rapid disease progression in the males (as indicated by an AIDS diagnosis occurring at or before the time of recruitment of the couple) is associated with sexual transmission of HIV to the female (represented by the binary indicator hivp). In addition to observed counts, the table includes proportions of the outcome by AIDS diagnosis in the male partners, and the measures of association described above. The table also presents the results of Fisher's exact test. Similar to the

Table 3.6. Female Partner's HIV Status by AIDS Diagnosis of Male Partner

```
. cs hivp aids, or exact

                 | AIDS diag. in male |
                 | [1=yes/0=no]       |
                 |  Exposed   Unexposed |     Total
-----------------+--------------------+----------
        Cases |      3          4 |       7
     Noncases |      2         22 |      24
-----------------+--------------------+----------
        Total |      5         26 |      31
                 |                    |
         Risk |     .6    .1538462 | .2258065
                 |                    |
                 | Point estimate    | [95\% Conf. Interval]
                 |--------------------+----------------------
Risk difference |     .4461538      | -.0050928    .8974005
     Risk ratio |          3.9      | 1.233644    12.32933
Attr. frac. ex. |     .7435897      |  .1893933    .9188926
Attr. frac. pop |     .3186813      |
     Odds ratio |         8.25      | 1.200901    57.1864  (Cornfield)
                 +----------------------------------------
                   1-sided Fisher's exact P = 0.0619
                   2-sided Fisher's exact P = 0.0619
```

χ^2 test, the Fisher test addresses the hypothesis of independence of outcome and predictor. However, the P-value is computed exactly, conditioning on the observed marginal totals. The P-value for the χ^2 test applied to the data in Table 3.6 (not shown) is 0.029. Similarly, the lower 95% confidence limits for the RR and OR exclude the value one, also indicating a statistically significant association. By contrast, the (two-sided) P-value for the Fisher's exact test for Table 3.6 is 0.062, indicating failure to reject the hypothesis of independence at the 5% level.

A commonly cited rule-of-thumb is that the Fisher's exact test should be used whenever any of the expected cell counts are less than 5. Note that Fisher's exact test applies to tables formed by variables with more than two categories. Although it can almost always be used in place of the χ^2 test, the associated computations can be lengthy for large sample sizes, especially for tables with dimensions larger than 2×2. Given the increased speed of modern desktop computers and the availability of more computationally efficient algorithms, we recommend using the exact P-value whenever it can easily be computed (i.e., in a matter of minutes) or is provided, and especially in cases where either actual or expected minimum cell counts are less than 5.

3.4.3 Predictors With Multiple Categories

In the WCGS study discussed above, one potentially important predictor of CHD risk is age at entry into the study. Despite the fact that this can be considered as a continuous variable for the purpose of analyses, we might begin investigating the relationship by grouping age into multiple categories and summarizing CHD risk in the resulting groups. Table 3.7 shows the results obtained by dividing subjects into five-year age intervals using a constructed five-level categorical variable AGEC. With the exception of the first two columns,

Table 3.7. CHD Events by Age in WCGS Cohort

```
. tabulate chd69 agec, col chi2

              |                        agec
    CHD event |    35-40    41-45    46-50    51-55    56-60 |    Total
--------------+--------------------------------------------------+----------
           no |      512    1,036      680      463      206 |    2,897
              |    94.29    94.96    90.67    87.69    85.12 |    91.85
--------------+--------------------------------------------------+----------
          yes |       31       55       70       65       36 |      257
              |     5.71     5.04     9.33    12.31    14.88 |     8.15
--------------+--------------------------------------------------+----------
        Total |      543    1,091      750      528      242 |    3,154
              |   100.00   100.00   100.00   100.00   100.00 |   100.00

          Pearson chi2(4) =   46.6534    Pr = 0.000
```

the estimated percentages of individuals with CHD in the second row of the table clearly increase with increasing age. In addition, the accompanying χ^2 test indicates that age and CHD risk are associated.

As mentioned above, the conclusion of association based on the χ^2 test does not reveal anything about the nature of the relationship between these variables. More insight could be gained by computing measures of association between age and CHD risk. However, unlike the two-by-two table case, the fact that age is represented with five levels means that a single measure will not suffice here. In fact, odds ratios can be computed to compare any two age groups. For example, the ER, RR, and OR comparing CHD risk in 56 to 60-year-olds with that in 35 to 40-year-olds are calculated by applying the formulas in (3.14) as follows:

$$ER = (36/242) - (31/543) = 0.092$$
$$RR = \frac{36/242}{31/543} = 2.606$$
$$OR = \frac{\frac{36/242}{206/242}}{\frac{31/543}{512/543}} = 2.886. \tag{3.15}$$

The results in Table 3.8 further reinforce our observation that CHD risk is increasing with increasing age. The odds ratios in the table are all computed using the youngest age group as the reference category. The pattern of increase in estimated odds ratios mirrors that seen in Table 3.7. Note that each odds ratio in the table is accompanied by a 95% confidence interval and associated hypothesis test. In addition, two global tests providing additional information are provided: The `Test of homogeneity` addresses the null hypothesis that odds ratios do not differ across age categories. In this case, the P-value indicates rejection, confirming the observed difference in the odds ratios mentioned above. Since age can be viewed as a continuous variable, and the categorical version considered here is ordinal, more specific alternatives to non-homogeneity of odds are of greater scientific interest. The `Score test for trend` in Table 3.8 addresses the alternative hypothesis that there is a linear trend in the odds of CHD with increasing age categories. The statistically significant results indicate support for this hypothesis, and represent a stronger conclusion than non-homogeneity. Note that this test is not applicable to nominal categorical variables.

Table 3.8. Odds Ratios for CHD Events by Age Group

```
. tabodds chd69 agec, or

       ---------------------------------------------------------------------
       agec |  Odds Ratio         chi2        P>chi2     [95% Conf. Interval]
       ------------+--------------------------------------------------------
      35-40 |    1.000000          .             .             .         .
      41-45 |    0.876822         0.32         0.5692      0.557454   1.379156
      46-50 |    1.700190         5.74         0.0166      1.095789   2.637958
      51-55 |    2.318679        14.28         0.0002      1.479779   3.633160
      56-60 |    2.886314        18.00         0.0000      1.728069   4.820876
       ---------------------------------------------------------------------
Test of homogeneity (equal odds): chi2(4)  =     46.64
                                  Pr>chi2   =    0.0000

Score test for trend of odds:     chi2(1)  =     40.76
                                  Pr>chi2   =    0.0000
```

Despite the useful information gained from the analysis in Tables 3.7 and 3.8, we may be concerned that our conclusions depend on the arbitrary choice of grouping age into five categories. Increasing the number of age categories may provide more information on how risk varies with age, but will also reduce the number of individuals in each category and lead to more variable estimates of risk in each group. This dilemma is one of the primary motivations for introducing a regression model for the dependence of outcome risk on a continuous predictor variable. Another motivation (which will be explored briefly below and more fully in Chapter 6) arises when we consider the joint effects on risk of multiple (categorical and/or continuous) predictor variables.

3.4.4 Analyses Involving Multiple Categorical Predictors

A common feature of observational clinical and epidemiological studies is that investigators do not experimentally control the distributions of characteristics of interest among participants in the sample. Unlike randomized trials in which random allocation serves to balance the distributions of characteristics across treatment arms, observational data are usually characterized by differing distributions across subgroups defined by predictors of primary interest. For example, observational studies of the relationship between dietary factors and cancer typically adjust for age since it is frequently related to both diet and cancer risk. A fundamental part of drawing inferences regarding the relationship between the outcome and key predictors in observational studies is to consider the potential influence of these other characteristics. This topic will be covered in detail from regression models in Chapter 5. Here we give a brief introduction for binary outcomes and categorical predictors.

Consider the cross-tabulation of a binary indicator 20-year mortality and self-reported smoking presented in Table 3.9. These data represent women

Table 3.9. Twenty-Year Vital Status by Smoking Behavior

```
. cs vstatus smoker [freq = nn], or

                 | smoker                    |
                 | Exposed    Unexposed   |    Total
-----------------+------------------------+----------
         Cases | 139             230    |      369
      Noncases | 443             502    |      945
-----------------+------------------------+----------
         Total | 582             732    |     1314
               |                        |
          Risk | .2388316   .3142077    | .2808219
               |                        |
               | Point estimate         | [95% Conf. Interval]
               |------------------------+----------------------
Risk difference | -.075376               | -.1236536   -.0270985
     Risk ratio | .7601076               |  .6347365    .9102415
 Prev. frac. ex. | .2398924              |  .0897585    .3652635
 Prev. frac. pop | .1062537              |
     Odds ratio | .6848366               |  .5354784    .8758683  (Cornfield)
               +-------------------------------------------------
                        chi2(1) =    9.12  Pr>chi2 = 0.0025
```

participating in a health survey in Whickham, England in 1972–1974 (Vanderpump et al., 1996). Deaths were ascertained via follow-up of participants over a 20-year period. The results indicate a statistically significant negative association between smoking and mortality (where **Cases** denote deceased women).

Before concluding that this somewhat unintuitive inverse relationship between smoking and mortality may reflect a real association in the population being studied, we need to consider the possibility that it may be due to the influence of other characteristics of women in the sample. The standard approach for controlling for the influence of additional categorical predictors in contingency tables is via a *stratified* analysis, where a relationship of interest is examined in subgroups defined by a additional variable (or variables).

Table 3.10 presents the same analysis stratified by a three-level categorical variable **agegrp** representing three categories of participant age (as ascertained in the original survey). The age-specific odds ratios and associated

Table 3.10. Twenty-year Vital Status by Smoking Behavior, Stratified by Age

```
. cs vstatus smoker [freq = nn], or by(agegrp)

         agegrp |       OR      [95% Conf. Interval]    M-H Weight
----------------+-------------------------------------------------
        18-44 |   1.776666    .8727834    3.615113    5.568471 (Cornfield)
        45-64 |   1.320359    .8728567    1.997089   19.55856 (Cornfield)
          64+ |   1.018182    .4240727    2.43359     4.772727 (Cornfield)
----------------+-------------------------------------------------
        Crude |    .6848366   .5354784    .8758683
   M-H combined |  1.357106    .9710409   1.896662
----------------------------------------------------------------
Test of homogeneity (M-H)      chi2(2) =     0.945  Pr>chi2 = 0.6234

                 Test that combined OR = 1:
                       Mantel--Haenszel chi2(1) =       3.24
                                    Pr>chi2 =        0.0719
```

95% confidence intervals indicate a positive (but not statistically significant) association between smoking and vital status in two of the three age groups. The crude odds ratio reproduces the result obtained in Table 3.9, while the age-adjusted (M-H combined, or *Mantel–Haenszel*) estimate is computed via a weighted average of the the age-specific estimates, where the stratum-specific weights are given in the right table margin (M-H Weight). Because this estimate is based on separate estimates made in each age stratum, the weighted average adjusts for the influence of age.

Comparison of the crude estimate with the adjusted estimate reveals that adjusting for age reverses the direction (and alters the significance) of the unadjusted result. Considering that none of the stratum-specific estimates indicate reduced risk associated with smoking, the crude estimate is surprising. This seemingly paradoxical result is often referred to as *Simpson's paradox*. To aid in further interpretation, Table 3.10 also includes results from two hypothesis tests of properties of the stratum-specific and combined odds ratios. The *test of homogeneity* addresses the null hypothesis that the three age-specific odds ratios are identical. Rejection of this hypothesis would provide evidence that the stratum-specific odds ratios differ, and may indicate a differential effect of smoking on mortality across different age groups. This phenomenon is also known as *interaction* or *effect modification*. In this case, the results indicate that the data do not support rejecting the null hypothesis in favor of the alternative hypothesis of differing age-specific odds ratios. We conclude that there is no strong evidence of interaction and that the age-specific odds ratios are similar.

The second test result presented in Table 3.10 addresses the null hypothesis that the true age-adjusted ("combined") odds ratio for the association between vital status and smoking is different than one. This hypothesis is meaningful if we have already failed to reject the hypothesis of homogeneity. In this case, we have already concluded that we do not have strong evidence that the age-specific odds ratios differ, and the results of the test for an age-adjusted association indicate failure to reject the null hypothesis at the 5% significance level. We conclude that the observed unadjusted negative association between vital status and smoking is at least partially explained by age adjustment. In fact, adjusting for age results in a positive association between smoking and vital status, that is more in accordance with our expectations that smokers may experience more health problems.

The results of the Whickham example are an instance of a more general phenomenon in observational studies known as *confounding*. In the example, the seemingly paradoxical finding of a positive association (albeit not statistically significant) after adjustment for age can be explained by differences between age groups in the proportion of women who were smokers (women in the intermediate age group were more likely to smoke than women in the other groups), and the fact that mortality was much higher in the older women. Of course, other measured or unmeasured factors may also influence the relationship between smoking and vital status. A complete analysis would consider these. Also, it would be a good idea to consider alternate measures of age and smoking if available (e.g. treating them as continuous variables in a regression model). The phenomena of confounding and interaction will be discussed extensively in the regression context in the remaining chapters of the book.

3.5 Basic Methods for Survival Analysis

In the previous section we considered binary outcomes – that is, whether or not an event has occurred. Survival data represents an extension in which we take into account the time until the event occurs – or until the end of follow-up, if the event has not yet occurred at that point. These more complex outcomes are studied using techniques collectively known as *survival analysis*. The term reflects the origin of these methods in demographic studies of life expectancy.

3.5.1 Right Censoring

To illustrate the special characteristics of survival data, we consider a study of 6-mercaptopurine (6-MP) as maintenance therapy for children in remission from acute lymphoblastic leukemia (ALL) (Freireich *et al.*, 1963). Forty-two patients achieved remission from induction therapy and were then randomized in equal numbers to 6-MP or placebo. The survival time studied was from randomization until relapse. At the time of the analysis, all 21 patients in the placebo group had relapsed, whereas only 9 of 21 patients in the 6-MP group had.

One crucial characteristic of these survival times is that for the 12 patients in the 6-MP group who remained in remission at the time of the analysis, the exact time to relapse was unobserved; it was only known to exceed the follow-up time. For example, one patient had only been under observation for six weeks, so we only know that the relapse time is longer than that. Such a survival time is said to be *right-censored* – "right" because on a graph the relapse time would lie somewhere to the right of the censoring time of six weeks.

> Definition: A survival time is said to be *right-censored* at time t if it is only known to be greater than t.

Table 3.11 displays follow-up times in the leukemia study. Asterisks mark the right-censored remission times.

Table 3.11. Weeks in Remission Among Leukemia Patients

```
Placebo:  1,1,2,2,3,4,4,5,5,8,8,8,8,11,11,12,
          12,15,17 22,23

6-MP:     6,6,6,6*,7,9*,10,10*,11*,13,16,17*,
          19*,20*,22,23,25*,32*,32*,34*,35*
```

Because of the censoring, we could not validly estimate the effects of 6-MP on time to relapse simply by comparing average follow-up times in the two groups (say, with a t-test). This simple approach would not work because the right-censored follow-up times in the 6-MP group are shorter, possibly much shorter, than the actual unobserved times to relapse for these patients. Furthermore, five of the right-censored values in the 6-MP group exceed the largest follow-up time in the placebo group; to ignore this would be throwing away valuable evidence for the effectiveness of the treatment. Survival analysis makes it possible to analyze right-censored data like these without bias or losing information contained in the length of the follow-up times.

3.5.2 Kaplan–Meier Estimator of the Survival Function

Suppose we would like to describe the probability of remaining in remission during each of the first 10 weeks of the leukemia study. This probability is called the *survival function*.

> Definition: The *survival function* at time t, denoted $S(t)$, is the probability of being event-free at t; equivalently, the probability that the survival time is greater than t.

We will first show how the survival function can be estimated for the 21 placebo patients. Because there is no right-censoring in the placebo group, we could simply estimate the survival function by the sample proportion in remission for each week. However, we will use a more complicated method because it accommodates right-censored data. This method depends on writing the survival function in any given week as a chain of conditional probabilities.

In Table 3.12 the placebo data are summarized by consecutive one-week intervals. The number of subjects who remain both in remission and in follow-up at the start of the week is given in the second column. The third and fourth columns list the numbers who relapse and who are censored during the week, respectively. Since none are censored, the number in follow-up is reduced only during weeks when a patient relapses. From the table, we see that in the first

Table 3.12. Follow-Up Table for Placebo Patients in the Leukemia Study

Week of follow-up	No. followed	No. relapsed	No. censored	Conditional prob. of remission	Survival function
1	21	2	0	$19/21 = 0.91$	0.91
2	19	2	0	$17/19 = 0.90$	$0.90 \times 0.91 = 0.81$
3	17	1	0	$16/17 = 0.94$	$0.94 \times 0.81 = 0.76$
4	16	2	0	$14/16 = 0.88$	$0.88 \times 0.76 = 0.67$
5	14	2	0	$12/14 = 0.86$	$0.86 \times 0.67 = 0.57$
6	12	0	0	$12/12 = 1.00$	$1.00 \times 0.57 = 0.57$
7	12	0	0	$12/12 = 1.00$	$1.00 \times 0.57 = 0.57$
8	12	4	0	$8/12 = 0.67$	$0.67 \times 0.57 = 0.38$
9	8	0	0	$8/8 = 1.00$	$1.00 \times 0.38 = 0.38$
10	8	0	0	$8/8 = 1.00$	$1.00 \times 0.38 = 0.38$

week, 19 of 21 patients remained in remission, so a natural estimate of the probability of being in remission in the first week is $19/21 = 0.91$. In the second week, 2 of the 19 placebo patients still in remission in the first week relapsed, and the remaining 17 remained in remission. Thus the probability of not relapsing in the second week, conditional on not having relapsed in the first, is estimated by $17/19 = 0.90$. It follows that the overall probability of remaining in remission in the second week is estimated by $19/21 \times 17/19 = 17/21 = 0.81$. Likewise, the probability of remaining in remission in the third week is estimated by $19/21 \times 17/19 \times 16/17 = 16/21 = 0.76$. In this case where there is no censoring, our chain of conditional probabilities reduces to the overall sample proportion in remission at the end of every week. You can easily verify that after ten weeks, the survival function estimate given by the chain of conditional probabilities is equal to the sample proportion still in remission.

Now we show how the survival function estimate based on the chain of conditional probabilities accommodates the censoring in the 6-MP group, as shown in Table 3.13. The problem we have to address is that two 6-MP subjects are censored prior to week 10. Since it is unknown whether they would have relapsed before the end of that week, we can no longer estimate the survival function at week 10 by the sample proportion still in remission at that point.

Table 3.13. Follow-Up Table for 6-MP Patients in the Leukemia Study

Week of follow-up	No. followed	No. relapsed	No. censored	Condition. prob. of remission	Survival function
1	21	0	0	$21/21 = 1.00$	1.00
2	21	0	0	$21/21 = 1.00$	$1.00 \times 1.00 = 1.00$
3	21	0	0	$21/21 = 1.00$	$1.00 \times 1.00 = 1.00$
4	21	0	0	$21/21 = 1.00$	$1.00 \times 1.00 = 1.00$
5	21	0	0	$21/21 = 1.00$	$1.00 \times 1.00 = 1.00$
6	21	3	1	$18/21 = 0.86$	$0.86 \times 1.00 = 0.86$
7	17	1	0	$16/17 = 0.94$	$0.94 \times 0.86 = 0.81$
8	16	0	0	$16/16 = 1.00$	$1.00 \times 0.81 = 0.81$
9	16	0	0	$16/16 = 1.00$	$1.00 \times 0.81 = 0.81$
10	16	0	1	$16/16 = 1.00$	$1.00 \times 0.81 = 0.81$

The rows of Table 3.13 for weeks 6 and 7 show how the method works with right-censored data. In week 6, three patients are observed to relapse, and one is censored (by assumption at the end of the week). Thus the probability of remaining in remission in week 6, conditional on having remained in remission in week 5, is $18/21 = 0.86$. Then we estimate the probability of remaining in remission in week 7, conditional on having remained in remission in week 6, as $16/17$: in short, the patient censored during week 6 has disappeared from the denominator, and does not contribute to the calculations for any subsequent week. Using this method for dealing with the censored observations, the conditional probabilities can still be estimated. As a result, we obtain a valid estimate of the probability of remaining in remission at the end of week 10, even though it is unknown whether the two censored patients remained in remission at that time.

In essence we have estimated the survival functions in the placebo and 6-MP groups using the well-known Kaplan–Meier estimator to deal with right censoring. In this example, the follow-up times have been grouped into weeks, but the method also applies to cases where they are observed more exactly. In Sect. 7.5.4 we examine the important assumption of *independent censoring* which underlies these procedures.

3.5.3 Interpretation of Kaplan–Meier Curves

Plots of the Kaplan–Meier estimates of $S(t)$ for the 6-MP and placebo groups in the leukemia study are shown in Fig. 3.2. Note that the curves drop at observed relapse times and are flat in the intervening periods. As a result, we can infer periods of high risk, when the survival curve descends rapidly, as well as periods of lower risk, when it remains relatively flat. In particular, placebo patients appear to be at high risk of relapse in the first five weeks.

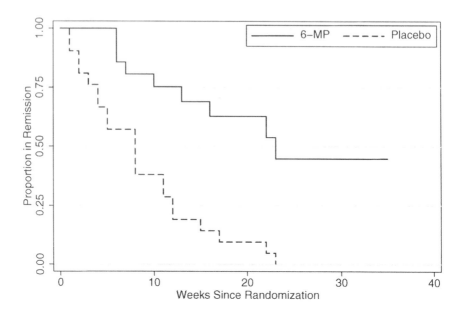

Fig. 3.2. Survival Curves by Treatment for Leukemia Patients

In addition, the estimated survival function for the 6-MP group is above the placebo curve over the entire follow-up period, giving evidence for higher probability of remaining in remission, or equivalently longer times in remission and lower risk of relapse in patients treated with 6-MP. In Sect. 3.5.6 below we show how to test the null hypothesis that the survival functions are the same in the two groups.

3.5.4 Median Survival

The Kaplan–Meier results may also be used to obtain estimates of the median survival time, defined as the time at which half the relevant population has experienced the outcome event. In the absence of censoring, with every survival time observed exactly, the median survival time could be simply estimated by the sample median of survival times: that is, the earliest time at which half the study participants have experienced the event. From Table 3.12 we can see that median time to relapse is eight weeks in the placebo group – the first week in which at least half the sample (12/21) have relapsed.

In the presence of censoring, however, we need to use the Kaplan–Meier estimate $\hat{S}(t)$ to estimate the median. In this case, the median survival time is estimated by the earliest time at which the Kaplan–Meier curve dips below 0.50. In the leukemia example, Fig. 3.2 shows that estimated median time to relapse is 23 weeks for 6-MP group, as compared to eight weeks for placebo – more evidence for the effectiveness of 6-MP as maintenance therapy for ALL.

By extension, other quantiles of the distribution of survival times can be obtained from the Kaplan–Meier estimate $\hat{S}(t)$. The pth quantile is estimated as the earliest time at which the Kaplan–Meier curve drops below $1 - p$. For instance, the lower quartile (i.e., the 0.25 quantile) is the earliest time at which the curve drops below $1 - 0.25 = 0.75$. The lower quartiles for the 6-MP and placebo groups are 13 and 4 weeks, respectively. However, a limitation of the Kaplan–Meier estimate is that when the curve does not reach $1 - p$, the pth percentile cannot be estimated. For example, Fig. 3.2 makes it clear that for the 6-MP group, quantiles of the distribution of remission times larger than the 0.6th cannot be estimated using the Kaplan–Meier method.

Note that while we can estimate the median and other quantiles of the distribution of survival times using the Kaplan–Meier results, we are unable to estimate the mean of the distribution in the typical case, as in the 6-MP group, where the longest follow-up time is censored (Problem 3.7).

A final note: graphs are useful for giving overall impressions of the survival function, but it can difficult to read quantities from them (e.g., median survival time or $\hat{S}(t)$ for some particular t). To obtain precise values, the results in Tables 3.12 and 3.13 can be printed in Stata using the sts list and stsci commands.

3.5.5 Cumulative Incidence Function

Another useful summary of survival data is the probability of having experienced the outcome event by time t. In terms of our leukemia example, this would mean estimating the probability of having relapsed by the end of each week of the study.

> <u>Definition</u>: The *cumulative incidence function* at time t, denoted $F(t)$, is the probability that the event has occurred time by t, or equivalently, the probability that the survival time is less than or equal to t. Note that $F(t) = 1 - S(t)$.

The cumulative incidence function is estimated by the complement of the Kaplan–Meier estimate of the survival function: that is, $\hat{F}(t) = 1 - \hat{S}(t)$. If t has the same value τ for all study participants, then $F(\tau)$ is interpretable as the outcome risk discussed in Sect. 3.4 on contingency table methods for binary outcomes. The cumulative incidence plots shown in Fig. 3.3 are also easily obtained in Stata.

Note that parametric methods can also be used to estimate survival distributions, as well quantities that are not immediately available from the Kaplan–Meier approach (e.g., the mean and specified quantiles). However, because they rest on explicit assumptions about the form of these distributions, they are somewhat less robust than the methods presented here. For

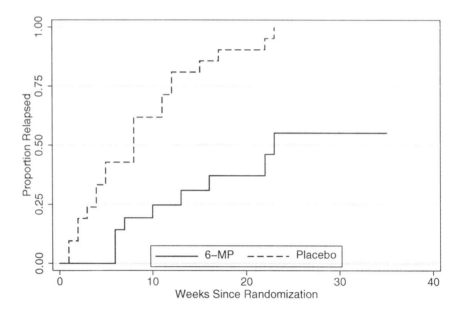

Fig. 3.3. Cumulative Incidence Curves by Treatment for Leukemia Patients

example, the mean can be poorly estimated in situations where a large proportion of the data are censored, with the result that the right tail of the survival function is only "known" by extrapolation.

3.5.6 Comparing Groups Using the Logrank Test

The Kaplan–Meier estimator provides an interpretable description of the survival experience of two treatment groups in the study of 6-MP as maintenance therapy for ALL. With those descriptions in hand, how do we go on to compare differences in relapse between the treatments?

The primary tool for the comparison of the survival experience of two or more groups is the *logrank test*. The null hypothesis for this test is that the survival distributions being compared are equal at all follow-up times. In the leukemia example, this implies that the population survival curves for 6-MP and placebo coincide. The alternative hypothesis is that the two survival curves differ at one or more points in time. Like the Kaplan–Meier estimator, the logrank test accommodates right-censoring. It works by comparing observed numbers of events in each group to the number expected if the survival functions were the same. The comparison accounts for differences in length of follow-up in calculating the expected numbers of events. Results are shown in Table 3.14.

Table 3.14. Logrank Test for Leukemia Example

```
Log-rank test for equality of survival functions
-------------------------------------------------
           |   Events          Events
group      |  observed        expected
-----------+-------------------------
6 MP       |      9            19.25
Placebo    |     21            10.75
-----------+-------------------------
Total      |     30            30.00

              chi2(1) =        16.79
              Pr>chi2 =        0.0000
```

There are a total of 30 events in the sample, 21 in the placebo group and 9 in the 6-MP group. The column labeled `Events expected` gives the expected number of events in the two groups under the null hypothesis of equal survival functions. In the leukemia data, average follow-up was considerably shorter in the placebo group and hence fewer events would be expected in that group. Clearly there were many more events than expected among placebo participants, and many fewer than expected in the 6-MP group. The resulting χ^2 statistic of 16.8 is statistically significant ($P < 0.00005$), in accord with our informal earlier impression that 6-MP is effective maintenance therapy for patients with ALL.

The logrank test is easily generalized to the comparison of more than two groups. The logrank test statistic for $K > 2$ groups follows an approximate χ^2 distribution with $K - 1$ degrees of freedom. In this more general case, the null hypothesis is

$$H_0 : S_1(t) = \ldots = S_K(t) \quad \text{for all } t \tag{3.16}$$

where $S_k(t)$ is the survival function for the kth group at time t. In analogy to the F-test discussed in Sect. 4.3.3, the alternative hypothesis is that some or all of the survival curves differ at one or more points in time.

When the null hypothesis is rejected, visual inspection of the Kaplan–Meier plots can help to determine where the important differences arise. Another common procedure for understanding group differences is to conduct pairwise logrank tests. This requires cautious interpretation; see Sect. 4.3.4 for approaches to handling potential difficulties with multiple comparisons.

Like some other nonparametric methods reviewed earlier in this chapter, and as its name implies, the logrank test only uses information about the *ranks* of the survival times rather than their actual values. The semi-parametric Cox proportional hazards model covered in Chapter 7 also works this way. In every instance, the nonparametric approach reduces the need for making restrictive

and sometimes hard-to-verify assumptions, with a view toward making estimates more robust.

There is an extensive literature on testing differences in survival between
groups. These tests have varying levels of similarity to the logrank test. The
most popular are extensions of the Wilcoxon test for censored data; these
tests can be viewed as a weighted versions of the logrank test. Such weighting
can make sense, for example, if early events are judged to be particularly
important.

Chapter 7 covers censoring and other types of missing data in greater
depth, and also presents more comprehensive methods of analysis for survival
data, including the multipredictor Cox proportional hazards regression model.

3.6 Bootstrap Confidence Intervals

Bootstrapping is a widely applicable method for obtaining standard errors and
confidence intervals in cases where approximate methods for computing valid
confidence intervals have been developed but not conveniently implemented
in statistical packages; other situations where development of such methods
has turned out to be intractable; and data sets where the assumptions underlying the established methods are badly enough violated that the resulting
confidence intervals would be unreliable.

In general, standard errors and confidence intervals reflect the sampling
distribution of statistics of interest, such as regression coefficient estimates:
that is, their relative frequency if we repeatedly drew independent samples of
the same size from the source population, and recalculated the statistics in
each new sample. In standard problems such as linear regression, the sampling
distribution of the regression coefficient estimates is well known on theoretical
grounds, provided the data meet underlying assumptions.

Bootstrap procedures approximate the sampling distribution of statistics
of interest by a *resampling* procedure. Specifically, the actual sample is treated
as if it were the source population, and bootstrap samples are repeatedly
drawn from it. Bootstrap samples of the same size as the actual sample – a
key determinant of precision – are obtained by resampling *with replacement*,
so that in a given bootstrap sample some observations appear more than once,
some once, and some not at all. We use the sample to represent the population
and hence resampling from the actual data mimics drawing repeated samples
from the source population.

Then, from each of a large number of bootstrap samples, the statistics of
interest are computed. For example, if our focus was on the difference between
the coefficient estimates for a predictor of interest before and after adjustment
for a covariate, the two models would be estimated in each bootstrap sample,
and the difference between the two coefficient estimates tabulated across samples. The result would be the bootstrap distribution of the difference, which
can in turn be regarded as an estimate of its actual sampling distribution.

Confidence intervals for the statistic of interest would then be computed from the bootstrap distribution. Stata calculates bootstrap confidence intervals using three procedures:

- *Normal approximation* If the bootstrap distribution of the statistic of interest is reasonably normal, it may be enough to compute its standard deviation, then compute a conventional confidence interval centered on the observed statistic, simply substituting the bootstrap SD for the usual model-based standard error of the statistic. The bootstrap SD is a relatively stable estimate of the standard error, since it is based on the complete set of bootstrap samples, so a relatively small number of bootstrap samples may suffice. However, we often resort to the bootstrap precisely because the sampling distribution of the statistic of interest is unlikely to be normal, particularly in the tails. Thus this method is less reliable for constructing confidence intervals than for estimating the standard error of the statistic.
- *Percentile Method* The confidence interval for the statistic of interest is constructed from the relevant quantiles of the bootstrap distribution. Because the extreme percentiles of a sample are very noisy estimates of the corresponding percentiles of a population distribution, a much larger number of bootstrap samples is required. If 1,000 samples were used, then a 95% CI for the statistic of interest would span the 25th to 975th largest bootstrap estimates.
- *Bias-Corrected Percentile Method* The percentile-based confidence interval is shifted to account for bias, as evidenced by a difference between the observed statistic and the median of the bootstrap estimates. Again, a relatively large number of bootstrap samples is required.

Table 3.15 shows Stata output for the simple linear regression model for SBP shown earlier in Table 3.4, now with a bootstrap confidence interval. In this instance, all three bootstrap results are fairly consistent with the parametric 95% CI (0.73–0.81 mmHg). See Sects. 4.5.3, 6.5.1, 7.5.1, and 8.6.1 for other examples where bootstrap confidence intervals are computed.

3.7 Interpretation of Negative Findings

Confidence intervals obtained either by standard parametric methods or by the bootstrap play a particularly important role when the data do not enable us to reject a null hypothesis of interest. It is easy to overstate such negative findings. Recall that $P > 0.05$ does not prove the null hypothesis; it only indicates that the observed result could have arisen by chance, not that it necessarily did. A negative result worth discussing is best interpreted in terms

Table 3.15. Bootstrap Confidence Interval for Association of Age With SBP

```
. reg SBP age

      Source |       SS       df       MS              Number of obs =     276
-------------+------------------------------           F(  1,   274) =    5.58
       Model | 2179.70702       1  2179.70702          Prob > F      =  0.0188
    Residual | 106991.347     274  390.47937           R-squared     =  0.0200
-------------+------------------------------           Adj R-squared =  0.0164
       Total | 109171.054     275  396.985652          Root MSE      =  19.761


------------------------------------------------------------------------------
         sbp |      Coef.   Std. Err.      t    P>|t|     [95% Conf. Interval]
-------------+----------------------------------------------------------------
         age |   .4405286    .186455     2.36   0.019     .0734621    .8075952
       _cons |    105.713   12.40238     8.52   0.000      81.2969     130.129
------------------------------------------------------------------------------

. bootstrap '"reg SBP age"' _b, reps(1000)

command:     reg SBP age
statistics:  b_age      = _b[age]

Bootstrap statistics                      Number of obs     =        276
                                          Replications      =       1000


------------------------------------------------------------------------------
Variable     | Reps  Observed      Bias  Std. Err. [95% Conf. Interval]
-------------+----------------------------------------------------------------
       b_age | 1000 .4405287 -.0078003 .1744795    .0981403     .782917   (N)
             |                                      .0655767    .7631486   (P)
             |                                      .0840077    .7690148  (BC)
------------------------------------------------------------------------------

Note:  N  = normal
       P  = percentile
       BC = bias-corrected
```

of the point estimate and confidence interval. In the following example, we can distinguish four possible cases, in increasing order of the strength of the negative finding. Suppose that a 20% reduction risk of recurrent heart attacks would justify the risks and costs of a possible new treatment, but that a risk reduction of only 5% would not meet this standard. The four cases are–

- The estimated risk reduction was large enough to be substantively important, but the confidence interval spanned the null value and was thus too wide to provide strong evidence for effectiveness. Example: treatment reduced recurrence risk an estimated 20% (95% CI –1% to 37%). In this case we might conclude that the study gives inconclusive evidence *for* the potential importance of the treatment; but it would be also important to note that the confidence interval includes effects too small to be worthwhile.

- The estimated risk reduction was too small to be important, but the confidence interval extended to values that could be important. Example: treatment reduced recurrence risk an estimated 5% (95% CI –15% to 22%). In this case the point estimate provides little support for the importance of the treatment, but the confidence interval does not clearly rule out a potentially important effect.

- The estimated risk reduction was too small to be important, and while the confidence interval did not include the null (i.e., $P < 0.05$), it did exclude values that could be important. Example: treatment reduced recurrence risk an estimated 3% (95% CI: 1% to 5%). In this case, we can definitively say that the treatment does not have a clinically important benefit, even though we can also rule out no effect.

- The estimated risk reduction was too small to be important, and the confidence interval both included the null and excluded values that could be important. Example: treatment reduced recurrence risk an estimated 1% (95% CI –2% to 4%). Again, we can definitively say that the treatment does not have a clinically important benefit.

This approach using the point estimate and confidence interval is preferable to interpretations based on *ex post facto* power calculations, which are driven by assumptions about the true effect size, and often inappropriately based on treating the observed effect size as if it were the true population value (Hoenig and Heisey, 2001). A variant of this approach is to suggest that with a larger sample, the observed effect would have been statistically significant. But of course the confidence interval for most negative findings tells us that the true effect size may well be nil or worse, which a larger sample might also firmly establish. In contrast to these problematic interpretations, the point estimate and confidence interval can together be used to summarize what the data at hand have to tell us about the strength of the association and the precision of our information about it.

3.8 Further Notes and References

Among the best introductory statistics books are Freedman *et al.* (1991), Devore and Peck (1986), and Pagano and Gavreau (1993). Consult these for more complete coverage of basic statistical inference, analysis of variance, and linear regression. Good references on methods for the analysis of contingency tables include Fleiss *et al.* (2003) and Jewell (2004). Two applied survival analysis texts with a biomedical orientation are Miller *et al.* (1981) and Marubini and Valsecchi (1995). Finally, for a review of bootstrap methods, see Efron and Tibshirani (1986, 1993).

3.9 Problems

Problem 3.1. An alternative to OLS is least absolute deviation (LAD) regression, in which the regression line is selected to minimize the sum of the absolute vertical differences (rather than squared differences) between the line and the data. Explain how this might reduce sensitivity to outliers.

Problem 3.2. To create a new age variable age10 in units of ten years, we would divide the original variable age (in years) by ten, so that a woman of age 67 would have age10 = 6.7. Similarly, the standard deviation of age10 is changed by the same factor: that is, the SD of age is 6.38, so the SD of age10 is 0.638. Suppose we want to estimate the effect of age in SD units, as is commonly done. How do we compute the new variable and what is *its* SD?

Problem 3.3. Using (3.12) and a statistical analysis program, demonstrate with your own data that the slope coefficient in a univariate linear model with continuous predictor and outcome is a rescaled transformation of the sample correlation between predictor and outcome.

Problem 3.4. The correlation coefficient is a measure of *linear* association. Suppose x takes on values evenly over the range from –10 to 10, and that $E[y|x] = x^2$. In this case the correlation of x and y is zero, even though there is clearly a systematic relationship. What does this suggest about the need to test model assumptions? Using a statistical package, generate a random sample of 100 values of x uniformly distributed on $[-10, 10]$, compute $E[y|x]$ for each value of x, add randomly generated standard normal errors to get the 100 values of y, and check the sample correlation of x and y.

Problem 3.5. Verify the estimates for the excess risk, relative risk, and odds ratio for the HIV example presented in Table 3.6.

Problem 3.6. The data presented below are from a case-control study of esophageal cancer. (The study and data are described in more detail in Sect. 6.3.)

```
. tabulate case ditob

      Case |
    status |
   (1=case, |          tobacco
  0=control) | 0-9 g/day  10+ g/day |    Total
  -----------+----------------------+----------
         0 |     255        520 |      775
         1 |       9        191 |      200
  -----------+----------------------+----------
     Total |     264        711 |      975
```

The rows (labeled according to Case status) represent 200 cancer cases and 775 cancer-free controls selected from the same population as the cases. The columns represent a binary indicator of reported consumption of more than ten grams of tobacco per day.

Compute the odds ratio comparing the risk of cancer in individuals who report consuming more than ten grams of tobacco per day with the the corresponding risk in the group reporting less or no consumption. Next, compute the odds ratio comparing the proportion of individuals reporting higher levels of consumption among cases with that among the controls. Comment.

Problem 3.7. Suppose we could estimate the value of the survival function $S(t)$ for every possible survival time from $t = 0$ onward. Clearly $S(t) \rightarrow 0$ as t becomes large. It can be shown that the mean survival time is equal to the area under this "complete" survival curve. Why are we unable to estimate mean survival from the Kaplan–Meier result when the largest follow-up time is censored? To gain insight, contrast the survival curves for the 6-MP and placebo groups in Fig. 3.2.

Problem 3.8. In the leukemia study, the probability of being relapse-free at 20 weeks, conditional on being relapse-free at 10 weeks, can be estimated by the Kaplan–Meier estimate for 20 weeks, divided by the corresponding estimate for 10 weeks. In the placebo group, those estimates are 0.38 and 0.10 respectively. Verify that the estimated conditional probability of remission at week 20, conditional on being in remission at week 10, is 0.25. In the 6-MP group, estimated probabilities of remaining in remission are 0.81, 0.63, and 0.45 at 10, 20, and 30 weeks, respectively. Use these values to estimate the probabilities of remaining in remission at 20 and 30 weeks, conditional on being in remission at 10 weeks.

3.10 Learning Objectives

1. Be familiar with the t-test (including versions for paired and unequal-variance data), one-way ANOVA, the correlation coefficient r, and some nonparametric alternatives.
2. Describe the assumptions and mechanics of the simple linear model for continuous outcomes, and interpret the results.
3. Define the basic measures of association (i.e., excess risk, relative risk, and odds ratio) for binary outcomes.
4. Be familiar with standard contingency table approaches to evaluating associations between binary outcomes and categorical predictors, including the χ^2 test and the Mantel–Haenszel approach to estimating odds ratios adjusted for the confounding influence of additional predictors.
5. Define right-censoring.
6. Interpret Kaplan–Meier survival and cumulative incidence curves.
7. Calculate median survival from an estimated survival curve.
8. Interpret the results of a logrank test.

4

Linear Regression

Post-menopausal women who exercise less tend to have lower bone mineral density (BMD), putting them at increased risk for fractures. But they also tend to be older, frailer, and heavier, which may explain the association between exercise and BMD. People whose diet is high in fat on average have higher low-density lipoprotein (LDL) cholesterol, a risk factor for coronary heart disease (CHD). But they are also more likely to smoke and be overweight, factors which are also strongly associated with CHD risk. Increasing body mass index (BMI) predicts higher levels of hemoglobin Hb_{a1c}, a marker for poor control of glucose levels; however, older age and ethnic background also predict higher Hb_{a1c}.

These are all examples of potentially complex relationships in observational data where a continuous outcome of interest, such as BMD, SBP, and Hb_{a1c}, is related to a risk factor in analyses that do not take account of other factors. But in each case the risk factor of interest is associated with a number of other factors, or potential *confounders*, which also predict the outcome. So the simple association we observe between the factor of interest and the outcome may be explained by the other factors.

Similarly, in experiments, including clinical trials, factors other than treatment may need to be taken into account. If the randomization is properly implemented, treatment assignment is on average not associated with any prognostic variable, so confounding is usually not an issue. However, in stratified and other complex study designs, multipredictor analysis is used to ensure that confidence intervals, hypothesis tests, and P-values are valid. For example, it is now standard practice to account for clinical center in the analysis of multi-site clinical trials, often using the random effects methodology to be introduced in Chapter 8. And with continuous outcomes, stratifying on a strong predictor in both design and analysis can account for a substantial proportion of outcome variability, increasing the efficiency of the study. Multipredictor analysis may also be used when baseline differences are apparent between the randomized groups, to account for potential confounding of treatment assignment.

Another way the predictor–outcome relationship can depend on other factors is that an association may not be the same in all parts of the population. For example, the association of lipoprotein(a) levels with risk of CHD events appears to vary by ethnicity. Hormone therapy has a smaller beneficial effect on LDL levels among post-menopausal women who are also taking statins, and its effect on BMD may be greater in younger post-menopausal women. These are examples of *interaction*, where the association of a factor of primary interest with a continuous outcome is modified by another factor.

The problem of sorting out complex relationships is not restricted to continuous outcomes; the same issues arise with the binary outcomes covered in Chapter 6, survival times in Chapter 7, and repeated measures in Chapter 8. A general statistical approach to these problems is needed.

The topic of this chapter is the multipredictor linear regression model, a flexible and widely used tool for assessing the joint relationships of multiple predictors with a continuous outcome variable. We begin by illustrating some basic ideas in a simple example (Sect. 4.1). Then in Sect. 4.2 we present the assumptions of the multipredictor linear regression model and show how the simple linear model reviewed in Chapter 3 is extended to accommodate multiple predictors. Sect. 4.3 shows how categorical predictors with multiple levels are coded and interpreted. Sect. 4.4 describes how multipredictor regression models deal with confounding; in particular Sect. 4.4.1 uses a *counterfactual* view of *causal effects* to show how and under what conditions multipredictor regression models might be used to estimate them. These themes recur in Sects. 4.5 and 4.6 on mediation and interaction, respectively. Sect. 4.7 introduces some simple methods for assessing the fit of the model to the data and how well the data conform to the underlying assumptions of the model. In Chapter 5 we discuss the difficult problem of which variables and how many to include in a multipredictor model.

4.1 Example: Exercise and Glucose

Glucose levels above 125 mg/dL are diagnostic of diabetes, while levels in the range from 100 to 125 mg/dL signal increased risk of progressing to this serious and increasingly widespread condition. So it is of interest to determine whether exercise, a modifiable lifestyle factor, would help people reduce their glucose levels and thus avoid diabetes.

To answer this question definitively would require a randomized clinical trial, a difficult and expensive undertaking. As a result, research questions like this are often initially looked at using observational data. But this is complicated by the fact that people who exercise differ in many ways from those who do not, and some of the other differences might explain any unadjusted association between exercise and glucose level.

Table 4.1 shows a simple linear model using a measure of exercise to predict baseline glucose levels among 2,032 participants without diabetes in the HERS

Table 4.1. Unadjusted Regression of Glucose on Exercise

```
. reg glucose exercise if diabetes == 0

      Source |       SS        df       MS              Number of obs =     2032
-------------+------------------------------           F(  1,  2030) =    14.97
       Model |  1412.50418      1  1412.50418           Prob > F      =   0.0001
    Residual |  191605.195   2030  94.3867954           R-squared     =   0.0073
-------------+------------------------------           Adj R-squared =   0.0068
       Total |  193017.699   2031  95.0357946           Root MSE      =   9.7153


     glucose |      Coef.   Std. Err.      t    P>|t|     [95% Conf. Interval]
-------------+----------------------------------------------------------------
    exercise |  -1.692789   .4375862    -3.87   0.000    -2.550954   -.8346243
       _cons |   97.36104   .2815138   345.85   0.000     96.80896    97.91313
```

clinical trial of hormone therapy (Hulley *et al.*, 1998). Women with diabetes are excluded because the research question is whether exercise might help to prevent progression to diabetes among women at risk, and because the causal determinants of glucose may be different in that group. Furthermore, glucose levels are far more variable among diabetics, a violation of the assumption of homoscedasticity, as we show in Sect. 4.7.3 below. The coefficient estimate (Coef.) for exercise shows that average baseline glucose levels were about 1.7 mg/dL lower among women who exercised at least three times a week than among women who exercised less. This difference is statistically significant ($t = -3.87, P < 0.0005$).

However, women who exercise are slightly younger, a little more likely to use alcohol, and in particular have lower average body mass index (BMI), all factors associated with glucose levels. This implies that the lower average glucose we observe among women who exercise could be due at least in part to differences in these other predictors. Under these conditions, it is important that our estimate of the difference in average glucose levels associated with exercise be "adjusted" for the effects of these potential confounders of the unadjusted association. Ideally, adjustment using a multipredictor regression model provides an estimate of the causal effect of exercise on average glucose levels, by *holding the other variables constant*. In Sect. 4.4 below, the rationale for estimation of causal effects using multipredictor regression models is explained in more detail.

From Table 4.2 we see that in a multiple regression model that also includes – that is, adjusts for – age, alcohol use (drinkany), and BMI, average glucose is estimated to be only about 1 mg/dL lower among women who exercise (95% CI 0.1–1.8, $P = 0.027$), holding the other three factors constant. The multipredictor model also shows that average glucose levels are about 0.7 mg/dL higher among alcohol users than among non-users. Average levels also increase by about 0.5 mg/dL per unit increase in BMI, and by 0.06 mg/dL for each additional year of age. Each of these associations is statistically significant after adjustment for the other predictors in the model. Furthermore, the

Table 4.2. Adjusted Regression of Glucose on Exercise

```
. reg glucose exercise age drinkany BMI if diabetes == 0;

      Source |       SS       df       MS              Number of obs =    2028
-------------+------------------------------           F(  4,  2023) =   39.22
       Model | 13828.8486       4  3457.21214          Prob > F      =  0.0000
    Residual | 178319.973    2023  88.1463042          R-squared     =  0.0720
-------------+------------------------------           Adj R-squared =  0.0701
       Total | 192148.822    2027  94.7946828          Root MSE      =  9.3886


     glucose |    Coef.   Std. Err.      t    P>|t|     [95% Conf. Interval]
-------------+----------------------------------------------------------------
    exercise | -.950441     .42873    -2.22   0.027    -1.791239   -.1096426
         age |  .0635495   .0313911     2.02   0.043     .0019872    .1251118
     drinkany |  .6802641   .4219569     1.61   0.107    -.1472513     1.50778
         BMI |   .489242   .0415528    11.77   0.000     .4077512    .5707328
       _cons |  78.96239   2.592844    30.45   0.000     73.87747    84.04732
```

association of each of the four predictors with glucose levels is adjusted for the effects of the other three, in the sense of taking account of its correlation with the other predictors and their adjusted associations with glucose levels. In summary, the multipredictor model for glucose levels shows that the unadjusted association between exercise and glucose is partly but not completely explained by BMI, age, and alcohol use, and that exercise remains a statistically significant predictor of glucose levels after adjustment for these three other factors – that is, when they are held constant by the multipredictor regression model.

Still, we have been careful to retain the language of association rather than cause and effect, and in Sect. 4.4 and Chapter 5 will suggest that adjustment for additional potential confounders would be needed before we could consider a causal interpretation of the result.

4.2 Multiple Linear Regression Model

Confounding thus motivates models in which the average value of the outcome is allowed to depend on multiple predictors instead of just one. Many basic elements of the multiple linear model carry over from the simple linear model, which was reviewed in Sect. 3.3. In Sects. 4.4.1–4.4.9 below, we show how this model is potentially suited to estimating causal relationships between predictors and outcomes.

4.2.1 Systematic Part of the Model

For the simple linear model with a single predictor, the regression line is defined by

$$E[y|x] = \text{average value of outcome } y \text{ given predictor value } x$$
$$= \beta_0 + \beta_1 x. \tag{4.1}$$

In the multiple regression model, this generalizes to

$$E[y|\mathbf{x}] = \beta_0 + \beta_1 x_1 + \beta_2 x_2 + \cdots + \beta_p x_p, \tag{4.2}$$

where $\mathbf{x}$ represents the collection of p predictors $x_1, x_2, \ldots x_p$ in the model, and $\beta_1, \beta_2, \ldots \beta_p$ are the corresponding regression coefficients.

The right-hand side of model (4.2) has a relatively simple form, a *linear combination* of the predictors and coefficients. Analogous linear combinations of predictors and coefficients, often referred to as the *linear predictor*, are used in all the other regression models covered in this book. Despite the simple form of (4.2), the multipredictor linear regression model is a flexible tool, and with the elaborations to be introduced later in this chapter, usually allows us to represent with considerable realism how the average value of the outcome varies systematically with the predictors. In Sect. 4.7, we will consider methods for examining the adequacy of this part of the model and for improving it.

Interpretation of Adjusted Regression Coefficients

In (4.2), the coefficient $\beta_j, j = 1, \cdots, p$ gives the change in $E[y|\mathbf{x}]$ for an increase of one unit in predictor x_j, holding other factors in the model constant; each of the estimates is adjusted for the effects of all the other predictors. As in the simple linear model, the intercept β_0 gives the value of $E[y|\mathbf{x}]$ when all the predictors are equal to zero; "centering" of the continuous predictors can make the intercept interpretable. If confounding has been persuasively ruled out, we may be willing to interpret the adjusted coefficient estimates as representing causal effects.

4.2.2 Random Part of the Model

As before, individual observations of the outcome y_i are modeled as varying by an error term ε_i about an average determined by their predictor values $\mathbf{x}_i$:

$$y_i = E[y_i|\mathbf{x}_i] + \varepsilon_i$$
$$= \beta_0 + \beta_1 x_{1i} + \beta_2 x_{2i} + \cdots + \beta_p x_{pi} + \varepsilon_i, \tag{4.3}$$

where x_{ji} is the value of predictor variable x_j for observation i. We again assume that $\varepsilon_i \sim$ i.i.d $\mathcal{N}(0, \sigma_\varepsilon^2)$; that is, ε is normally distributed with mean zero and the same standard deviation σ_ε at every value of $\mathbf{x}$, and that its values are statistically independent.

Fitted Values, Sums of Squares, and Variance Estimators

From (4.2) it is clear that the fitted values $\hat{y}_i$, defined for the simple linear model in Equation (3.4), now depend on all p predictors and the corresponding regression coefficient estimates, rather than just one predictor and two coefficients. The resulting sums of squares and variance estimators introduced in Sect. 3.3 are otherwise unchanged in the multipredictor model.

   In the glucose example, the residual standard deviation, shown as Root MSE, declines from 9.7 in the unadjusted model (Table 4.1) to 9.4 in the model adjusting for age, alcohol use, and BMI (Table 4.2).

Variance of Adjusted Regression Coefficients

Including multiple predictors does affect the variance of $\hat{\beta}_j$, which now depends on an additional factor r_j, the multiple correlation of x_j with the other predictors in the model. Specifically,

$$\hat{\mathrm{Var}}(\hat{\beta}_j) = \frac{s_{y|\mathbf{x}}^2}{(n-1)s_{x_j}^2(1-r_j^2)}. \tag{4.4}$$

where, as before, $s_{y|\mathbf{x}}^2$ is the residual variance of the outcome and $s_{x_j}^2$ is the variance of x_j; r_j is equivalent to $r = \sqrt{R^2}$ from a multiple linear model in which x_j is regressed on all the other predictors. The term $1/(1-r_j^2)$ is known as the *variance inflation factor*, since $\mathrm{Var}(\hat{\beta}_j)$ is increased to the extent that x_j is correlated with other predictors in the model.

   However, inclusion of other predictors, especially powerful ones, also tends to decrease $s_{y|\mathbf{x}}^2$, the residual or unexplained variance of the outcome. Thus the overall impact of including other predictors on $\mathrm{Var}(\hat{\beta}_j)$ depends on both the correlation of x_j with the other predictors and how much additional variability they explain. In the glucose example, the standard error of the coefficient estimate for exercise declines slightly, from 0.44 to 0.43, after adjustment for age, alcohol use, and BMI. This reflects the reduction in residual standard deviation previously described, as well as a variance inflation factor in the adjusted model of only 1.03.

t-**Tests and Confidence Intervals**

The t-tests of the null hypothesis H_0: $\beta_j = 0$ and confidence intervals for β_j carry over almost unchanged for each of the βs estimated by the model, only using (4.4) rather than (3.11) to compute the standard error of the regression coefficient, and comparing the t-statistic to a t-distribution with $n - (p+1)$ degrees of freedom (p is the number of predictors in the model, and an extra degree of freedom is used in estimation of the intercept β_0).

   However, there is a substantial difference in interpretation, since the results are now adjusted for other predictors. Thus in rejecting the null hypothesis

H_0: $\beta_j = 0$ we would be making the stronger claim that, in the population, x_j predicts y, holding the other factors in the model constant. Similarly, the confidence interval for β_j refers to the parameter which takes account of the other $p - 1$ predictors in the model.

We have just seen that $\text{Var}(\hat{\beta}_j)$ may not be increased by adjustment. However, in Sect. 4.4 we will see that including other predictors in order to control confounding commonly has the effect of attenuating the unadjusted estimate of the association of x_j with y. This reflects the fact that the population parameter being estimated in the adjusted model is often closer to zero than the parameter estimated in the unadjusted model, since some of the unadjusted association is explained by other predictors. If this is the case, then even if $\text{Var}(\hat{\beta}_j)$ is unchanged, it may be more difficult to reject H_0: $\beta_j = 0$ in the adjusted model. In the glucose example, the adjusted coefficient estimate for exercise is considerably smaller than the unadjusted estimate. As a result the t-statistic is reduced in magnitude from –3.87 to –2.22 – still statistically significant, but less highly so.

4.2.3 Generalization of R^2 and r

The coefficient of determination $R^2 = \text{MSS} / \text{TSS}$ retains its interpretation as the proportion of the total variability of the outcome that can be accounted for by the predictor variables. Under the model, the fitted values summarize all the information that the predictors supply about the outcome. Thus the multiple correlation coefficient $r = \sqrt{R^2}$ now represents the correlation between the outcome y and the fitted values $\hat{y}$. It is easy to confirm this identity by extracting the fitted values from a regression model and computing their correlation with the outcome (Problem 4.3). In the glucose example, R^2 increases from less than 1% in the unadjusted model to 7% after inclusion of age, alcohol use, and BMI, a substantial increase in relative if not absolute terms.

4.2.4 Standardized Regression Coefficients

In Sect. 3.3.9 we saw that the slope coefficient β_1 in a simple linear model is systematically related to the Pearson correlation coefficient (3.12); specifically, $r = \beta_1 \sigma_x / \sigma_y$, where σ_x and σ_y are the standard deviations of the predictor and outcome. Moreover, we pointed out that the scale-free correlation coefficient makes it easier to compare the strength of association between the outcome and various predictors across single-predictor models. In the context of a multipredictor model, *standardized regression coefficients* play this role. Obtained using the **beta** option to the **regress** command in Stata, the standardized regression coefficient $\hat{\beta}_j^s$ for predictor x_j is defined in analogy to (3.12) as

$$\hat{\beta}_j^s = \hat{\beta}_j \text{SD}(x_j)/\text{SD}(y), \tag{4.5}$$

where $SD(x_j)$ and $SD(y)$ are the sample standard deviations of predictor x_j and the outcome y. These standardized coefficient estimates are what would be obtained from the regression if the outcome and all the predictors were first rescaled to have standard deviation 1. Thus they give the change in standard deviation units in the average value of y per standard deviation increase in the predictor. Standardized coefficients make it easy to compare the strength of association of different continuous predictors with the outcome within the same model.

For binary predictors, however, the unstandardized regression coefficients may be more directly interpretable than the standardized estimates, since the unstandardized coefficients for such predictors simply estimate the differences in the average value of the outcome between the two groups defined by the predictor, holding the other predictors in the model constant.

4.3 Categorical Predictors

In Chapter 3 the simple regression model was introduced with a single continuous predictor. However, predictors in both simple and multipredictor regression models can be binary, categorical, or discrete numeric, as well as continuous numeric.

4.3.1 Binary Predictors

The exercise variable in the model for LDL levels shown in Table 4.1 is an example of a binary predictor. A good way to code such a variable is as an *indicator* or *dummy* variable, taking the value 1 for the group with the characteristic of interest, and 0 for the group without the characteristic. With this coding, the regression coefficient corresponding to this variable has a straightforward interpretation as the increase or decrease in average outcome levels in the group with the characteristic, with respect to the reference group.

To see this, consider the simple regression model for average glucose values:

$$E[\texttt{glucose}|x] = \beta_0 + \beta_1\texttt{exercise} \tag{4.6}$$

With the indicator coding of exercise ($1 =$ yes, $0 =$ no), the average value of glucose is $\beta_0 + \beta_1$ among women who do exercise, and β_0 among the rest. It follows directly that β_1 is the difference in average glucose levels between the two groups. This is consistent with our more general definition of β_j as the change in $E[y|\mathbf{x}]$ for a one-unit increase in x_j. Furthermore, the t-test of the null hypothesis $H_0\colon \beta_1 = 0$ is a test of whether the between-group difference in average glucose levels is statistically significant. In fact this unadjusted model is equivalent to a t-test comparing glucose levels in women who do and do not exercise. A final point: when coded this way, the average value of the exercise variable gives the proportion of women who exercise.

A commonly used alternative coding for binary variables is (1 = yes, 2 = no). With this coding, the coefficient β_1 retains its interpretation as the between-group difference in average glucose levels, but now among women who do not exercise as compared to those who do, a less intuitive way to think of the difference. Furthermore, with this coding the coefficient β_0 has no straightforward interpretation, and the average value of the binary variable is not equal to the proportion of the sample in either group. However, overall model fit, including fitted values of the outcome, standard errors, and P-values, are the same with either coding (Problem 4.1).

4.3.2 Multilevel Categorical Predictors

The 2,763 women in the HERS cohort also responded to a question about how physically active they considered themselves compared to other women their age. The five-level response, designated physact, ranged from "much less active" to "much more active," and was coded in order from 1 to 5. This is an example of an *ordinal* variable, as described in Chapter 2, with categories that are meaningfully ordered, but separated by increments that may not be accurately reflected in the numerical codes used to represent them. For example, responses "much less active" and "somewhat less active" may represent a larger difference in physical activity than "somewhat less active" and "about as active."

Multilevel categorical variables can also be *nominal*, in the sense that there is no intrinsic ordering in the categories. Examples include ethnicity, marital status, occupation, and geographic region. With nominal variables it is even clearer that the numeric codes often used to represent the variable in the database cannot be treated like the values of a numeric variable such as glucose.

Categories are usually set up to be mutually exclusive and exhaustive, so that every member of the population falls into one and only one category. In that case both ordinal and nominal categories define subgroups of the population.

Both types of categorical variables are easily accommodated in multi-predictor linear and other regression models, using indicator or dummy variables. As with binary variables, where two categories are represented in the model by a single indicator variable, categorical variables with $K \geq 2$ levels are represented by $K - 1$ indicators, one for each of level of the variable except a baseline or reference level. Suppose level 1 is chosen as the baseline level. Then for $k = 2, 3, \ldots, K$, indicator variable k has value 1 for observations belonging to the category k, and 0 for observations belonging to any of the other categories. Note that for $K = 2$ this also describes the binary case, in which the "no" response defines the baseline or reference group and the indicator variable takes on value 1 only for the "yes" group.

Stata automatically defines indicator variables using the xi: command prefix in conjunction with the i. variable prefix. By default it uses the level

Table 4.3. Coding of Indicators for a Multilevel Categorical Variable

physact	Indicator variables			
	_Iphysact_2	_Iphysact_3	_Iphysact_4	_Iphysact_5
Much less active	0	0	0	0
Somewhat less active	1	0	0	0
About as active	0	1	0	0
Somewhat more active	0	0	1	0
Much more active	0	0	0	1

with the lowest value as the reference group; for text variables this means using the first in alphabetic order. Following the Stata convention for the naming of the four indicator variables, Table 4.3 shows the values of the four indicator variables corresponding to the five response levels of physact. Each level of physact is defined by a unique pattern in the four indicator variables.

Furthermore, the corresponding βs have a straightforward interpretation For the moment, consider a simple regression model in which the five levels of physact are the only predictors. Then

$$E[\text{glucose}|\mathbf{x}] = \beta_0 + \beta_2\_\text{Iphysact\_2} + \cdots + \beta_5\_\text{Iphysact\_5}. \tag{4.7}$$

For clarity, the βs in (4.7) are indexed in accord with the levels of physact, so β_1 does not appear in the model. Letting the four indicators take on values of 0 or 1 as appropriate for the five groups defined by physact, we obtain

$$E[\text{glucose}|\mathbf{x}] = \begin{cases} \beta_0 & \text{physact} = 1 \\ \beta_0 + \beta_2 & \text{physact} = 2 \\ \beta_0 + \beta_3 & \text{physact} = 3 \\ \beta_0 + \beta_4 & \text{physact} = 4 \\ \beta_0 + \beta_5 & \text{physact} = 5. \end{cases} \tag{4.8}$$

From (4.8) it is clear that the intercept β_0 gives the value of $E[\text{glucose}|\mathbf{x}]$ in the reference or much less active group (physact $= 1$). Then it is just a matter of subtracting the first line of (4.8) from the second to see that β_2 gives the difference in the average glucose in the somewhat less active group (physact $= 2$) as compared to the much less active group. Accordingly the t-test of H_0: $\beta_2 = 0$ is a test of whether average glucose levels are the same in the much less and somewhat less active groups (physact $= 1$ and 2). And similarly for β_3, β_4, and β_5.

Four other points are to be made from (4.8).

- Without other predictors, or covariates, the model is equivalent to a one-way ANOVA (Problem 4.10). Also the model is said to be *saturated* and the population group means would be estimated under model (4.8) by the sample averages. With covariates, the estimated means for each group would be adjusted for between-group differences in the covariates included in the model.

- The parameters of the model can be manipulated to give the estimated mean in any group, using (4.8), or to give the estimated differences between any two groups. For instance, the difference in average outcome levels between the much more and somewhat more active groups is equal to $\beta_5 - \beta_4$ (why?). All regression packages make it straightforward to estimate and test hypotheses about these *linear contrasts*. This implies that choice of reference group is in some sense arbitrary. While a particular choice may be best for ease of presentation, possibly because contrasts with the selected reference group are of primary interest, alternative reference groups result in essentially the same model (Problem 4.4).
- The five estimated group means can take on almost any pattern with respect to each other, in either the adjusted or unadjusted model. In contrast, if physact were treated as a score with integer values 1 through 5, the estimated means would be constrained to lie on a straight regression line.

Table 4.4 shows results for the model with physact treated as a categorical variable, again using data for women without diabetes in HERS. In the regression output, $\hat{\beta}_0$ is found in the column and row labeled Coef. and _cons; we see that average glucose in the much less active group is approximately 98.4 mg/dL. The differences between the reference group and the two most active groups are statistically significant; for instance, the average glucose level in the much more active group (_Iphysact_5) is 3.3 mg/dL lower than in the much less active group ($t = -2.92$, $P = 0.003$).

Using (4.8), the first lincom command after the regression computes the estimated mean in the somewhat less active group, equal to the sum of $\hat{\beta}_0$ (_cons) and $\hat{\beta}_2$ (_Iphysact_2), or 97.6 mg/dL (95% CI 96.5–98.6 mg/dL). We can also use the lincom command to assess pairwise differences between two groups when neither is the referent. For example, the second lincom result in Table 4.4 shows that average glucose is 2.1 mg/dL lower in among women in the much more active (physact $= 5$) group as compared to those who are about as active (physact $= 3$), and that this difference is statistically significant ($t = -2.86$, $P = 0.004$). The last two results in the table are explained below.

4.3.3 The F-Test

Although every pairwise contrast between levels of a categorical predictor is readily available, the t-tests for these multiple comparisons provide no overall evaluation of the importance of the categorical variable, or more precisely a single test of the null hypothesis that the mean level of the outcome is the same at all levels of this predictor. In the example, this is equivalent to a test of whether any of the four coefficients corresponding to physact differ from zero. The testparm result in Table 4.4 ($F(4, 2027) = 4.43, P = 0.0014$) shows that glucose levels clearly differ among the groups defined by physact.

Table 4.4. Regression of Glucose on Physical Activity

```
. xi: reg glucose i.physact if diabetes == 0;
i.physact          _Iphysact_1-5       (naturally coded; _Iphysact_1 omitted)

      Source |       SS       df       MS              Number of obs =    2032
-------------+------------------------------          F(  4,  2027) =    4.43
       Model | 1673.09022        4  418.272554         Prob > F      =  0.0014
    Residual | 191344.609     2027  94.3979322         R-squared     =  0.0087
-------------+------------------------------          Adj R-squared =  0.0067
       Total | 193017.699     2031  95.0357946         Root MSE      =  9.7159


-----------------------------------------------------------------------------
     glucose |     Coef.   Std. Err.       t     P>|t|     [95% Conf. Interval]
-------------+---------------------------------------------------------------
  _Iphysact_2 | -.8584489   1.084152    -0.79    0.429    -2.984617    1.267719
  _Iphysact_3 | -1.226199   1.011079    -1.21    0.225     -3.20906    .7566629
  _Iphysact_4 | -2.433855   1.010772    -2.41    0.016    -4.416114   -.4515951
  _Iphysact_5 | -3.277704   1.121079    -2.92    0.003    -5.476291   -1.079116
       _cons |  98.42056   .9392676   104.78    0.000     96.57853    100.2626
-----------------------------------------------------------------------------

. lincom _cons + _Iphysact_2;
 ( 1)  _Iphysact_2 + _cons = 0

-----------------------------------------------------------------------------
     glucose |     Coef.   Std. Err.       t     P>|t|     [95% Conf. Interval]
-------------+---------------------------------------------------------------
         (1) |  97.56211   .5414437   180.19    0.000     96.50027    98.62396
-----------------------------------------------------------------------------

. lincom _Iphysact_5 - _Iphysact_3;
 ( 1) - _Iphysact_3 + _Iphysact_5 = 0

-----------------------------------------------------------------------------
     glucose |     Coef.   Std. Err.       t     P>|t|     [95% Conf. Interval]
-------------+---------------------------------------------------------------
         (1) | -2.051505    .717392    -2.86    0.004    -3.458407   -.6446024
-----------------------------------------------------------------------------

. testparm _I*;
 ( 1)  _Iphysact_2 = 0
 ( 2)  _Iphysact_3 = 0
 ( 3)  _Iphysact_4 = 0
 ( 4)  _Iphysact_5 = 0
       F(  4,  2027) =     4.43
             Prob > F =    0.0014

. test - _Iphysact_2 + _Iphysact_4 + 2 * _Iphysact_5 = 0;
 ( 1) - _Iphysact_2 + _Iphysact_4 + 2 _Iphysact_5 = 0
       F(  1,  2027) =    12.11
             Prob > F =    0.0005
```

4.3.4 Multiple Pairwise Comparisons Between Categories

When the focus is on the difference between a single pre-specified pair of subgroups, the overall F-test is of limited interest and the t-test for the single contrast between those subgroups can be used without inflation of the type-I error rate. All levels of the categorical predictor should still be retained in the analysis, however, because residual variance can be reduced, sometimes substantially, by splitting out the remaining groups. Furthermore, this avoids combining the remaining subgroups with either of the pre-specified groups, focusing the contrast on the comparison of interest.

However, it is frequently of interest to examine multiple pairwise differences between levels of a categorical predictor, especially when the overall F-test is statistically significant, and in some cases even when it is not. Examples include comparisons between treatments in a clinical trial with more than one active treatment arm, or in longitudinal data, to be discussed in Chapter 8, when between-treatment differences are evaluated at multiple points in time.

For this case, various methods are available for controlling the experiment-wise type-I error rate (EER) for the wider set of comparisons. These methods differ in the trade-off made between power and the breadth of the circumstances under which the type-I error rate is protected. One of the most straightforward is Fisher's *least significant difference* (LSD) procedure, in which the pairwise-comparisons are carried out using t-tests at the nominal type-I error rate, but only if the overall F-test is statistically significant; otherwise the null hypothesis is accepted for all the pairwise comparisons. This protects the EER under the *complete null hypothesis* that all the group-specific population means are the same. However, it is subject to inflation of the EER under *partial null hypotheses* – that is, when there are some real population differences between subgroups.

More conservative procedures that protect the EER under partial null hypotheses include setting the level of the pairwise tests required to declare statistical significance equal to α/k (Bonferroni) or $1 - (1 - \alpha)^{1/k}$ (Sidak), where α is the desired EER and k is the number of pre-planned comparisons to be made. The Sidak correction is slightly more liberal for small values of k, but otherwise equivalent. The Scheffé method is another, though very conservative, method in which differences can be declared statistically significant only when the overall F-test is also statistically significant. The Tukey *honestly significant difference* (HSD) and Tukey-Kramer methods are more powerful than the Bonferroni, Sidak, or Scheffé approaches and also perform well under partial null hypotheses.

A special case arises when only comparisons with a single reference group are of interest, as might arise in a clinical trial with multiple treatments and a single placebo control. In this situation, Dunnett's test achieves better power than alternatives designed for all pairwise comparisons, while still protecting the EER under partial null hypotheses. It also illustrates the general principle that controlling the EER for a smaller number of contrasts is less costly in terms of power, so that it makes sense to control only for the contrasts of interest. Compare this approach to Scheffé's, which controls the EER for all possible linear contrasts but at a considerable expense in power.

The previous alternatives provide simultaneous inference on all the pairwise comparisons considered. Various *step-down* and *step-up* multiple-stage testing procedures attempt to improve power using testing of cleverly sequenced hypotheses that only continues as long as the test results are statistically significant. The Duncan and Student-Newman-Keuls procedures fall in this class. However, neither protects the EER under partial null hypotheses.

As noted in Sect. 3.1.5, the Bonferroni, Sidak, and Scheffé procedures are available with the oneway ANOVA in Stata, but not in the regression regress command used for linear regression. Thus using these methods in examining estimates provided by a multipredictor linear model may require help from a statistician.

4.3.5 Testing for Trend Across Categories

The coefficient estimates for the categories of physact shown in Table 4.4 decrease in order, so a linear trend in physact might be an adequate representation of the association with glucose. Tests for linear trend across the values of physact are best performed using a *linear contrast* in the coefficients corresponding to the various levels of the categorical predictor. As compared to a simpler approach in which the numeric values of the categorical variable are treated as a score, this approach is more efficient, in that the model captures both trend and departures from it, reducing the residual variance that makes regression effects harder to detect.

Table 4.5. Linear Contrasts Used for Testing Trend

Number of levels	Linear contrast
3	$\beta_3 = 0$
4	$-\beta_2 + \beta_3 + 3\beta_4 = 0$
5	$-\beta_2 + \beta_4 + 2\beta_5 = 0$
6	$-3\beta_2 - \beta_3 + \beta_4 + 3\beta_5 + 5\beta_6 = 0$

Table 4.5 summarizes linear contrasts that would be used for testing trend when the categorical variable has 3–6 levels with evenly spaced numeric codes (e.g., 1, 2, 3, 4, 5), and the category with the lowest numeric code is treated as the reference. As in the physact example, β_k is the coefficient corresponding to the indicator for category k. These contrasts can be motivated as the slope coefficients from a regression in which the group means are modeled as linear in the sequential numeric codes for the categorical variable. Note that for a categorical variable with only three levels, the t-test for β_3, the coefficient for the category with the largest numeric code, provides the test for trend. These formulas are valid for all the other models in this book.

In the physact example, shown in Table 4.4, we tested the hypothesis H_0: $-\beta_2 + \beta_4 + 2\beta_5 = 0$. The result ($F(1, 2027) = 12.11, P = 0.0005$) leaves little doubt that there is a declining trend in glucose levels with increasing values of physact.

The pattern in average glucose across the levels of a categorical variable could be characterized by both a linear trend and a departure from trend. Given a statistically significant trend according to the test of the linear contrast, it is easy to test for such a departure. This test uses a model in which

Table 4.6. Model Assessing Departures from Linear Trend

```
. xi: reg glucose physact i.physact if diabetes == 0;
i.physact              _Iphysact_1-5        (naturally coded; _Iphysact_1 omitted)

        Source |       SS       df       MS              Number of obs =     2032
-------------+------------------------------            F(  4,  2027) =     4.43
       Model |  1673.09022      4  418.272554            Prob > F      =   0.0014
    Residual |  191344.609   2027  94.3979322            R-squared     =   0.0087
-------------+------------------------------            Adj R-squared =   0.0067
       Total |  193017.699   2031  95.0357946            Root MSE      =   9.7159


-----------------------------------------------------------------------------
     glucose |     Coef.   Std. Err.      t    P>|t|     [95% Conf. Interval]
-------------+---------------------------------------------------------------
     physact |  -.8194259   .2802698    -2.92   0.003    -1.369073    -.269779
 _Iphysact_2 |   -.039023   .9015677    -0.04   0.965    -1.807119    1.729073
 _Iphysact_3 |   .4126531   .6739888     0.61   0.540     -.90913    1.734436
 _Iphysact_4 |    .024423   .6366194     0.04   0.969    -1.224074     1.27292
 _Iphysact_5 |  (dropped)
       _cons |   99.23999   1.184013    83.82   0.000     96.91798     101.562
-----------------------------------------------------------------------------

. testparm _I*;
 ( 1)   _Iphysact_2 = 0
 ( 2)   _Iphysact_3 = 0
 ( 3)   _Iphysact_4 = 0
 ( 4)   _Iphysact_5 = 0
        Constraint 4 dropped
        F(  3,  2027) =     0.26
          Prob > F =     0.8511
```

the categorical variable appears both as a *score* (i.e., is treated as a continuous predictor) and as a set of indicators. In Table 4.6 the F-test for the joint effect of **physact** as a categorical variable ($F(3, 2027) = 0.26, P = 0.85$) shows that there is little evidence for departures from a linear trend in this case.

It is important to note that in Table 4.6, both the coefficient and the t-test for the effect of **physact** as a score ($\hat{\beta} = -0.82, t = -2.92, P = 0.003$) are not easily interpretable, because their values depend on which additional indicator is dropped from the model. The test for trend must be carried out using the linear contrast described earlier.

4.4 Confounding

In Table 4.1, the unadjusted coefficient for **exercise** estimates the difference in mean glucose levels between two subgroups of the population of women with heart disease. But this contrast ignores other ways in which those subgroups may differ. In other words, the analysis does not take account of confounding of the association we see. Although the unadjusted contrast may be useful for describing subgroups, it would be risky to infer any causal connection between exercise and glucose on this basis. In contrast, the adjusted coefficient for **exercise** in Table 4.2 takes account of the fact that women who exercise also have lower BMI and are slightly younger and more likely to report alcohol use, all factors which are associated with differences in glucose levels. While this adjusted model is clearly rudimentary, the underlying premise of multi-predictor regression analysis of observational data is that with a sufficiently refined model (and good enough data), we can estimate causal effects, free or almost free of confounding.

To understand what confounding means, and to see how and under what conditions a multipredictor regression model might be able to overcome it, requires that we first state more clearly what we mean by the causal effect of a predictor variable. What would it mean, in more precise terms, for exercise to have a causal effect on glucose?

4.4.1 Causal Effects and Counterfactuals

To simplify the discussion, we focus on a binary predictor, a generic "exposure." Now suppose that we could run an experiment in which every member of a population is exposed and the value of the outcome observed; then, turning back the clock, we observe the outcome in the absence of exposure for every member of the population. Because we can never really turn back the clock, one of the two experimental outcomes for every individual is an unobservable *counterfactual*. However, this counterfactual experiment is central to our definition of the causal effect of the exposure.

> Definition: The *causal effect of an exposure on a continuous outcome* is the difference in population mean values of the outcome in the presence as compared to the absence of exposure, when the actual and counterfactual outcomes are observed for every member of the population as if by experiment, holding all other variables constant. If the means differ, then the exposure is a *causal determinant* of the outcome.

Three comments:

- The causal effect is defined as a *difference in population means.* This does not rule out variation in the causal effects of exposure at the individual level, possibly depending on the values of other variables. It might even be the case that exposure increases outcome levels for some members of the population and decreases them for others, yet the population means under the two conditions are equal. That is, we could have individual causal effects in the absence of an overall population causal effect.
- In our counterfactual experiment, turning back the clock to observe the outcome for each individual under both conditions means that the individual characteristics and experimental conditions that help determine the outcome are held constant, except for exposure. Thus the exposed and unexposed population means represent averaging over the same distribution of individual characteristics

and experimental conditions. In other words, all other causal determinants of outcome levels are perfectly balanced in the exposed and unexposed populations.

- Holding other variables constant does not imply that other causal effects of exposure are held constant after the experiment is initiated. These other effects may include *mediators* of the causal effect of exposure on the outcome (Sect. 4.5).

4.4.2 A Linear Model for the Counterfactual Experiment

To gain insight into our counterfactual experiment, we can write down expressions for Y_1 and Y_0, the outcomes under exposure and in its absence, using notation introduced earlier. In the following, X_1 is the indicator of exposure, with $0 =$ unexposed and $1 =$ exposed. For simplicity, we also assume that all the other determinants of the outcome – that is, the personal characteristics and experimental conditions held constant within individuals when we turn back the clock in the counterfactual experiment – are captured by another binary variable, X_2, which also has a causal effect on the outcome in the sense of our definition. Thus, for individual i the outcome under exposure is

$$y_{1i} = \beta_0 + \beta_1^c + \beta_2^c x_{2i} + \varepsilon_{1i}. \tag{4.9}$$

In (4.9)

- β_0 represents the mean of the outcome when $X_1 = X_2 = 0$.
- β_1^c is the *causal effect* of X_1: that is, the difference in population mean values of the outcome in the counterfactual experiment where X_1 is varied and X_2 is held constant.
- β_2^c is the causal effect of X_2, defined analogously as the difference in population means in a second counterfactual experiment in which X_1 is held constant and X_2 is varied.
- Variable x_{2i} is the observed value of X_2 for individual i.
- The error term ε_1 has mean zero and is assumed not to depend on X_1 or X_2. It captures variation in the causal effects across individuals as well as error in the measurement of the outcome.

Thus the population mean value of the outcome under exposure is

$$\begin{aligned} \mathrm{E}[Y_1] &= \mathrm{E}[\beta_0 + \beta_1^c + \beta_2^c X_2 + \varepsilon_1] \\ &= \beta_0 + \beta_1^c + \beta_2^c \mathrm{E}[X_2], \end{aligned} \tag{4.10}$$

where $\mathrm{E}[X_2]$ is the mean value of X_2 across all members of the population. Similarly, the outcome for individual i in the absence of exposure is

$$y_{0i} = \beta_0 + \beta_2^c x_{2i} + \varepsilon_{0i}, \tag{4.11}$$

and the population mean outcome under this experimental condition is

$$E[Y_0] = E[\beta_0 + \beta_2^c X_2 + \varepsilon_0]$$
$$= \beta_0 + \beta_2^c E[X_2]. \tag{4.12}$$

Crucially, in the counterfactual experiment, X_2 has the same mean $E[X_2]$ under both the exposed and unexposed conditions, because it is held constant within individuals, each of whom contributes both an actual and counterfactual outcome. Subtracting (4.12) from (4.10), the difference in population means is

$$E[Y_1] - E[Y_0] = \beta_0 + \beta_1^c + \beta_2^c E[X_2] - \beta_0 - \beta_2^c E[X_2]$$
$$= \beta_1^c. \tag{4.13}$$

Thus the linear model reflects the fact that in the counterfactual experiment, the difference in population means is equal to β_1^c, the causal effect of X_1, even in the presence of the other causal effects represented by X_2.

To illustrate using our first example, suppose that β_0, the mean glucose value when $X_1 = X_2 = 0$, is 100 mg/dL; η_1^c, the causal effect of exercise is to lower glucose levels an average of 2 mg/dL; that β_2^c, the causal effect of X_2 (which may represent younger age, lower BMI, alcohol use, as well as other factors) is to lower glucose 4 mg/dL; and that $E[X_2]$, in this case the proportion of women with $X_2 = 1$, is 0.5. Now consider comparing the counterfactual population means. Using (4.10), mean glucose under the exercise condition would be

$$\beta_0 + \beta_1^c + (\beta_2^c \times 0.5) = 100 - 2 - (4 \times 0.5) = 96 \text{ mg/dL}. \tag{4.14}$$

In the absence of exercise, mean glucose would be

$$\beta_0 + (\beta_2^c \times 0.5) = 100 - (4 \times 0.5) = 98 \text{ mg/dL}. \tag{4.15}$$

Thus, using (4.13), or subtracting (4.15) from (4.14), the difference in the counterfactual means would be

$$\beta_1^c = -2 \text{ mg/dL}. \tag{4.16}$$

Now suppose we could sample randomly from this population of individuals and observe both actual and counterfactual outcomes for each, and that we used the simple linear model

$$E[Y|x] = \beta_0 + \beta_1 x_1 \tag{4.17}$$

to estimate the causal effect of exposure. Equation (4.13) implies that fitting the simple linear model (4.17) would result in an unbiased estimate of the causal effect β_1^c. By unbiased we mean that that over many repeated samples drawn from the population, the average or expected value of the estimates based on each sample would equal the population causal effect. Equivalently, using our notation for expected values,

$$E[\hat{\beta}_1] = \beta_1^c. \tag{4.18}$$

Thus if we could sample from the counterfactual experiment the difference in sample averages under the exposed and unexposed conditions would provide an unbiased estimate of the causal effect of exercise on glucose.

4.4.3 Confounding of Causal Effects

In reality, of course, causal effects cannot be estimated in counterfactual experiments. The outcome is generally observable for each individual under only one of the two conditions. In place of a counterfactual experiment, we usually have to compare mean values of the outcome in two distinct populations, one composed of exposed individuals and the other of unexposed. In doing so, there is no longer any guarantee that the mean values of X_2 would be equal in the exposed ($X_1 = 1$) and unexposed ($X_1 = 0$) populations. Note that this inequality would mean that X_1 and X_2 are correlated.

However, since both β_1^c and β_2^c represent causal effects, we can still use (4.10) and (4.12) to express the two population means. Letting $E_1[X_2]$ denote the mean of X_2 in the exposed, the mean outcome value in that population is

$$E[Y_1] = \beta_0 + \beta_1^c + \beta_2^c E_1[X_2]. \tag{4.19}$$

Similarly, with $E_0[X_2]$ denoting the mean of X_2 among the unexposed, the mean of the outcome in that population is

$$E[Y_0] = \beta_0 + \beta_2^c E_0[X_2]. \tag{4.20}$$

This implies that

$$
\begin{aligned}
E[Y_1] - E[Y_0] &= \beta_0 + \beta_1^c + \beta_2^c E_1[X_2] - \beta_0 - \beta_2^c E_0[X_2] \\
&= \beta_1^c + \beta_2^c (E_1[X_2] - E_0[X_2]).
\end{aligned} \tag{4.21}
$$

Thus the difference in population means is now arbitrarily different from the causal effect, depending on the difference between $E_1[X_2]$ and $E_0[X_2]$ and on the magnitude of β_2^c, the population causal effect of X_2. From this it follows that if we sampled randomly from the combined exposed and unexposed populations, an estimate of β_1 found using the simple linear model (4.17) ignoring X_2 would usually be biased for β_1^c, the causal effect of X_1 on Y. In short, our estimate of the causal effect of X_1 would be confounded by the causal effect of X_2.

Definition: *Confounding* is present when the difference in mean values of the outcome between populations defined by a potentially causal variable of interest is not equal to its causal effect on that outcome. In terms of our binary exposure, this can be expressed as $E[Y_1] - E[Y_0] \neq \beta_1^c$. As a consequence, the regression coefficient estimate for the causal variable given by fitting a simple linear model to a random sample of data from the combined population will be biased for the causal effect.

The effect of such confounding can be large and go in either direction. Returning to our first example, we again suppose that β_0, the mean glucose value in the population with $X_1 = X_2 = 0$ is 100 mg/dL; β_1^c, the causal effect of exercise is to lower glucose levels an average of 2 mg/dL; and that β_2^c, the causal effect of the potential confounder X_2 is to lower glucose 4 mg/dL. Now consider comparing populations where $E_1[X_2]$, the proportion with $X_2 = 1$ among women who exercise, is 0.8; but $E_0[X_2]$, the corresponding proportion among women who do not, is only 0.2. Then, using (4.19), mean glucose in the population of women who exercise would be

$$\beta_0 + \beta_1^c + (\beta_2^c \times 0.8) = 100 - 2 - (4 \times 0.8) = 94.8 \text{ mg/dL}. \tag{4.22}$$

In the population of women who do not exercise, mean glucose would be

$$\beta_0 + (\beta_2^c \times 0.2) = 100 - (4 \times 0.2) = 99.2 \text{ mg/dL}. \tag{4.23}$$

Thus, using (4.21), or subtracting (4.23) from (4.22), the difference in population means would be

$$\beta_1^c + \beta_2^c \times (0.8 - 0.2) = -2 - (4 \times 0.6) = -4.4 \text{ mg/dL}. \tag{4.24}$$

So the difference in population means would be considerably larger than the population causal effect of exercise. It follows that an unadjusted estimate of the causal effect using the simple linear model (4.17) would on average be substantially too large. In sum, under the plausible assumption that the other determinants of glucose have a real causal effect, (that is, $\beta_2^c \neq 0$), then only if the mean of X_2 were the same in both the exposed and unexposed populations – that is, $E_1[X_2] = E_0[X_2]$ – would the simple unadjusted comparison of sample averages – or population means – be free of confounding.

4.4.4 Randomization Assumption

The condition under which the difference in population means is equal to the causal effect can now be stated in terms of counterfactual outcomes: this equality will hold if the process determining whether individuals belong to the exposed or unexposed population is independent of their actual and counterfactual outcomes under those two conditions (exposure and its absence). In the glucose example, this would imply that exercising (or not) does not depend in any way on what glucose levels would be under either condition. This is known as the *randomization assumption*.

In general this assumption is met in randomized experiments, since in that setting, exposure – that is, treatment – is determined by a random process and does not depend on future outcomes. But in the setting of observational data where multipredictor regression models are most useful, this assumption clearly cannot be assumed to hold. In the HERS cohort, the randomization assumption holds for assignment to hormone therapy. However, in the glucose example, the randomization assumption is violated when the co-determinants of glucose differ according to exercise. Essentially this is because the other factors captured by X_2 are causal determinants of glucose levels (or proxies for such determinants) *and* correlated with exercise.

4.4.5 Conditions for Confounding of Causal Effects

There are two conditions under which a covariate X_2 may confound the difference in mean values of an outcome Y in populations defined by the primary causal variable X_1:

- *X_2 is a causal determinant of Y, or a proxy for such determinants.*
- *X_2 is a causal determinant of X_1, or they share a common causal determinant.*

We note that age is one commonly used proxy for underlying causal effects. Further, if X_1 is a causal determinant of X_2, rather than the opposite, then X_2 would *mediate* rather than confound the causal effects of X_1. Mediation is discussed in more detail below in Sect. 4.5. Finally, bi-directional causal pathways between X_1 and X_2 would require more complex methods beyond the scope of this book.

4.4.6 Control of Confounding

The key to understanding how a multiple regression model can control for confounding when the randomization assumption does not hold is the concept of *holding other causal determinants of the outcome constant*. This is easiest to see in our example where all the causal determinants of the outcome Y other than X_1 are captured by the binary covariate X_2. The underlying argument is that within levels of X_2, we should be able to determine the causal effect of X_1, since within those strata X_2 is the same for all individuals and thus cannot explain differences in mean outcome levels according to X_1. Under the two-predictor linear model

$$E[Y|\mathbf{x}] = \beta_0 + \beta_1 x_1 + \beta_2 x_2, \tag{4.25}$$

it is straightforward to write down the population mean value of the outcome for the four groups defined by X_1 and X_2. For purposes of illustration, we assume as in the previous example that $\beta_0 = 100$ mg/dL, $\beta_1^c = -2$ mg/dL, and $\beta_2^c = -4$ mg/dL. The results are shown in Table 4.7.

Examining the effect of X_1 while holding X_2 constant thus means comparing groups 1 and 2 as well as groups 3 and 4. It is easy to see that in both cases the between-group difference in $E[y|\mathbf{x}]$ is simply β_1^c, or -2 mg/dL. We have made it possible to hold X_2 constant by modeling its effect, β_2^c. Furthermore, under our assumption that all causal determinants of Y other than X_1 are captured by X_2, the randomization assumption holds within the

Table 4.7. Linear Model for Causal Effects of X_1 and X_2

Group	X_1	X_2	$E[y|\mathbf{x}]$	Population mean
1	0	0	β_0	100 mg/dL
2	1	0	$\beta_0 + \beta_1^c$	98 mg/dL
3	0	1	$\beta_0 + \beta_2^c$	96 mg/dL
4	1	1	$\beta_0 + \beta_1^c + \beta_2^c$	94 mg/dL

strata defined by X_2. As a result, the regression parameters β_1^2 and β_2^2 are interpretable as causal effects.

By extension from this simple example, the rationale for using multiple regression to control for confounding is the prospect of obtaining unbiased estimates of the causal effects of predictors of interest by modeling the effects of confounding variables. Furthermore, these arguments for the potential to control confounding using the multipredictor linear model can be extended, with messier algebra, to settings where there is more than one causal co-determinant of the outcome, where any or all of the predictor variables are continuous, counts, or multi-level categories, rather than binary, and where the outcome is binary or a survival time, as discussed in later chapters.

4.4.7 Range of Confounding Patterns

In our hypothetical causal example comparing distinct rather than counterfac-tual populations, the causal effect of X_1 is smaller than the simple difference between population means. We also saw this pattern in the estimate for the effect of exercise on glucose levels after adjustment for age, alcohol use, and BMI.

However, qualitatively different patterns can arise. We now consider a small hypothetical example where x_1, the predictor of primary interest, is binary and coded 0 and 1, and the potential confounder, x_2, is continuous. At one extreme, the effect of a factor of interest may be completely confounded by a second variable. In the upper left panel of Fig. 4.1, x_1 is shown to be strongly associated with y in unadjusted analysis, as represented in the scatterplot. However, the upper right panel shows that the unadjusted difference in y can be entirely explained by the continuous covariate x_2. The regression lines for x_2 are the same for both groups defined by x_1; in other words, there is no association with x_1 after adjustment for x_2.

At the other extreme, we may find little or no association in unadjusted analysis, because it is *masked* or *negatively confounded* by another predictor. The lower panels of Fig. 4.1 show this pattern. On the left, there is clearly no association between the binary predictor x_1 and y, but on the right the regression lines for x_2 are very distinct for the groups defined by x_1. In short, the association between x_1 and y is unmasked by adjustment for x_2. Negative confounding can occur under the following circumstances:

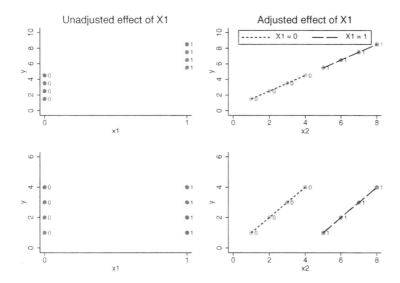

Fig. 4.1. Complete and Negative Confounding Patterns

- the predictors are inversely correlated, but have regression coefficients with the same sign.
- the two predictors are positively correlated, but have regression coefficients with the opposite sign.

The example shown in the lower panels of Fig. 4.1 is of the second kind.

4.4.8 Diagnostics for Confounding in a Sample

In Sects. 4.4.3 and 4.4.5 a definition and conditions for confounding were stated in terms of causal relationships defined by counterfactual differences in population means, which are clearly not verifiable. Randomized experiments provide the best approximation to these conditions, since the randomization assumption holds in that context. However, many epidemiologic questions about the causes of disease cannot be answered by experiments. In an observational sample, we do our best to control confounding by modeling the effects of potential confounders in multipredictor regression models.

In this context, we have to assess the potential for confounding in terms of associations between predictors and outcomes, an assessment best carried out within a hypothesized causal framework to help us distinguish potential confounders from mediators, defined below in Sect. 4.5. There are four useful diagnostics for potential confounding of the effect of a predictor of interest:

- The potential confounder must be associated with the outcome.

- The potential confounder must be associated with the predictor of interest.
- Adjustment for the potential confounder must affect the magnitude of the coefficient estimate for the predictor of interest. Note that this change could be in either direction, and may even involve change in sign; attenuation is the most common pattern, but increases in the absolute value of the coefficient are consistent with negative confounding.
- The potential confounder must make sense in terms of the hypothetical causal framework. In particular it should be plausible as a causal determinant of the predictor of interest, or as a proxy for such a determinant, and at the same time, it should clearly not represent a causal effect of the predictor of interest.

The first two diagnostics are the sample analogs of the conditions for confounding of causal effects given in Sect. 4.4.5. The third condition is the sample analog of a discrepancy between the causal effect of exposure, defined as the difference in mean values of the outcome between counterfactual populations, and the simple but potentially confounded difference between outcome means in distinct populations defined by exposure. If the fourth condition does not hold, we might see a similar pattern of associations and change in coefficients, but a different analysis is appropriate, as explained below in Sect. 4.5 on mediation.

4.4.9 Confounding Is Difficult To Rule Out

The problem of confounding is more resistant to multipredictor regression modeling than the simple two-predictor causal model in Sect. 4.4.6 might suggest. We assumed in that case that all causal determinants of Y other than X_1 were completely captured in the binary covariate X_2 – a substantial idealization. Of course, the multipredictor linear model (4.2) can (within limits imposed by sample size) include many more than two predictors, giving us considerable freedom to model the effects of other causal determinants. Nonetheless, for the multipredictor linear model to control confounding successfully and estimate causal effects without bias, all potential confounders must have been–

- recognized and assessed by design in the study,
- measured without error, and
- accurately represented in the systematic part of the model.

Logically, of course, it is not possible to show that all confounders have been measured, and in some cases it may be clear that they have not. Furthermore, the hypothetical causal framework may be uncertain, especially in the early stages of an investigating a research question. Also, measurement error in predictors is common; this may arise in some some cases because the study has only measured proxies for the causal variables which actually confound a predictor of interest. Finally, Sect. 4.7 will show that accurate modeling of systematic relationships cannot be taken for granted.

4.4.10 Adjusted vs. Unadjusted $\hat{\beta}$s

In Sect. 4.4.3 we emphasized that confounding induces bias in unadjusted (or inadequately adjusted) estimates of the causal effects that are commonly the focus of our attention. This implies that unadjusted parameter estimates are always biased and adjusted estimates less so. But there is a sense in which this is misleading. In fact the two estimate different population quantities. The observed difference in average glucose levels between women who do and do not exercise is clearly interpretable, though it almost surely does not have a causal interpretation. Thus it should not be expected to have the same value as the causal parameter.

4.4.11 Example: BMI and LDL

With a more formal definition of confounding now in hand, we turn to a relatively simple example, again using data from the HERS cohort. Body mass index (BMI) and LDL cholesterol are both established heart disease risk factors. It reasonable to hypothesize that BMI is a causal determinant of LDL. An unadjusted model for BMI and LDL is shown in Table 4.8. The unadjusted estimate shows that average LDL increases .42 mg/dL per unit increase in BMI (95% CI: 0.16–0.67 mg/dL, $P = 0.001$). However, age, ethnicity (nonwhite), smoking, and alcohol use (drinkany) may confound this unadjusted association. These covariates may either represent causal determinants of LDL or be proxies for such determinants, and are correlated with but almost surely not caused by BMI, and so may confound the BMI–LDL relationship. After adjustment for these four demographic and lifestyle factors, the estimated increase in average LDL is 0.36 mg/dL per unit increase in BMI, an association that remains highly statistically significant ($P = 0.007$). In addition, average LDL is estimated to be 5.2 mg/dL higher among nonwhite women, after adjustment for between-group differences in BMI, age, smoking, and alcohol use. The association of smoking with higher LDL is also statistically significant, and there is some evidence for lower LDL among older women and those who use alcohol.

In this example, smoking is a negative confounder, because women with higher BMI are less likely to smoke, but both are associated with higher LDL. Negative confounding is further evidenced by the fact that the adjusted coefficient for BMI is *larger* (0.36 vs. 0.32 mg/dL) in the fully adjusted model shown in Table 4.8 than in a model adjusted for age, nonwhite, and drinkany but not for smoking (reduced model not shown).

The covariates in the adjusted model shown in Table 4.8 can all be shown to meet sample diagnostic criteria for potential confounding of the effect of

Table 4.8. Unadjusted and Adjusted Regressions of LDL on BMI

```
. reg LDL bmi
```

Source	SS	df	MS		Number of obs =	2747
					F(1, 2745) =	10.14
Model	14446.0223	1	14446.0223		Prob > F =	0.0015
Residual	3910928.63	2745	1424.74631		R-squared =	0.0037
					Adj R-squared =	0.0033
Total	3925374.66	2746	1429.48822		Root MSE =	37.746

LDL	Coef.	Std. Err.	t	P>|t|	[95% Conf. Interval]	
BMI	.4151123	.1303648	3.18	0.001	.1594894	.6707353
_cons	133.1913	3.7939	35.11	0.000	125.7521	140.6305

```
. reg LDL bmi age nonwhite smoking drinkany
```

Source	SS	df	MS		Number of obs =	2745
					F(5, 2739) =	5.97
Model	42279.1877	5	8455.83753		Prob > F =	0.0000
Residual	3881903.3	2739	1417.27028		R-squared =	0.0108
					Adj R-squared =	0.0090
Total	3924182.49	2744	1430.09566		Root MSE =	37.647

LDL	Coef.	Std. Err.	t	P>|t|	[95% Conf. Interval]	
BMI	.3591038	.1341047	2.68	0.007	.0961472	.6220605
age	-.1897166	.1130776	-1.68	0.094	-.4114426	.0320095
nonwhite	5.219436	2.323673	2.25	0.025	.6631081	9.775764
smoking	4.750738	2.210391	2.15	0.032	.4165363	9.08494
drinkany	-2.722354	1.498854	-1.82	0.069	-5.661351	.2166444
_cons	147.3153	9.256449	15.91	0.000	129.165	165.4656

BMI. For example, LDL is 5.2 mg/dL higher and average BMI 1.7 kg/m^2 higher among nonwhite women, and the adjusted effect of BMI is 13% smaller than the unadjusted estimate. Note that while the associations of ethnicity with both BMI and LDL are statistically significant in this example, ethnicity might still meaningfully confound BMI even if the differences were not nominally signficant. Evidence for this would still be provided by the substantial ($\geq$ 10%) change in the coefficient for BMI after adjustment for ethnicity, according to a useful (albeit ultimately arbitrary) rule of thumb (Greenland, 1989). Recommendations for inclusion of potential confounders in multipredictor regression models are given in Chapter 5.

Fig. 4.2 shows the unadjusted regression line for LDL and BMI, together with the adjusted lines specific to the white and nonwhite women, holding the other variables constant at their respective means. Two comments about Fig. 4.2:

- Some of the upward slope of the unadjusted regression line reflects the fact that women with higher BMI are more likely to be nonwhite, younger, and not to use alcohol – all factors associated with

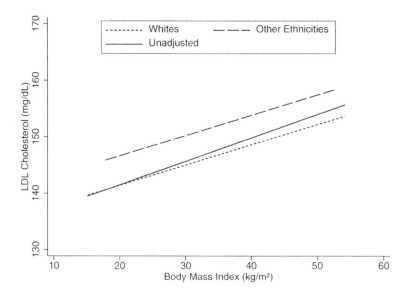

Fig. 4.2. Unadjusted and Adjusted Regression Lines

higher LDL. Despite the negative confounding by smoking, when these all these effects are accounted for using the multipredictor regression model, the slope for BMI is attenuated.

- The adjusted regression lines for white and nonwhite women are parallel, both with the same slope of 0.36 mg/dL per unit increase in BMI. Similar patterns are assumed to hold for adjusted regression lines specific to subgroups defined by smoking and alcohol use. Accordingly, the lines are separated by a vertical distance of 5.2 mg/dL at every value of BMI – the adjusted difference in average LDL by ethnicity. This pattern reflects the fact that the model does not allow for interaction between BMI and ethnicity. We assume that the slope for BMI is the same in both ethnic groups, and, equivalently, that the difference in LDL due to ethnicity is the same at every value of BMI. Testing the no-interaction assumption will be examined in Sect. 4.6 below.

4.5 Mediation

In Sect. 4.4.5 we presented conditions under which a covariate X_2 may confound the difference in mean values of an outcome Y in populations defined by the primary causal variable X_1:

- X_2 *is a causal determinant of* Y, *or a proxy for such determinants.*

- X_2 is a causal determinant of X_1, or they share a common causal determinant.

However, if X_1 is a causal determinant of X_2, then X_2 would not confound X_1 even if the first condition held; rather this would be an instance of *mediation* of the causal effects of X_1 on Y via its causal effects on X_2. That is, X_2 is affected by X_1 and in turn affects Y. For example, statin drugs reduce low-density LDL cholesterol levels, which in turn appear to reduce risk of heart attack; in this model, reductions in LDL mediate the protective effect of statins. The causal pathway from increased abdominal fat to development of diabetes and heart disease may operate through – that is, be mediated by – chemical messengers made by fat cells. The protective effect of bisphosponate drugs against fracture is mediated in part by the increases in bone mineral density (BMD) achieved by the drugs.

Definition: A *mediating variable* is a predictor hypothesized to lie on the causal pathway between a predictor of interest and the outcome, and thus to mediate the predictor's effects.

With both mediation and confounding, the mediator/confounder is associated with both the predictor of interest and the outcome, and adjustment for it typically attenuates the estimated association of the primary predictor with the outcome. At the extreme, a mediating variable based on continuous monitoring of a common pathway at a point near the final outcome may almost completely remove the effects of antecedent predictors; an example is heart failure and death from coronary heart disease.

However, in contrast to confounding, the coefficient for the primary predictor before adjustment for the proposed mediator has the more direct interpretation as the overall causal effect, while the coefficient adjusted for the mediator represents its direct causal effect via other pathways that do not involve the mediator. In this instance, the adjusted analysis is used to estimate the direct effect of the primary predictor via other pathways, its indirect effect via the mediator, and the degree of mediation. In the context of clinical trials, the relative change in the coefficient for treatment after adjustment for a mediator is sometimes referred to as the *proportion of treatment effect explained*, or PTE (Freedman *et al.*, 1992). A new approach to estimation of PTE has been developed by Li *et al.* (2001).

4.5.1 Modeling Mediation

If a potential mediator is identified on *a priori* grounds, then a series of models can be used to examine whether–

- the predictor of interest also predicts the mediator;
- the mediator predicts the outcome in a model controlling for the predictor of interest;

- addition of the mediator to a multipredictor model for the outcome attenuates the estimated coefficient for the predictor of interest.

If all three elements of this pattern are present, then the data are consistent with the mediation hypothesis. However, because this pattern also reflects what is typically seen with confounding, the two causal models must be distinguished on non-statistical grounds.

Estimation of the overall and direct effects of the predictor of interest, as well as its indirect effects via the proposed mediator, has many potential difficulties. For example, longitudinal data would clearly provide stronger support the hypothesized causal model by potentially showing that changes or differences in the predictor of interest are associated with subsequent changes in the mediator, which in turn predict the outcome still later in time. However, as discussed in Sect. 7.3.1, longitudinal analyses set up to examine such temporal patterns can be misleading if the mediator also potentially confounds the association between the primary predictor and outcome (Hernan *et al.*, 2001). Furthermore, bias in estimation of the direct effects of the primary predictor can arise from uncontrolled confounding of the association between the mediator and the outcome (Robins and Greenland, 1992; Cole and Hernan, 2002) – even in clinical trials where the primary predictor is randomized treatment assignment.

4.5.2 Confidence Intervals for Measures of Mediation

In principle, confidence intervals for PTE or for the difference in coefficient estimates for the same predictor before and after adjustment for a mediator are straightforward to compute, particularly for linear models; they have also been developed for evaluating mediation using logistic (Freedman *et al.*, 1992; Li *et al.*, 2001) and Cox proportional hazards models (Lin *et al.*, 1997). However, unlike simple comparisons of coefficients estimated in the same model, assessing mediation involves comparing coefficient estimates from two *different* models estimated using the *same* data. As a result, the two estimates are correlated, making confidence intervals more difficult to compute. Since standard statistical packages generally do not provide them, this would require the analyst to carry out computations requiring moderately advanced programming skills. An alternative is provided by *bootstrap* procedures, which were introduced in Sect. 3.6.

4.5.3 Example: BMI, Exercise, and Glucose

In Sect. 4.1 we saw that the association of exercise and glucose levels among women at risk for diabetes was substantially confounded by age, alcohol use, and BMI. In that model, BMI was shown to be a powerful predictor of glucose levels, with each kg/m^2 increase in BMI associated with a 0.49 mg/dL increase in average glucose (95% CI 0.41–0.57, $P < 0.0005$). In fact, most of the attenuation of the coefficient for exercise in the adjusted model was due to controlling for BMI, as is easily demonstrated by re-fitting the adjusted model omitting BMI.

In treating BMI as a confounder of exercise, we implicitly assumed that higher BMI makes women less likely to exercise: in short, BMI is a causal determinant of exercise. Of course, exercise might also be a determinant of BMI, which would considerably complicate the picture. However, exercise vigorous enough to result in weight loss was very uncommon in this cohort of older post-menopausal women with heart disease; furthermore, exercise was weakly associated ($P = 0.06$) with a small *increase* in BMI of 0.12 kg/m^2 over the first year of the study, after adjusting for age, ethnicity, smoking, and self-report of poor or fair health. Thus the potential causal pathway from exercise to decreased BMI appears negligible in this population.

Accordingly, we examined the extent to which the effects of BMI on glucose levels might be mediated through its effects on likelihood of exercise. In implementing the series of models set out in Sect. 4.5.1, we first used a multipredictor logistic regression model (Chap. 6) to show that each kg/m^2 increase in BMI is associated with an 8% decrease in the odds of exercise (95% CI 4–10%, $P < 0.0005$). We have already observed that exercise is associated with a decrease in average glucose of about 1 mg/dL (95% CI 0.1–1.9, $P = 0.027$), after adjusting for BMI as well as age and alcohol use. However, the coefficient for BMI is only slightly attenuated when exercise is added to the model, from 0.50 to 0.49 mg/dL per kg/m^2 increase in BMI, a decrease of only 2.9%. As shown in Table 4.9, a bias-corrected bootstrap confidence interval for the percentage decrease in the BMI coefficient due to mediation by exercise, the equivalent of PTE, was 0.3–6.0%, showing that the attenuation was not just due to chance. Nonetheless, this analysis suggests that only a very small part of the effect of BMI on glucose levels is mediated by its effects on likelihood of exercising. Note that a short program had to be "defined" in order to fit the nested models before and after adjustment for exercise in each bootstrap sample and then compute PTE.

Of note, there was little evidence of bias in the estimate of PTE in this example, since bias correction did not affect the percentile-based confidence interval. However, the interval based on the normal approximation was somewhat different from either percentile-based CI, running from 0.1 to 5.7% and thus indicating some departure from normality in the sampling distribution of PTE; this is not uncommon with ratio estimates. Of course, the qualitative interpretation would be unchanged.

4.6 Interaction

In Sect. 4.4.5 we outlined the conditions under which a two-predictor linear model could be successfully used to eliminate confounding of the effects of a primary predictor X_1 by a confounder X_2. We presented the two-predictor

Table 4.9. Bootstrap Confidence Interval for PTE

```
. program define mediate, rclass
  1. version 8.0
  2. tempname bg0
  3. reg glucose BMI age10 drinkany if diabetes == 0
  4. scalar 'bg0' = _b[BMI]
  5. reg glucose exercise BMI age10 drinkany if diabetes == 0
  6. return scalar pte = ('bg0' - _b[BMI]) / 'bg0' * 100
  7.   end
. bootstrap "mediate" r(pte), reps(1000)
command:      mediate
statistic:    _bs_1    = r(pte)

Bootstrap statistics                    Number of obs   =   2028
                                        Replications    =   1000


-------------------------------------------------------------------
Variable   |  Reps  Observed    Bias  Std. Err. [95% Conf. Interval]
-----------+-------------------------------------------------------
    _bs_1 |  1000  2.900681  .0546537  1.406885    .1398924  5.661469  (N)
          |                                        .3417959  6.031512  (P)
          |                                        .3424993  6.032569  (BC)
-------------------------------------------------------------------
Note:  N   = normal
       P   = percentile
       BC  = bias-corrected
```

model (4.25) under the tacit assumption that causal effect of X_1 on Y was the same within both strata defined by X_2. However, this may not hold. In this section we show how linear models can be used to model the interaction and thereby estimate causal effects that differ according to the level of a covariate.

4.6.1 Causal Effects and Interaction

In Sect. 4.4.2 we recognized that individual causal effects of exposure might vary across individuals from the overall population causal effect. In that context we assumed that the two differ for each individual by an amount that has mean zero and does not depend on either X_1 or X_2. However, suppose that in our counterfactual experiment, the population causal effect of X_1 does differ systematically according to the value of X_2. We could continue to compare the actual and counterfactual outcomes for each individual, thus holding all other variables constant, but in this case find that the causal effects, defined as the difference in population mean values of the outcome under exposure as compared to its absence, now vary across the strata of the population defined by X_2.

Definition: *Interaction* means that the causal effect of a predictor on the outcome differs according to the level of another predictor.

Interaction is also referred to as *effect modification* or *moderation*, and must be distinguished from mediation (Baron and Kenny, 1986).

4.6.2 Modeling Interaction

Continuing with our example of a primary predictor X_1 and a single covariate X_2, it is straightforward to model the interaction between X_1 and X_2 using a three-predictor linear model. As before, the randomization assumption must hold within the two strata defined by X_2, so that the stratum-specific difference in population means is equal to the causal effect of X_1 within each stratum. But in this case we do not assume that the causal effects of X_1 are the same in both strata. To allow for the interaction, we use the following three-predictor linear model:

$$E[Y|\mathbf{x}] = \beta_0 + \beta_1 x_1 + \beta_2 x_2 + \beta_3 x_1 x_2, \tag{4.26}$$

where $x_1 x_2$ simply denotes the product of the two predictors, equal to one only in the case where $X_1 = X_2 = 1$. It is again straightforward to write down the population mean values of the outcome for the four groups defined by X_1 and X_2. We assume as in the previous example that $\beta_1^c = -2$ mg/dL, $\beta_2^c = -4$ mg/dL, and $\beta_0 = 100$ mg/dL. But now in addition we assume that $\beta_3^c = -2$ mg/dL. The results are shown in Table 4.10.

Table 4.10. Interaction Model for Causal Effects of X_1 and X_2

Group	X_1	X_2	$X_1 X_2$	$E[y|\mathbf{x}]$	Population mean
1	0	0	0	β_0	100 mg/dL
2	1	0	0	$\beta_0 + \beta_1$	98 mg/dL
3	0	1	0	$\beta_0 + \beta_2$	96 mg/dL
4	1	1	1	$\beta_0 + \beta_1 + \beta_2 + \beta_3$	92 mg/dL

Examining the effect of X_1 while holding X_2 constant again means comparing groups 1 and 2 as well as groups 3 and 4. Now we do so not only to eliminate confounding, but also because the causal effects differ. In this case, when $X_2 = 0$, the between-group difference in $E[y|\mathbf{x}]$ is simply β_1, or –2 mg/dL. However, when $X_2 = 1$, the difference is $\beta_1 + \beta_3$, or –4 mg/dL. We hold X_2 constant by modeling its effect with the parameter β_2, and allow for the interaction by modeling the difference in the causal effects of X_1 with the parameter β_3. Again, assuming that the causal determinants of Y other than X_1 are captured by X_2, the randomization assumption holds within the strata defined by X_2. As a result, $\beta_1 = \beta_1^c$ and $\beta_3 = \beta_3^c$, so the regression parameters remain interpretable as causal effects.

4.6.3 Overall Causal Effect in the Presence of Interaction

It is important to point out that even when the causal effect of X_1 differs according to the level of X_2, the overall causal effect of X_1 remains well-defined in the counterfactual experiment as the difference in population mean values of the outcome in the presence as compared to the absence of exposure defined by X_1. In fact this overall causal effect is simply the weighted average of its causal effects within the strata defined by X_2, with weights defined by the proportions of the total population with $X_2 = 0$ and 1. This is not very different from averaging over the individual causal effects, except that in this case X_2 has a systematic effect on them. Furthermore, an estimate of β_1 using the two-predictor linear model (4.25) would be unbiased for this overall causal effect, provided the four groups in Table 4.7 are sampled in proportion to their relative sizes in the population.

However, in settings where an important interaction operates – especially where the causal effects differ in direction across strata – the overall causal effect is sometimes difficult to interpret. In addition, estimation and comparison of the stratum-specific causal effects will usually be of greater interest.

4.6.4 Example: Hormone Therapy and Statin Use

As an example of interaction, we examined whether the effect of hormone therapy (HT) on LDL cholesterol differs according to baseline statin use, using data from HERS. Suppose both assignment to hormone therapy and use of statins at baseline are coded using indicator variables. Then the product term for assessing interaction is also an indicator, in this case with value 1 only for the subgroup of women who reported using statins at baseline and were randomly assigned to hormone therapy. Now consider the regression model

$$E[\text{LDL}|\mathbf{x}] = \beta_0 + \beta_1\text{HT} + \beta_2\text{statins} + \beta_3\text{HTstatins}, \qquad (4.27)$$

where HT is the indicator of assignment to hormone therapy, statins the indicator of baseline statin use, and HTstatins the product term.

Table 4.11. Model for Interaction of HT and Statins

Group	HT	statins	HTstatins	$E[\text{LDL}|\mathbf{x}]$
1	0	0	0	β_0
2	1	0	0	$\beta_0 + \beta_1$
3	0	1	0	$\beta_0 + \beta_2$
4	1	1	1	$\beta_0 + \beta_1 + \beta_2 + \beta_3$

Table 4.11 shows the values of (4.27) for each of the four groups of women defined by HT and statins. The difference in $E[y|\mathbf{x}]$ between groups 1 and 2 is β_1, the effect of HT among women not using statins. Similarly, the difference in $E[y|\mathbf{x}]$ between groups 3 and 4 is $\beta_1 + \beta_3$, the effect of HT among statin users. So the interaction term β_3 gives the difference in treatment effects in these two groups. Accordingly, a t-test of H_0: $\beta_3 = 0$ is a test for the equality

of the effects of HT among statin users as compared to non-users. Note that within the strata defined by baseline statin use, the randomization assumption can clearly be assumed to hold for HT, the indicator for random treatment assignment.

Taking analogous differences between groups 1 and 3 or 2 and 4 would show that β_2 gives the difference in average LDL among statin users as compared to non-users among women assigned to placebo, while $\beta_2 + \beta_3$ gives the analogous difference among women assigned to HT. However, in this case the randomization assumption does not hold, implying that that unbiased estimation of the causal effects of statin use would require careful adjustment for *confounding by indication* – that is, for the prognostic factors that lead physicians to prescribe this treatment.

Table 4.12. Interaction of Hormone Therapy and Statin Use

```
. gen HTstatins = HT * statins
. reg LDL1 HT statins HTstatins

      Source |       SS       df       MS              Number of obs =    2608
-------------+------------------------------           F(  3,  2604) =   52.68
       Model |  227141.021      3  75713.6735          Prob > F      =  0.0000
    Residual |  3742707.78   2604  1437.29177          R-squared     =  0.0572
-------------+------------------------------           Adj R-squared =  0.0561
       Total |  3969848.80   2607  1522.76517          Root MSE      =  37.912


------------------------------------------------------------------------------
        LDL1 |     Coef.   Std. Err.      t    P>|t|    [95% Conf. Interval]
-------------+----------------------------------------------------------------
          HT | -17.72836   1.870629    -9.48   0.000   -21.39643   -14.06029
     statins | -13.80912   2.15213     -6.42   0.000   -18.02918   -9.589065
   HTstatins |  6.244416   3.076489     2.03   0.042    .2118044    12.27703
       _cons |  145.1567   1.325549   109.51   0.000    142.5575    147.756
------------------------------------------------------------------------------

. lincom HT + HTstatins
 ( 1)  HT + HTstatins = 0.0

------------------------------------------------------------------------------
        LDL1 |     Coef.   Std. Err.      t    P>|t|    [95% Conf. Interval]
-------------+----------------------------------------------------------------
         (1) | -11.48394   2.442444    -4.70   0.000   -16.27327   -6.694615
------------------------------------------------------------------------------
```

Table 4.12 shows that there is some evidence for a smaller effect of HT on LDL among women reporting statin use at study baseline. The coefficient for HT, or $\hat{\beta}_1$, shows that among women who did not report statin use at baseline, average cholesterol at the first annual HERS visit was almost 18 mg/dL lower in the HT arm than in placebo, a statistically significant subgroup treatment effect. To obtain the estimate of the effect of HT among baseline statin users, we sum the coefficients for HT and HTstatins (that is, $\hat{\beta}_1 + \hat{\beta}_3$) using the lincom command. This shows that the treatment effect among baseline statin users was only –11.5 mg/dL, although this was also statistically significant. The difference ($\hat{\beta}_3$) of 6.2 mg/dL between the two treatment effects was also statistically significant ($t = 2.03, P = .042$). Finally, the results for variable statins indicate that among women assigned to placebo, baseline statin use is a statistically significant predictor of LDL levels at the first annual visit.

4.6.5 Example: BMI and Statin Use

While it is often hard to obtain unbiased estimates of the causal effects of treatments like statins using observational data, a more tractable question of interest is whether the causal relationships of variables related to statin use may be modified by use of these drugs. Or it may be of interest simply to find out whether other risk factors differentially predict outcomes of interest according to use of related medications.

For example, the association between BMI and baseline LDL cholesterol levels was shown in Sect. 4.4.11 to be statistically significant after adjustment for demographics and lifestyle factors. However, treatment with statins may modify this association, possibly by interrupting the causal pathway between higher BMI and increased LDL. This would imply that BMI is less strongly associated with increased average LDL among statin users than among non-uses.

In examining this interaction, centering the continuous predictor variable BMI about its mean value of 28.6 kg/m^2 makes the parameter estimate for statin use more interpretable, as we show below. Then, to implement the analysis, we would first compute the product term statcBMI = statins × cBMI, where cBMI is the new centered BMI variable. Note that because statins is an indicator variable coded 1 for users and 0 for non-users, the product variable statcBMI is by definition equal to cBMI in statin users, but equal to zero for non-users. We then fit a multipredictor regression model including all these three predictors, as well as the potential confounders adjusted for previously. The resulting model for baseline LDL is

$$\mathrm{E}[\mathrm{LDL}|\mathbf{x}] = \beta_0 + \beta_1\mathtt{statins} + \beta_2\mathtt{cBMI} + \beta_3\mathtt{statcBMI}$$
$$+\beta_4\mathtt{age} + \beta_5\mathtt{nonwhite} + \beta_6\mathtt{smoking} + \beta_7\mathtt{drinkany}. \quad (4.28)$$

Thus among women who do not use statins,

$$\mathrm{E}[\mathrm{LDL}|\mathbf{x}] = \beta_0 + \beta_2\mathtt{cBMI}$$
$$+\beta_4\mathtt{age} + \beta_5\mathtt{nonwhite} + \beta_6\mathtt{smoking} + \beta_7\mathtt{drinkany}, \quad (4.29)$$

and the slope associated with cBMI in this group is β_2. In contrast, among statin users

$$\mathrm{E}[\mathrm{LDL}|\mathbf{x}] = \beta_0 + \beta_1\mathtt{statins} + \beta_2\mathtt{cBMI} + \beta_3\mathtt{statcBMI}$$
$$+\beta_4\mathtt{age} + \beta_5\mathtt{nonwhite} + \beta_6\mathtt{smoking} + \beta_7\mathtt{drinkany}$$
$$= \beta_0 + \beta_1\mathtt{statins} + (\beta_2 + \beta_3)\mathtt{cBMI}$$
$$+\beta_4\mathtt{age} + \beta_5\mathtt{nonwhite} + \beta_6\mathtt{smoking} + \beta_7\mathtt{drinkany}. \quad (4.30)$$

In this group, the slope associated with BMI is $\beta_2 + \beta_3$; so clearly the interaction parameter β_3 gives the difference between the two slopes.

The model also posits that the difference in average LDL between statin users and non-users depends on BMI. Subtracting (4.29) from (4.30), the difference in average LDL in statin users as compared to non-users is $\beta_1 + \beta_3 \text{cBMI}$. However, we may be reluctant to interpret this result as an unbiased estimate of the causal effects of statin use in view of the potential for uncontrolled confounding by indication.

Table 4.13. Interaction Model for BMI and Statin Use

```
. reg LDL cBMI statins statcBMI age nonwhite smoking drinkany
```

Source	SS	df	MS		Number of obs =	2745
					F(7, 2737) =	22.85
Model	216681.484	7	30954.4978		Prob > F =	0.0000
Residual	3707501	2737	1354.58568		R-squared =	0.0552
					Adj R-squared =	0.0528
Total	3924182.49	2744	1430.09566		Root MSE =	36.805

LDL	Coef.	Std. Err.	t	P>|t|	[95% Conf. Interval]	
statins	-16.25301	1.468788	-11.07	0.000	-19.13305	-13.37296
cBMI	.5821275	.160095	3.64	0.000	.2682082	.8960468
statcBMI	-.701947	.2693752	-2.61	0.009	-1.230146	-.1737478
age	-.1728526	.1105696	-1.56	0.118	-.3896608	.0439556
nonwhite	4.072767	2.275126	1.79	0.074	-.3883702	8.533903
smoking	3.109819	2.16704	1.44	0.151	-1.139381	7.359019
drinkany	-2.075282	1.466581	-1.42	0.157	-4.950999	.8004354
_cons	162.4052	7.583312	21.42	0.000	147.5356	177.2748

```
. lincom cBMI + statcBMI;
( 1)  cBMI + statcBMI = 0
```

LDL	Coef.	Std. Err.	t	P>|t|	[95% Conf. Interval]	
(1)	-.1198195	.2206807	-0.54	0.587	-.5525371	.3128981

Table 4.13 shows the results of the interaction model for statin use and BMI. The estimated coefficients have the following interpretations:

- statins: Among women with cBMI = 0, or equivalently, with BMI = 28.6 kg/m^2, statin use was associated with LDL levels that were more than 16 mg/dL lower on average. Note that if we had not first centered BMI, this coefficient would be an estimate of the statin effect in women with BMI = 0.
- cBMI: Among women who do not use statins, the increase in average LDL is 0.58 mg/dL per unit increase in BMI. The association is statistically signficant (t=3.64, $P < 0.0005$).
- statcBMI: The slopes for the average change in LDL per unit increase in BMI differ by approximately –0.70 mg/dL according to

baseline statin use. That is, the increase in average LDL associated with increases in BMI is much less rapid among women who use statins. Moreover, the interaction is statistically significant ($t = -2.61, P = 0.009$).

- `lincom` is used to estimate the slope for BMI among statin users, equal to the sum of the slope among non-users plus the estimated difference in slopes. The estimate of -.12 mg/dL per unit increase in BMI is not statistically significant ($t = -0.54, P = 0.59$), but the 95% confidence interval (–0.55 to 0.31 mg/dL per unit increase in BMI) is consistent with effects comparable in magnitude to the point estimate for non-users.

Fig. 4.3 shows the estimated regression lines in the two groups, demonstrating that the parallel lines assumption is no longer constrained to hold in the interaction model. In summary, the analysis suggests that any adverse causal effects of higher BMI on LDL may be blocked by statin use.

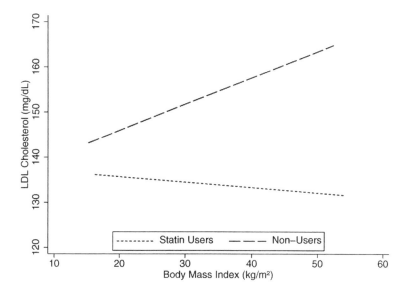

Fig. 4.3. Stratum-Specific Regression Lines

4.6.6 Interaction and Scale

Interaction models like (4.26) are often distinguished from simpler *additive* models typified by (4.25), which do not included product terms such as $x_1 x_2$. Moreover, the simpler additive model is generally treated as the default in predictor selection, with a product term being added only if there is more-or-less persuasive evidence that it is needed. It is important to recognize, however, that the need for interaction terms is dependent on the scale on which the outcome is measured (or, in the models discussed in later chapters, the scale on which its mean is modeled).

In Sects. 4.7.2 and 4.7.3 below we examine changes of the scale on which the outcome is measured to address violations of the linear model assumptions of normality and constant variance. Log transformation of the outcome, among the most commonly used changes of scale, effectively means modeling the average value of the outcome on a relative rather than absolute scale, as we show in Sect. 4.7.5 below. Similarly, in the analysis of before-and-after measurements of a response to treatment, we have the option of modeling percent rather than absolute change from baseline.

The issue of the dependence of interaction on scale arises in a similar but subtly different way with the other models discussed later in this book. For example, in logistic regression (Chap. 6) the *logit* transformation of $E[Y|\mathbf{x}]$ is modeled, while in some generalized linear models (GLMs; Chap. 9), including the widely used Poisson model, the log of $E[Y|\mathbf{x}]$ is modeled. Note that modeling $E[\log(Y)|\mathbf{x}]$, as we might do in a linear model, is different from modeling $\log(E[Y|\mathbf{x}])$ in the Poisson model. In these cases, the default model is *additive on a multiplicative scale*, as explained in Chapters 6, 7, and 9.

The need to model interaction depends on outcome scale because the simpler additive model can only hold exactly on one such scale, and may be an acceptable approximation on some scales but not others. This is in contrast to confounding; if X_2 confounds X_1, then it does so on every outcome scale. In the case of the linear model, the dependence of interaction on scale means that transformation of the outcome will sometimes succeed in eliminating an interaction.

4.6.7 Example: Hormone Therapy and Baseline LDL

The effect of hormone therapy on LDL cholesterol in the HERS trial was dependent on baseline values of LDL, with larger reductions seen among women with higher baseline values. An interaction model for absolute change in LDL from baseline to the first annual visit is shown in Table 4.14. Note that baseline LDL is centered in this model in order to make the coefficient for hormone therapy (HT) easier to interpret. The coefficients in the model have the following interpretations:

- HT: Among women with the average baseline LDL level of 135 mg/dL, the effect of HT is to lower LDL an average of 15.5 mg/dL over the first year of the study.
- cLDL0: Among women assigned to placebo, each mg/dL increase in baseline LDL is associated with a 0.35 mg/dL greater decrease in LDL over the first year. That is, women with higher baseline LDL

Table 4.14. Interaction Model for HT Effects on Absolute Change in LDL

```
. reg LDLch HT cLDL0 HTcLDL0

      Source |       SS       df       MS              Number of obs =    2597
-------------+------------------------------           F( 3,  2593) =  258.81
       Model |  721218.969     3  240406.323           Prob > F      =  0.0000
    Residual |  2408575.51  2593  928.876015           R-squared     =  0.2304
-------------+------------------------------           Adj R-squared =  0.2295
       Total |  3129794.48  2596  1205.62191           Root MSE      =  30.477


-------------------------------------------------------------------------------
       LDLch |      Coef.   Std. Err.      t    P>|t|     [95% Conf. Interval]
-------------+-----------------------------------------------------------------
          HT | -15.47703   1.196246   -12.94   0.000    -17.82273   -13.13134
       cLDL0 | -.3477064   .0225169   -15.44   0.000    -.3918593   -.3035534
     HTcLDL0 | -.0786871   .0316365    -2.49   0.013    -.1407226   -.0166517
       _cons | -4.888737   .8408392    -5.81   0.000    -6.537522   -3.239953
-------------------------------------------------------------------------------
```

experience greater decreases in the absence of treatment; this is in part due to regression to the mean and in part to greater likelihood of starting use of statins.

- HTcLDL0: The effect of HT is to lower LDL an additional 0.08 mg/dL for each additional mg/dL in baseline LDL. In short, larger treatment effects are seen among women with higher baseline values. The interaction is statistically significant ($P = 0.013$).

Table 4.15. Interaction Model for HT Effects on Percent Change in LDL

```
. reg LDLpctch HT cLDL0 HTcLDL0

      Source |       SS       df       MS              Number of obs =    2597
-------------+------------------------------           F( 3,  2593) =  165.33
       Model |  233394.163     3  77798.0542           Prob > F      =  0.0000
    Residual |  1220171.82  2593  470.563756           R-squared     =  0.1606
-------------+------------------------------           Adj R-squared =  0.1596
       Total |  1453565.98  2596  559.925263           Root MSE      =  21.692


-------------------------------------------------------------------------------
    LDLpctch |      Coef.   Std. Err.      t    P>|t|     [95% Conf. Interval]
-------------+-----------------------------------------------------------------
          HT | -10.79035   .8514335   -12.67   0.000    -12.45991   -9.120789
       cLDL0 | -.2162436   .0160265   -13.49   0.000    -.2476697   -.1848176
     HTcLDL0 |  .0218767   .0225175     0.97   0.331    -.0222773    .0660307
       _cons | -1.284976   .5984713    -2.15   0.032    -2.458506   -.1114456
-------------------------------------------------------------------------------
```

Inasmuch as the reduction in LDL caused by HT appears to be greater in proportion to baseline LDL, it is reasonable to ask whether the HT effect on *percent change* in LDL might be constant across baseline LDL levels. In that case, modeling an interaction between HT and the baseline value would not be necessary. This turns out to be the case, as shown in Table 4.15. In particular, the interaction term HTcLDL0 is no longer statistically significantly $(P = 0.331)$ and could be dropped from the model. Note that the coefficient for HT now estimates the average *percent* change in LDL due to treatment, among women at the average baseline level. In summary, analyzing percent rather than absolute change in LDL eliminates the interaction between HT and baseline LDL.

4.6.8 Details

There are several other more general points to be made about dealing with interaction in multipredictor regression models.

- Interactions between two multilevel categorical predictors require extra care in coding and interpretation. Simple computation of product terms involving a categorical predictor will almost always give mistaken results. The xi: command prefix and i. variable prefix in Stata handle this situation, but must be used with care. Furthermore, if one of the predictors has R levels and the other S levels, then the F-test for interaction would have $(R - 1)(S - 1)$ degrees of freedom. Many different patterns are subsumed by the alternative hypothesis of interaction, only a few of which may be of interest or biologically plausible.
- Interactions between two continuous variables are also tricky, especially if the two predictors are highly correlated. Both main effects in this case are hard to interpret. "Centering" of both variables on their respective sample means (Problem 4.7) resolves the interpretative problem only in part, since the coefficient for each predictor still refers only to the case where the value of other predictor is at its sample mean. Both the linearity of the interaction effect and the need for higher order interactions would need to be checked.
- In examining interactions, it is not enough to show that the predictor of primary interest has a statistically significant association with the outcome in a subgroup, especially when it is not a statistically significant predictor overall. So-called subgroup analysis of this kind can severely inflate the type-I error rate, and has a justifiably bad reputation in the analysis of clinical trials. Showing that the subgroup-specific regression coefficients are statistically different by testing for interaction sets the bar higher, is less prone to type-I error, and thus more persuasive (Brookes *et al.*, 2001).
- Methods have been developed (Gail and Simon, 1985) for assessing *qualitative interaction*, in which the sign of the coefficient for the predictor of interest differs across subgroups. This was nearly the case in the interaction of BMI and statin use. A more specific alternative of this kind is often easier to detect.

- Interaction can be hard to detect if the interacting variables are highly correlated. For example, it would be difficult to assess the interaction between two types of exposure if they occurred together either little or most of the time. This was not the case in the second HERS example, because statin use was reported by 36% of the cohort at baseline, and was uncorrelated with assignment to HT by virtue of randomization. However, in an observational cohort it might be much less common for women to report use of both medications. In that case, oversampling of dual users might be used if the interaction were of sufficient interest.

4.7 Checking Model Assumptions and Fit

In the simple linear model (4.1) as well as the multipredictor linear model (4.2), it has been assumed so far that $E[y|\mathbf{x}]$ changes linearly with each continuous predictor, and that the error term ε has a normal distribution with mean zero and constant variance for every value of the predictors. We have also implicitly assumed that model results are not unduly driven by any small subset of observations. Violations of these assumptions have the potential to bias regression coefficient estimates and undermine the validity of confidence intervals and P-values.

In this section, we show how to assess the validity of the linearity assumption for continuous predictors and suggest modifications to the model which can make it more reasonable. We also discuss assessments of normality, how to transform the outcome in order to make this assumption approximately hold, and discuss conditions under which it may be relaxed. We then discuss departures from the assumption of constant variance and methods for addressing them. All these procedures rely heavily on the transformations of both predictor and outcome that were introduced in Chapter 2. Finally, we show how to deal with *influential points*. Throughout, we emphasize the *severity* of departures, since model assumptions rarely hold exactly, and small departures are often benign, especially in large data sets. Nonetheless, careful attention to meeting model assumptions can prevent us from being seriously misled, and sometimes increase the efficiency of our analysis into the bargain.

4.7.1 Linearity

In modeling the effect of BMI on LDL, we have assumed that the regression is a straight line. However, this may not be an adequate representation of the true relationship. For example, we might find that average LDL stops increasing, or increases more slowly, among women with BMI in the upper reaches of its range – a *ceiling effect*. Analogously, the inverse relationship between BMI and HDL ("good") cholesterol may depart from linearity, with floor effects among very heavy women.

Component-Plus-Residual (CPR) Plots

In unadjusted analysis, checks for departures from linearity could be carried out using LOWESS, the nonparametric scatterplot smoother introduced in Chapter 2. This smoother approximates the regression line under the weaker assumption that it is smooth but not necessarily linear, with the degree of smoothness under our control, via the bandwidth. If the linear fit were satisfactory, the LOWESS curve would be close to the model regression line; that is, the nonparametric estimate found under the weaker assumption of smoothness would agree with the estimate found when linearity is assumed.

However, the direct approach of adding a LOWESS smooth to a scatterplot of predictor versus outcome is only effective for simple linear models with a single continuous predictor. For multipredictor regression models the analogous plot would have to accommodate $p + 1$ dimensions, where p is the number of predictors in the model – hard to imagine even for $p = 2$. Moreover, nonparametric smoothers work less well in higher dimensions.

Fortunately, the residuals from a regression model make it possible to examine the linearity of the adjusted association between a given predictor and the outcome, after taking account of the other predictors in the model. The basic idea is to plot the residuals versus each continuous predictor in the model; then a nonparametric smoother is used to detect departures from a linear trend in the average value of the residuals across the values of the predictor. This is a *residual versus predictor* (RVP) plot, obtained in Stata using the `rvpplot` command. However, for doing this check in Stata, we recommend the closely related *component plus residual* (CPR) plot, mainly because the `cprplot` command allows LOWESS smooths, which we find more informative and easier to control than the smooths available with `rvpplot`.

Fig. 4.4 shows CPR plots for multipredictor regression models for LDL and HDL, each adjusting the estimated effect of BMI for age, ethnicity, smoking, and alcohol use. If the linear fits for BMI were satisfactory, then there would be no nonlinear pattern across values of BMI in the component-plus-residuals. For LDL, shown on the left, the linear and LOWESS fits agree quite well, but for HDL, there is a substantial divergence. Thus the linearity assumption is rather clearly met by BMI in the model for LDL, but not in the model for HDL. The curvature in the relationship between BMI and HDL can be approximated by adding a quadratic term in BMI to the multipredictor linear model. The augmented model is then

$$E[\text{HDL}|\mathbf{x}] = \beta_0 + \beta_1 \text{BMI} + \beta_2 \text{BMI2}$$
$$+\beta_3 \text{age} + \beta_4 \text{nonwhite} + \beta_5 \text{smoking} + \beta_6 \text{drinkany}, \quad (4.31)$$

where `BMI2` is the square of `BMI`. A CPR plot for the relationship between BMI and HDL in this model is shown in Fig. 4.5. Except at the extremes of the range of `BMI`, where the LOWESS smooth would usually be unreliable, the quadratic fit is clearly an improvement on the simpler model. Moreover,

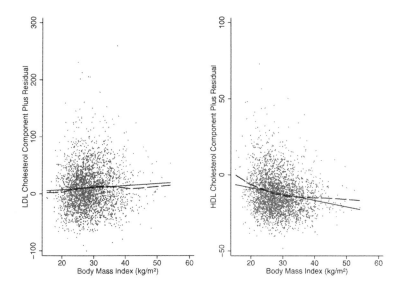

Fig. 4.4. CPR Plots for Multiple Regressions of LDL and HDL on BMI

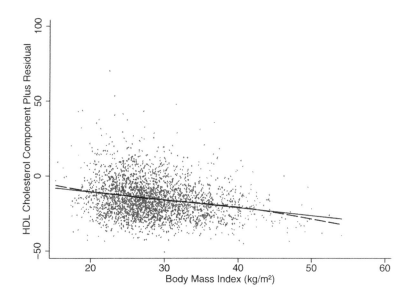

Fig. 4.5. CPR Plot for HDL Model with Quadratic Term in BMI

both the linear and quadratic terms in BMI are statistically significant (both $P < 0.0005$), and R^2 increases from 0.074 to 0.081, a gain of 9%.

Smooth Transformations of the Predictors

In the example of HDL and BMI, the departure from linearity was approximately addressed by adding a quadratic term in BMI to the model. This solution is often useful when the regression line estimated by the LOWESS smooth is convex or concave, and especially if the line becomes steeper at either side of the CPR plot.

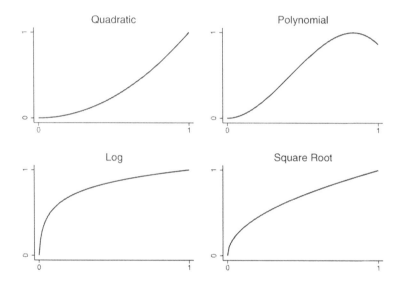

Fig. 4.6. Linearizing Predictor Transformations

However, other transformations of the predictor may sometimes be more successful and should be considered. Fig. 4.6 shows some of the predictor transformations commonly used to linearize the association between the predictor and the outcome. The upper left panel shows the typical curvature captured by adding a quadratic term in the predictor to the model. On the upper right, both quadratic and cubic terms have been included; in general such higher order polynomial tranformations are useful for S-shapes. A drawback is that these lines often fit badly in the tails of the predictor distribution if the data there are sparse. The lower panels show the log and square root transformations, which are useful in situations where the regression line increases more slowly with increasing values of the predictor, as we might expect in cases of floor or ceiling effects, and more generally where the slope becomes less steep. In Sect. 4.7.5 below, we discuss interpretation of the regression coefficients for a log-transformed predictor. Each of these transformations would work just as well for modeling the mirror image of the nonlinear shape, reversed top-to-bottom.

Categorizing the Predictor

Another transformation useful in exploratory analysis is to categorize the continuous predictor, either at cutpoints selected *a priori* or at percentiles that ensure adequate representation in each category. Then the model is estimated using indicators for all but the reference category of the transformed predictor, as in the `physact` example in Sect. 4.3. Clearly the transformed variable is ordinal in this case. This method models the association between the ordinal categories and the outcome as a *step function* (Fig. 4.7). Although this approach is unrealistic in not providing a smooth estimate of the regression line, and also less efficient, it has the advantage of flexibility, in that each step can be of any height. Such transformations are also easy to understand, especially when the categories are defined by familiar clinical cutpoints. In contrast, smooth transformations, in particular polynomials, are harder to motivate, present, and interpret.

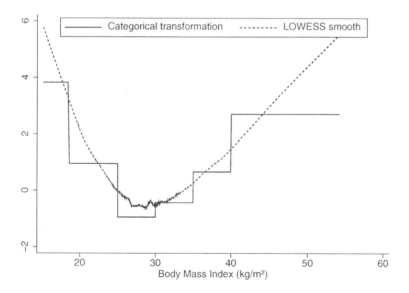

Fig. 4.7. Categorical Transformation of BMI

A final note: while diagnostics for nonlinearity using RVP and CPR plots do not carry over to the logistic, Cox, and generalized linear models presented in later chapters, departures from linearity can be addressed using quadratic terms as well as smooth and categorical transformations in all of these settings.

Evaluation

The choice of transformation will in a few cases be suggested by an understanding of mechanism or to make results more interpretable, but more often it will be made on the basis of what appears to fit best. Comparison of the LOWESS smooth in CPR plots with the transformations in Fig. 4.6 can help identify the best candidate transformations. After the revised model is estimated, repeating the diagnostic using a new CPR plot then provides an initial check on the adequacy of the transformation: there should be no remaining pattern in the residuals, and the smooth should be close to the linear fit. In cases where a quadratic or quadratic plus cubic term is added to the model, we can use t- or F-tests to evaluate the statistical significance of the addition to the model. This works because the original model is "nested" in the final model, in the sense that the predictors in the smaller model are a subset of those in the larger model. In other cases, for example, when we substitute the log-transformed for the untransformed predictor, the original and final models are not nested, so this testing procedure does not apply, although alternatives are available (Vuong, 1989). In both cases, however, we can check whether R^2 improves substantially with the transformation.

4.7.2 Normality

In Sect. 4.1 we stated that in the multipredictor linear model, the error term ε is assumed to have a normal distribution. Confidence intervals for regression coefficients and related hypothesis tests are based on the assumption that the coefficient estimates have a normal distribution. If ε has a normal distribution, and other assumptions of the multipredictor linear model are met, then ordinary least squares estimates of the regression coefficients can be shown to have a normal distribution, as required.

However, it can be shown that the regression coefficients are approximately normal in larger samples even if ε does not have a normal distribution. In that case, characterizing the distribution of the residuals is helpful for assessing whether the sample is large enough to trust the confidence intervals and hypothesis tests, since larger samples are required for this approximation to hold when departures from the normality of the errors are relatively serious. As with the t-test reviewed in Sect. 3.1, outliers are the principal worry with such departures, with the potential to erode the power of the model to detect real effects.

Residual Plots

Various graphical methods introduced in Chapter 2 are useful for assessing the normality of ε. In using these tools, it is important to distinguish between the distribution of the outcome y and the distribution of the residuals, which are the sample analogue of ε. The point here is that the residuals may be normally distributed when y is not, and conversely. Since our assumptions concern the distribution of ε, it is important to apply the diagnostic tools to the residuals rather than to the outcome variable itself.

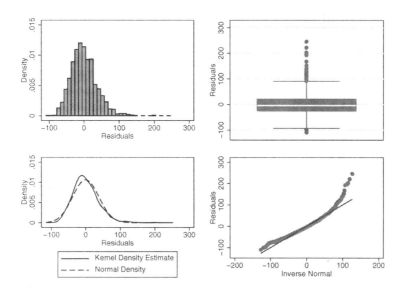

Fig. 4.8. Residuals With Untransformed LDL

Fig. 4.8 shows four useful graphical tools for assessing the normality of the residuals, in this case from our multipredictor regression model for LDL. In the upper panels the histogram and boxplot both suggest a somewhat long tail on the right. The lower left panel presents a nonparametric estimate of the distribution of the residuals obtained using the `kdensity, normal` command in Stata. For comparison, the solid line in that panel shows the normal distribution with the same mean and standard deviation. Comparing these two curves suggests some skewing to the right, with a long right and short left tail; but overall the shapes are quite close. Finally, as explained in Chapter 2, the upward curvature of the normal quantile-quantile (Q-Q) plot on the lower right is also diagnostic of right-skewness.

Interpretation of the results shown in Fig. 4.8 depends on the sample size. With 2,763 observations, there is little reason for concern about the moderate right-skewness. Given such a large data set, the distribution of the parameter estimates is likely to be well approximated by the normal, despite the mild departure from normality in the residuals. However, in a small data set, say, with 50 or fewer observations, the long right tail might be reason for concern, in part because it could make parameter estimates less precise and tests less powerful.

Testing for Departures From Normality

Various statistical tests are available for assessing the normality of the residuals, but have the drawback of being sensitive to sample size, often failing to reject the null hypothesis of normality in small samples where meeting this assumption is most important, and conversely rejecting it even for small violations in large data sets where inferences are relatively robust to departures from normality. For this reason, we do not recommend use of these tests; instead, the graphical methods just described should be used to judge the potential seriousness of the violation in the light of the sample size.

Log, Power, and Other Transformations of the Outcome

Transforming the outcome is often successful for reducing the skewness of residuals. The rationale is that the more extreme values of the outcome are usually the ones with large residuals (defined as $r_i = y_i - \hat{y}_i$); if we can "pull in" the outcome values in the tail of the distribution toward the center, then the corresponding residuals are likely to be smaller too.

One such transformation is to replace the outcome y with $\log(y)$. A constant can be added to an outcome variable with negative or zero values, so that all values are positive, though this may complicate interpretation. The log tranformation is now conventionally used to analyze viral load in studies of HIV and hepatitis infections, triglyceride levels in studies of cardiovascular disease, and in many other contexts. Fig. 4.9 shows that after log transformation of LDL, there is no more evidence of right-skewness; in fact there is slight evidence of too long a tail on the left. It should also be noted that there is no qualitative change in inferences for BMI. In Sect. 4.7.5 below, we discuss interpretation of regression coefficients in models where the outcome is log-transformed.

Power transformations are a flexible alternative to the log transformation. In this case, y is replaced by y^k. Smaller values of k "pull in" the right tail more strongly. As an example, square $(k = 1/2)$ and cube $(k = 1/3)$ root transformations were commonly used in analyzing CD4 lymphocyte counts in studies of HIV infection, since the distribution is very long-tailed on the right. Adding a constant so that all values of the outcome are non-negative will sometimes be necessary in this case too. The `ladder` command in Stata systematically searches for the power transformation of the outcome which is closest to normality.

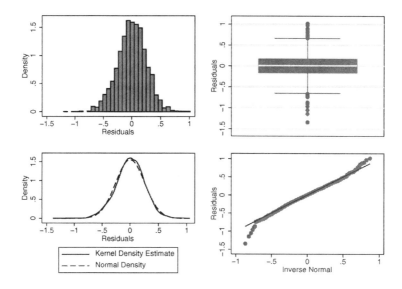

Fig. 4.9. Residuals With Log-Transformed LDL

A more difficult problem arises if both tails of the distribution of the residuals are too long, since neither log nor fractional power transformations will fix both tails. In this case one solution is the rank transformation, in which each outcome is replaced by its rank in the ordering of all the outcomes, as in the computation of the Spearman correlation coefficient (Sect. 3.2); this does not achieve normality but may reduce the loss of power. Another possibility is trimming the tails; for example, "Winsorizing" the outcome involves replacing outcome values more than 2 or 3 standard deviations from the average by that limiting value.

Generalized Linear Models (GLMs)

Some outcome variables cannot be satisfactorily transformed, or there may be compelling reasons to analyze them on the original scale. A good alternative is provided by the generalized linear models (GLMs) discussed in Chapter 9. A less efficient alternative is to dichotomize the outcome and analyze it using logistic models; alternatively, more than two outcome categories can be analyzed using proportional-odds or continuation-ratio models (Ananth and Kleinbaum, 1997; Greenland, 1994), as briefly described in Chapter 6.

4.7.3 Constant Variance

An additional assumption concerning ε is *homoscedasticity*, meaning that its variance σ_ε^2 is constant across observations. When this assumption is violated, the validity of confidence intervals and P-values can be affected. In particular, between-group contrasts can be misleading if σ_ε^2 differs substantially across the subgroups being compared, especially if the subgroups differ in size. Furthermore, in contrast to violations of the assumption that the residuals are normally distributed, heteroscedasticity is no less a problem in large samples than in small ones. Finally, while violations do not make the coefficient biased, some precision can be lost.

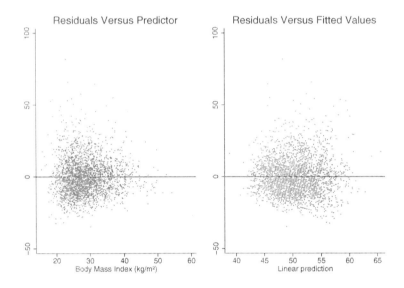

Fig. 4.10. Checking for Constant Residual Variance

Residual Plots

Diagnostics for violations of the constant variance assumption also use the residual versus predictor (RVP) plots used to check linearity of response to continuous predictors, as well as analogously defined residual versus fitted (RVF) plots. If the constant variance assumption is met, then the vertical spread of the residuals should be similar across the ranges of the predictors and fitted values; in contrast, heteroscedasticity is signaled by horizontal funnel shapes. Since the residuals of the LDL analysis gave no evidence of trouble, we examined the residuals from the companion model for HDL, which was shown in Sect. 4.7.1 to need a quadratic term in BMI to meet the linearity assumption.

Fig. 4.10 shows scatterplots of the residuals of the regression of HDL on BMI and its square, as well as age, ethnicity, smoking, and alcohol use. The plot against BMI shows somewhat wider range on the left, although this may partly be due to the fact that there are more observations on the left, and so more likely a few large residuals purely by chance. This evidence for non-constant variance is mirrored in the slightly wider spread on the right in the facing plot of the residuals against the fitted values.

Sub-Sample Variances

Constancy of variance across levels of categorical predictor can be checked by comparing the sample variance of the residuals for each category. In this example, the variance was essentially identical across groups defined by ethnicity, smoking, and alcohol use.

In contrast, in our analysis of the influence of exercise on glucose levels in Sect. 4.1, violation of the assumption of constant variance was one of several motivations for excluding women with diabetes. If they had been included, the variance of the residuals would have varied between this group of 734 women and the remainder of the HERS cohort by a factor of 26 (2,332 vs. 90). Even after log transformation of glucose, the variance would still have differed by a factor of 10 (0.097 vs. 0.0094). This pattern reflects the fact that diabetes is characterized by loss of control over glucose levels, and also variation in the use of medications that control them. These large differentials in residual variance would call into question inferences drawn from comparisons between women with and without diabetes.

Testing for Departures From Constant Variance

Statistical methods available for testing the assumption of homoscedasticity share the sensitivity to sample size described earlier for tests of normality. The resulting potential for giving false reassurance in small samples leads us to recommend against the use of these formal tests. Instead, we need to examine the severity of the violation.

When Departures May Cause Trouble

Violations of the assumption of constant variance should be addressed in cases where the variance of the residuals–

- changes by a factor of 2 or more across the range of the fitted values or a continuous predictor, judging from the LOWESS smooth of the squared residuals;
- differs by a factor of 2 or more between subgroups that differ in size by a factor of 2 or more;
- differs by a factor of 3 or more between subgroups that differ in size by a factor of less than 2.

Note that smaller differences in the *standard deviation* of the residuals would give reason for transformation.

Variance-Stabilizing Outcome Transformations

In simple cases where multiple predictors do not need to be taken into account, we could use t-tests with the `unequal` option to compare subgroups, allowing for the unequal variances. However, multipredictor modeling is often crucial; furthermore, use of a t-test with unequal variances would not address smooth dependence of σ_ε^2 either on $E[y|\mathbf{x}]$ or on a continuous predictor. In that case, non-constant variance can sometimes be addressed using a *variance-stabilizing* transformation of the outcome, including the log and square root transformations. As shown in Fig. 4.11, log transformation of HDL reduces, though it does not completely eliminate, the evidence for non-constant variance we found in Fig. 4.10. However, in this case our qualitative conclusions would be unchanged by log transformation of HDL.

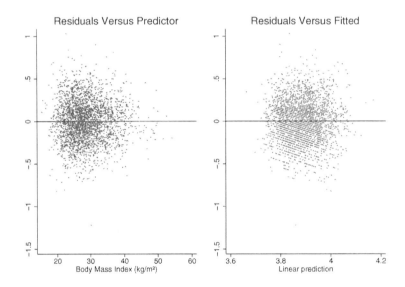

Fig. 4.11. Rechecking Constant Variance After Log-Transforming HDL

GLMs

The square root transformation has been widely used to stabilize the variance of counts. However, this has now been largely supplanted by GLMs such as the Poisson and negative binomial regression models (Chap. 9). As in other GLMs, including the logistic model (Chap. 6), the variance of the Poisson outcome is modeled as a function of its mean. In particular, this would potentially be useful in cases where a LOWESS smooth of the squared residuals, an alternative diagnostic for heteroscedasticity, increased in proportion to the fitted values. GLMs represent the primary alternative when transformation of the outcome fails to rectify substantial violations of the assumption of constant variance.

4.7.4 Outlying, High Leverage, and Influential Points

We have already pointed out that outlying observations with relatively large residuals can cause trouble, in part by inflating the variance of coefficient estimates, making it harder to detect statistically significant effects. In this section we consider *high-leverage* points, which could be described as x-outliers, since they tend to have extreme values of one or more predictors, or represent an unusual combination of predictor values. The importance of high-leverage points is that they are also potentially *influential*, in the sense that one or more of the coefficient estimates would change by an unduly large amount if the influential points were omitted from the data set. This can happen when a high-leverage point also has a large residual.

> Definitions: *High leverage points* are x-outliers with the potential to exert undue influence on regression coefficient estimates. *Influential points* are points that have exerted undue influence on the regression coefficient estimates.

Ultimately, our concern is that changes in coefficient estimates resulting from the omission of one or a few influential points could qualitatively affect the conclusions drawn from the analysis. This could arise if associations that were clearly statistically significant become clearly non-significant, or vice versa, including interaction and quadratic terms, or if associations change substantially in magnitude or direction. We would have good reason to mistrust substantive conclusions that were dependent on a few observations in this way. Similarly, in regression models oriented to prediction of future outcomes (Sect. 5.2), prediction error might be substantially affected.

Outlying, high leverage, and influential points are illustrated in Fig. 4.12. In all three of these small samples ($n = 26$), a problematic data point, marked with an X, is included. The solid and dashed lines in each plot show the regression lines estimated with and without the point, as a graphical measure of influence. The sample shown on the upper left includes an outlier with a very large positive residual. However, the leverage of the outlier is minimal, because it is in the center of the distribution of x. Accordingly, the slope estimate is unaffected by omission of this data point, Note that the point is influential for the intercept estimate, but this parameter may be of less direct interest.

In the upper right panel, the point at the extreme right has high leverage, but because this data point is fairly consistent with the prediction based on the other 25 data points, its influence is limited, and the estimated slope and

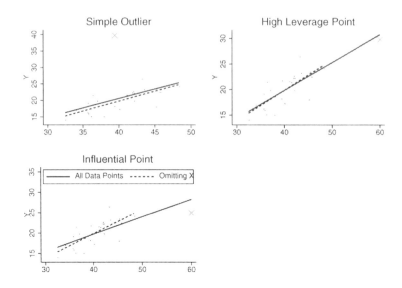

Fig. 4.12. Outlying, High-Leverage, and Influential Points

its statistical significance are almost unchanged by by omission of the high-leverage point. Certainly our qualitative interpretation of the slope would be unaffected.

In contrast, the point at the extreme right in the lower left panel has the same leverage as the point in the upper right panel, but in this case its influence is very strong, moving the slope estimate by more than 2 standard errors. The slope remains positive and statistically significant in this instance, so our qualitative interpretation would be similar, but in some circumstances omission of such a data point could make a non-significant result highly statistically significant, or vice versa. In part this reflects the small sample size, since a high leverage point is has a better chance of outweighing a relatively small number of other observations.

DFBETAs

To check for sensitivity of the conclusions of an analyis to a small number of high-leverage observations, we first need to identify potentially influential points. Of the various statistics for quantifying influence that have been defined, we recommend using DFBETA statistics, which quantify how much each of the coefficients would change if each observation were omitted from the data set. In linear regression, these statistics are exact; for logistic and Cox models, accurate approximations are available. DFBETA statistics are in standard error units – effectively on the same scale as the t-statistic, which is equal to $\hat{\beta}$ divided by its standard error. If the analysis is focused on one predictor of primary interest, then clearly the DFBETAs for that predictor are of central concern.

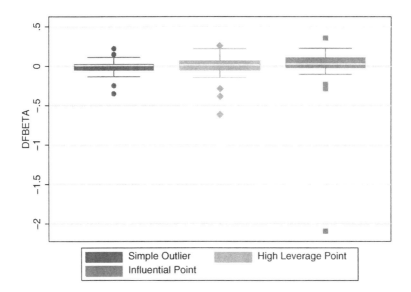

Fig. 4.13. DFBETAs for Data Sets Shown in Fig. 4.12

Boxplots are convenient for identifying a small set of extreme outliers among the DFBETA values for each predictor. DFBETAs often have a very small inter-quartile range, so that a substantial set of observations may lie beyond the whiskers of the plot. Thus we need to look for a small number of extreme values that are set off from the rest. Fig. 4.13 shows boxplots of the DFBETA statistics for the single predictor in the three data sets shown in Fig. 4.12. These plots clearly indicate the single influential point.

If a small set of observations meeting diagnostic criteria for undue influence is identified, the accuracy of those data points should first be checked and clearly erroneous observations corrected, or if this is impossible, deleted. Then if any of the apparently influential points are retained, a final step is sensitivity analyses in which the final model is rerun omitting some or all of the retained influential points. For example, suppose we have identified ten influential points that are not due to data errors, and that these include two observations with absolute DFBETAs greater than 2, three observations with values between 1 and 2, and five more with values between 0.5 and 1. Then a convenient *ad hoc* procedure would be to delete the two worst observations, then the worst five, and finally all ten potentially influential points. In each model, we would check whether the important conclusions of the analysis were affected. In prediction models, sensitivity would be assessed in terms of estimated prediction error (Sect. 5.2). In summary, we emphasize the underlying theme of sensitivity to the omission of a *small* number of points, relative to sample size; if we omit 10% or 20% of the data and the conclusions change, this would probably not indicate undue sensitivity.

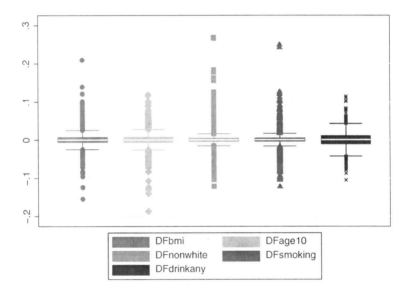

Fig. 4.14. DFBETAs for LDL Model

Fig. 4.14 below shows boxplots of DFBETAs for the multiple regression of LDL on BMI, age, ethnicity, smoking, and alcohol use. As compared to the clearly influential point shown in Fig. 4.13, the largest DFBETAs are much less extreme. Examination of the four observations with DFBETAs > 0.2 identified women with high LDL values (between 346 and 393 mg/dL).

Table 4.16. Sensitivity of LDL Model to Omission of Four Most Influential Points

Predictor variable	All observations			Omitting four observations		
	$\hat{\beta}$	95% CI	P-Value	$\hat{\beta}$	95% CI	P-Value
BMI	0.36	0.10, 0.62	0.007	0.34	0.08, 0.60	0.010
Age	−1.89	−4.11, 0.32	0.090	−1.86	−4.03, 0.31	0.090
Nonwhite	5.22	0.66, 9.78	0.025	4.19	−0.27, 8.66	0.066
Smoking	4.75	0.42, 9.08	0.032	3.78	−0.47, 8.03	0.072
Alcohol Use	−2.72	−5.66, 0.22	0.069	−2.64	−5.51, 0.23	0.072

The sensitivity of model results to the omission of these four points is summarized in Table 4.16.The changes are mostly minor, in particular for BMI, the predictor of primary interest. The P-values for ethnicity and smoking shift from nominally statistically significant to borderline significant, but these are not variables of primary interest and in any case our conclusions should not be unduly influenced by small shifts of this kind.

A potential weakness of these procedures is that DFBETAs capture the influence of omitting one observation at a time, but do not tell us how the omission of various *sets* of points, some of which may have small DFBETAs, will affect our conclusions. Unfortunately, user-friendly diagnostics for checking sensitivity to omission of sets of observations have not been developed, in part because the computational burden is too great.

Addressing Influential Points

If substantive conclusions are qualitatively affected by omission of influential points in the sensitivity analysis, *this should be reported.* In addition, it is often worthwhile to consider in substantive terms why these points have high leverage and are influential. For example, the WCGS data include an influential point with an extreme but accurately recorded cholesterol level of 645 mg/dL, which resulted from familial hypercholesterolemia, a rare condition. For research questions concerning the effects of cholesterol levels in the usual range determined by common risk factors, it would be reasonable to delete this point. But in many circumstances, deletion of influential points is hard to justify persuasively.

In that case, it may also be worth considering a more complex model that better accommodates the influential points. In Fig. 4.12, for example, a quadratic term would almost certainly reduce the influence of the observation causing trouble. Alternatively, interaction terms might accommodate influential data points characterized by an unusual combination of two predictor values. Nonetheless, changing the model in such a substantial way to accommodate one or a few data points should be undertaken with caution, with attention to the plausibility of the modified model, and the results clearly presented as data-driven, sensitive to influential points, and hypothesis-generating.

4.7.5 Interpretation of Results for Log-Transformed Variables

In Sect. 4.7 we discussed log-transforming predictors to achieve linearity, and proposed log transformation of the outcome as a means of normalizing the residuals or stabilizing their variance. Even if substantive interpretation and P-values are often not much changed, these transformations have a substantial effect on the estimated regression coefficients and their literal interpretation.

For both predictors and outcomes, log transformation changes the focus from absolute to relative or percentage change. Recall that for a predictor and outcome on their measured scale, the regression coefficient is interpretable as the change in the average value of the outcome for every unit increase in the predictor; for both predictor and outcome, we mean change on the measured, or absolute, scale.

Log Transformation of the Predictor

First consider log transformation of the predictor. In this case, the regression coefficient multiplied by $\log(1.01)$ can be interpreted as the change in the average value of the outcome for every 1% increase in the predictor. This is valid whether we use the natural log or logarithms with other bases. In a linear model using the natural log (ln) transformation of weight to predict systolic blood pressure (SBP), the estimated coefficient for ln weight is 3.004517. Thus we estimate that average SBP increases $3.004517 \times \ln(1.01) \approx 0.03$ mmHg for each 1% increase in weight. Similarly, if we multiply $\hat{\beta}$ by $\ln(1.05)$ or $\ln(1.1)$ we obtain the estimates that average SBP increases 0.15 mmHg for each 5% increase in weight and 0.29 mmHg for each 10% increase.

Within limits, we can approximate these results without using a calculator. Specifically, if the predictor is natural log-transformed, we can estimate the increase in the average value of the outcome per 1% increase in the predictor simply by $\hat{\beta}/100$. This follows because $\ln(1.01) \approx 0.01$. But this shortcut is not valid for logarithms with other bases, and analogous calculations for larger percentage increases in the predictor get progressively less accurate and should not be attempted by this means.

Log Transformation of the Outcome

Similarly, with natural log transformation of the outcome, $100(e^{\hat{\beta}} - 1)$ is interpretable as the *percentage* increase in the average value of the outcome per unit increase in the predictor. If base-10 logs were used to transform the outcome, then $100(10^{\hat{\beta}} - 1)$ has this interpetation. The coefficient for BMI in a linear model for the natural log transformation of triglyceride (TGL) is 0.0133487, so the model predicts a $100(e^{0.0133487} - 1) = 1.34\%$ increase in TGL per unit increase in BMI.

Again, we can approximate these results without a calculator under some circumstances. When the outcome is natural log-transformed, we can approximate the percentage change in the average value of the outcome per unit increase in the predictor by $100\hat{\beta}$. But this is acceptably accurate only if $\hat{\beta}$ is smaller than 0.1 in absolute value, and is again not valid using log transformations with other bases.

Log Transformation of Both Predictor and Outcome

If both predictor and outcome are transformed using natural logs, then $100(e^{\hat{\beta}\ln(1.01)} - 1)$ can be interpreted as the percentage increase in the average value of the outcome per 1% increase in the predictor. With the $\log_{10}$

transformation, $100(10^{\hat{\beta}\log_{10}(1.01)} - 1)$ has this interpretation. In this case, the back-of-the-envelope approximation for the percent increase in outcome for each 1% increase in the predictor is simply $\hat{\beta}$; this is accurate if both predictor and outcome are natural log-transformed and $\hat{\beta}$ is smaller than 0.1 in absolute value.

4.7.6 When to Use Transformations

Our graphical diagnostics for linearity, normality, and constant variance do not provide clearcut decision rules analogous to $P < 0.05$, and we do not recommend formal statistical tests in this context. Furthermore, addressing these violations will in many cases involve using transformations of predictors or outcomes that may make the results harder to interpret. A natural criterion for assessing the necessity for transformation is whether important substantive results differ qualitatively before and after transformation. If not, it may be reasonable not to use the transformations. Our example using BMI and diabetes to predict HDL is probably a case in point: while log transformation of HDL corrected departures from both normality and constant variance, the conclusions were unchanged. But if substantial differences do arise, then using transformed variables to meet model assumptions more closely helps us to avoid misleading results.

4.8 Summary

The multipredictor linear model is a straightforward extension of the simple linear model for continuous outcomes. Inclusion of multiple predictors in the model makes it possible to adjust for confounding variables, examine mediation, check for and model interactions, and increase efficiency, especially in experiments, by accounting for design factors. It is important to check the assumptions of the linear model and to use transformations of predictor and outcome variables as necessary to meet them more closely, especially in small samples. It is also important to recognize common data types where linear regression is not appropriate; these include binary, time-to-event, count, and repeated measures or clustered outcomes, and are addressed in subsequent chapters.

4.9 Further Notes and References

For more detailed information on the linear regression model, first-rate books include Weisberg (1985) and Draper and Smith (1981). Jewell (2004), in particular Chapter 8, gives an excellent introduction – to which we are indebted – to issues of causality in observational studies; Rothman and Greenland (1998) also address these issues in some detail. A cutting-edge book in this area, unfortunately of considerable difficulty, is van der Laan and Robins (2003). A standard book on regression diagnostics is Belsey *et al.* (1980), while Cleveland (1985) covers graphical methods for model checking in detail. See Breiman (2001) for a skeptical view of the sensitivity of the methods presented here for detecting lack of fit.

Splines and Generalized Additive Models

The Stata package implements a convenient and often more biologically plausible alternative to the categorical transformations presented in Sect. 4.7.1 for addressing departures from linearity, called the *linear spline*. We again specify cutpoints, usually called *knots* in this context. The resulting fitted regression line is continuous at each of the knots and linear in the intervals between them. The `mkspline` command in Stata can be used to set up the transformed predictor variables, one for each interval defined by the cutpoints. As with the categorical transformation, selection of the knots is a non-trivial problem.

Methods have also been developed for fitting linear as well as logistic (Chap. 6) and other generalized linear models (Chap. 9) in which the adjusted response to each predictor can be flexibly modeled as a smooth (piecewise cubic rather than piecewise linear) spline, or alternatively using a LOWESS curve. In both cases the degree of smoothness is under the control of the analyst. Known as *generalized additive models* (Hastie and Tibshirani, 1986, 1999), implementations in the R statistical package make it easy to model and test the statistical significance of departures from linearity. Implementations in R of smooth spline transformations of predictors are also available for the Cox model, discussed in Chapter 7.

4.10 Problems

Problem 4.1. Using the Western Collaborative Group Study (WCGS) data for middle-aged men at risk for heart disease, fit a multipredictor model for total cholesterol (`chol`) that includes the binary predictor `arcus`, which is coded 1 for the group with *arcus senilis*, a milky ring in the iris associated with high cholesterol levels, and 0 for the reference group. Save the fitted values. Now refit the model with the code for the reference group changed to 2. Compare the coefficients, standard errors, P-values, and fitted values from the two models. The WCGS data are available at `http://www.biostat.ucsf.edu/vgsm`.

Problem 4.2. Using (4.2), show that β_j gives the difference in $E[y|\mathbf{x}]$ for a one-unit increase in x_j, no matter what the values of x_j or the other predictors. *Hint:* Write the value of (4.2) for $x_j = x$ and then for $x_j = x+1$, for arbitrary (unspecified) values of the other predictors, all of which are held fixed, and subtract the first value from the second.

Problem 4.3. Using the WCGS data referenced in Problem 4.1, extract the fitted values from the multipredictor linear regression model for cholesterol and show that the square of the sample correlation between the fitted values and the outcome variable is equal to R^2. In Stata the following code saves the predicted values from the regression model in Table 4.2 to a new variable yhat:

```
. reg glucose exercise BMI smoking drinkany
. predict yhat
```

Then use the pwcorr and display commands to get the correlation between yhat and the predictor and square it.

Problem 4.4. Give an alternative coding for the unadjusted model predicting glucose from the five-level physical activity variable in which no intercept parameter is included in the model. In this case, there is no reference group, and all five group-specific indicators are included in the model. What is the interpretation of the βs in this model? How could the Stata lincom command be used to compare groups?

Problem 4.5. Use the test command in Stata or an equivalent command in another statistical package to show that $F = t^2$ for a pairwise contrast between any other level of a categorical predictor and the reference group used in the model.

Problem 4.6. In the model including an interaction between BMI and statin use, define a second new BMI variable so that estimates for BMI specific to women who do and do not use statins can be obtained directly from the regression coefficients, rather than having to compute sums of the coefficients for one of these groups. Define the values of the new BMI variable in the two groups, and then write down the regression equations analogous to (4.28), (4.29), and (4.30). Explain why the statin use variable needs to be included in this model.

Problem 4.7. If we "center" age – that is, replace it with a new variable defined as the deviation in age from the sample mean, what would be the interpretation of the intercept in the model for SBP (3.2)? If BMI had *not* been centered, how would the interpretation of the statin use variable change in the model in Sect. 4.6.5 allowing for interaction in predicting LDL?

Problem 4.8. Consider the associations between exercise and glucose levels among women without diabetes. What are the interpretations of the coefficient for exercise–

- in a simple linear model for glucose levels
- in a multipredictor linear regression model for glucose adjusting for all known confounders of the exercise association

Suppose factor X had been identified as a mediator of the exercise/glucose association. What would be the interpretation of the exercise coefficient in a multipredictor regression model that also adjusted for factor X, supposing that the exercise coefficient remained statistically significantly different from zero?

Problem 4.9. Suppose that in a clinical trial of the effects of a new treatment on glucose levels, the randomization is stratified on diabetes, an important predictor of this outcome. By virtue of randomization, the treatment is uncorrelated with diabetes. Using (4.4), explain why including diabetes in the analysis should provide a more efficient estimate of the treatment effect. Would it be a good idea to check for interaction between treatment and diabetes in this analysis? Why?

Problem 4.10. Using Stata (or another statistical package) and the WCGS data set referenced above in Problem 4.1 (or your own data set), verify that you get equivalent results from

- a t-test and a simple linear model with one binary predictor
- one-way ANOVA and a linear model with one multilevel categorical predictor.

Problem 4.11. What is the difference between showing that an interaction is statistically significant and showing that an association is statistically significant in one group but not in the other? Describe a pattern where the second condition holds but there would clearly be no interaction. Is that pattern of clinical interest?

Problem 4.12. Consider a predictor of interest for an important outcome in your field of expertise. Are there other predictors that might be hypothesized *a priori* to interact with the predictor of interest? Why?

Problem 4.13. Suppose a quadratic term in BMI is added to the model for HDL to rectify the departure from linearity and improve fit. How would you summarize this more complex association in presentations or a paper?

Problem 4.14. Consider a right-skewed outcome variable that could be adequately normalized using an unfamiliar fractional power transformation (say, the cube root). A simpler alternative is just to dichotomize the variable. Why would you expect this to be a costly choice in terms of efficiency? Now consider birth weights. Why might analysis of an indicator of low birth weight be worth the loss of efficiency in this case?

Problem 4.15. Suppose you fit a model with an influential point. With the point, the association of interest is just statistically significant, and without it, it is clearly not. What would you do?

4.11 Learning Objectives

1. Describe situations in which multipredictor analysis is needed. Given an analysis situation, decide if linear regression is appropriate.
2. Translate research questions appropriate for a regression model into specific questions about model parameters.
3. Use linear regression models to test hypotheses about relationships between variables, including confounding, mediation, and interaction.
4. Describe the linear regression model, its key assumptions, and their implications.
5. Explain why the estimates are called least squares estimates.
6. Define regression line, fitted value, residual, and influence.
7. State the relationships between
    - correlation and regression coefficients
    - the two-sample t-test and a regression model with one binary predictor
    - ANOVA and a regression model with categorical predictors.
8. Know how a statistical package is used to estimate the parameters in a regression model and make diagnostic plots to assess how well model assumptions are met.
9. Interpret regression model output output including regression parameter estimates, hypothesis tests, confidence intervals, and statistics which quantify the fit of the model.
10. Interpret regression coefficients when the predictor, outcome, or both are log-transformed.

Predictor Selection

Walter *et al.* (2001) developed a model to identify older adults at high risk of death in the first year after hospitalization, using data collected for 2,922 patients discharged from two hospitals in Ohio. Potential predictors included demographics, activities of daily living (ADLs), the APACHE-II illness-severity score, and information about the index hospitalization. A "backward" selection procedure with a restrictive inclusion criterion was used to choose a multipredictor model, using data from one of the two hospitals. The model was then validated using data from the other hospital. The goal was to select a model that best predicted future events, with a view toward identifying patients in need of more vigorous monitoring and intervention.

Grodstein *et al.* (2001) evaluated the efficacy of hormone therapy (HT) for secondary prevention of coronary heart disease (CHD), using observational data for 2,489 women with previous myocardial infarction or documented coronary artery disease in the Nurse's Health Study (NHS), a prospective cohort followed from 1976 forward. In addition to measures of the use of HT, a set of known CHD risk factors were controlled for, including age, body mass index (BMI), smoking, hypertension, LDL cholesterol levels, parental heart disease history, diet, and physical activity. The goal of predictor selection was to obtain a minimally confounded estimate of the effect of HT on risk of CHD events.

The Heart and Estrogen/Progestin Replacement Study (HERS), a randomized clinical trial addressing the same research question, was conducted among 2,763 post-menopausal women with clinically evident heart disease (Hulley *et al.*, 1998). As in the NHS, a wide range of predictors were measured at study entry. Yet in the pre-specified analysis of the main HERS outcome, the only predictor was treatment assignment. The goal was to obtain a valid test of the null hypothesis as well as an unbiased estimate of the effectiveness of assignment to HT.

Orwoll *et al.* (1996) examined independent predictors of axial bone mass using data from the Study of Osteoporotic Fractures (SOF). SOF was a large ($n = 9{,}704$) observational cohort study designed to address multiple research questions about osteoporosis and fractures among ambulatory women age 65 and up. Predictors considered by Orwoll had been identified in previous studies, and included weight, use of medications such as HT and diuretics, smoking history, alcohol and caffeine use, calcium intake, physical activity, and various measures of physical function and strength. All variables that were statistically significant at $P < .05$ in models adjusting for age were included in the final multipredictor linear regression model. The goal was to identify all important predictors of bone mass.

In each of these examples, many more potential predictor variables had been measured than could reasonably be included in a multivariable regression model. The difficult problem of how to select predictors was resolved differently, to serve three distinct inferential goals:

1. *Prediction.* Here the primary issue is minimizing prediction error rather than causal interpretation of the predictors in the model. The prediction error of the model selected by Walter *et al.* (2001) was evaluated using an independent data set from a second hospital.

2. *Evaluating a predictor of primary interest.* In pursuing this inferential goal, a central problem in observational data is confounding, which relatively inclusive models are more likely to minimize. Predictors necessary for *face validity* as well as those that behave like confounders should be included in the model. Randomized experiments like HERS represent a special case where the predictor of primary interest is the intervention; confounding is not usually an issue, but covariates are sometimes included in the model for other reasons.

3. *Identifying the important independent predictors of an outcome.* This is the most difficult of the three inferential goals, and one in which both causal interpretation and statistical inference are most problematic. Pitfalls include false-positive associations, the potential complexity of causal pathways, and the difficulty of identifying a single best model. We also endorse inclusive models in this context, and recommend a selection procedure that affords increased protection against false-positive results. Cautious interpretation of weak associations is key to this approach.

In summary–

Definition: *Predictor selection* is the process of choosing appropriate predictors for inclusion in a multipredictor regression model. A good model should be substantively motivated, appropriate to the inferential goal and sample size, interpretable, and persuasive.

5.1 Diagramming the Hypothesized Causal Model

Many potential confounders, effect modifiers, and mediators can be identified *a priori* from previous research. It can be useful to formalize this prior knowledge in a diagram of the hypothesized causal model. While most important for models oriented to causal interpretation, diagrams can also help select variables which are likely to be most useful for prediction.

Fig. 5.1 shows some conventional ways of diagraming causal relationships, illustrating the case with a predictor of primary interest. Causal links between predictors, or between predictor and outcome, are shown by arrows pointing from cause to effect. An arrow may represent negative (e.g., HT reduces LDL) and well as positive causal pathways. A statistical association between variables not directly linked as cause and effect is shown by two-headed arrows; in many cases this association reflects a common causal antecedent. Confounders are linked by an arrow to the outcome, and to the variables they confound by single or two-headed arrows (Fig. 5.1A, B). An interaction is represented by an arrow from the effect modifier which intersects the arrow for the modified causal effect (Fig. 5.1C). Mediation is represented by an arrow from the primary predictor to the mediator, and another arrow from mediator to outcome; where more than one causal pathway from predictor to outcome is hypothesized, multiple arrows can be used, some of them passing through mediators, some not (Fig. 5.1D).

Fig. 5.2 shows a hypothetical causal model relating HT use in a population and CHD events. Most CHD risk factors are shown as causally related to CHD events, but as correlates of HT use, since the links between voluntary use of HT seem unlikely to be direct. Some of the correlations are shown with question marks, and could easily be evaluated in the data. There are also many interrelationships among the risk factors. For example, BMI and age increase risk of diabetes, which in turn increases lipid (cholesterol) levels and hypertension; furthermore, all these factors are shown as potentially affecting CHD risk directly, as well as through the hypothesized mediating relationships. BMI is shown as possibly causing decreased exercise, as well as responding to it, a complexity that would be difficult to sort out using observational data.

Even though just a subset of the possible causal pathways, mediating relationships, and associations among known risk factors, HT, and CHD risk are shown in Fig. 5.2, the diagram suggests that the causal pathways leading to heart disease events are complex. The utility of causal diagrams is in helping to think through hypotheses in advance of selecting predictors for a multipredictor regression model. In the following sections, we examine how the relationships shown in a causal diagram can be implemented in predictor selection for three inferential goals: prediction of new events, assessing a predictor of primary interest, and identifying the important independent predictors of an outcome.

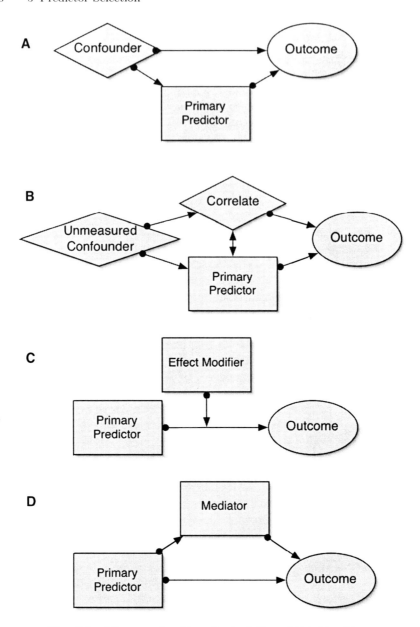

Fig. 5.1. Diagramming Hypothesized Causal Relationships

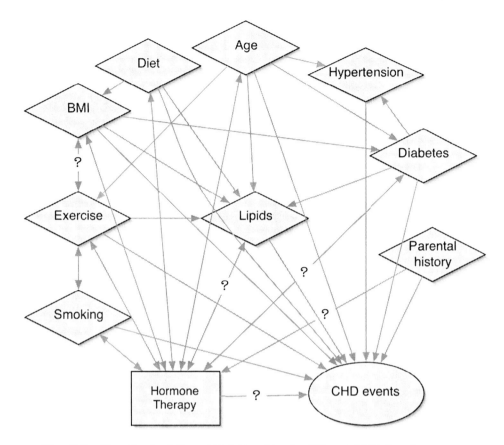

Fig. 5.2. Hypothesized Causal Relationships for HT and CHD Events

5.2 Prediction

In selecting a good prediction model, candidate predictors should be compared in terms of prediction error.

Definition: *Prediction error (PE)* measures how well the model is able predict the outcome for a new, randomly selected observation that was not used in estimating the parameters of the prediction model.

5.2.1 Bias–Variance Trade-off

Inclusive models that minimize confounding may not work as well for prediction as models with smaller numbers of predictors. This can be understood in terms of the *bias–variance trade-off*. Bias in predictions is often reduced when more variables are included in the model, provided they are measured and modeled adequately. But as less important covariates are added to the model, precision may start to erode, without commensurate decreases in bias. The larger models may be *overfitted* to the idiosyncrasies of the data at hand, and thus less able to predict new, independent observations. The bottom line is that a smaller model which gives slightly biased estimates of the regression coefficients may predict the outcome for new observations better than a larger but less precisely estimated model. The model which optimizes the bias–variance trade-off is by definition the model which minimizes prediction error, the natural criterion for selecting among candidate prediction models.

5.2.2 Estimating Prediction Error

R^2, interpretable as the proportion of variance explained by a linear regression model, increases with each additional covariate, even if it adds minimal information about the outcome. At the extreme, $R^2 = 1$ in a model with one predictor for each observation. Thus the model that maximizes R^2 is unlikely to minimize PE, essentially because the same observations are used to estimate the model and assess its predictiveness.

Better estimates of PE are based on *generalized cross-validation* (GCV), a class of methods that work by using different, independent sets of observations to estimate the model and to evaluate PE. The most straightforward example of cross-validation is the learning set/test set (LS/TS) approach, in which the parameter estimates are obtained from the learning set and PE is evaluated in the test set. In linear regression, computing PE is straightforward, using $\hat{\beta}$ from the learning set to compute the predicted value $\hat{y}$ and corresponding residual for each observation in the test set. In some implementations, the learning and test sets are obtained by splitting a single data set, often with two-thirds of the observations randomly assigned to the learning set. Other implementations, as in Walter's analysis of post-hospitalization mortality among high-risk older adults, use an independent sample as the test set. This may give more generalizable estimates of PE, since the test set is not sampled under exactly the same circumstances as the learning set. LS/TS is less efficient than some alternatives but easier to implement.

An alternative to LS/TS is leave-one-out or jackknife methods, in which all but one observation are used to estimate the model, and then PE is evaluated for the omitted observation. In linear regression models, this estimate of PE can be computed and averaged over every observation with minimal extra computation. The resulting predicted residual sum of squares (PRESS) is available from most regression packages. In logistic and Cox models, exact leave-one-out estimates of PE are time-consuming to compute, but fast one-step approximations are available. The best prediction model minimizes PRESS.

Midway between LS/TS and the jackknife is *h-fold cross-validation* (hCV). The data set is divided into h mutually exclusive subsets and a measure of PE is evaluated in each subset, using parameter estimates obtained from the remaining observations. A global estimate of PE is then found by averaging over the h subset estimates. Setting $h = 10$ is often recommended. This short-cut generally gives better estimates of PE than the leave-one-out jackknife method.

Alternative measures theoretically motivated by PE include adjusted R^2, which works by penalizing R^2 for the number of predictors in the model. Thus when a variable is added, adjusted R^2 increases only if the increment in R^2 outweighs the added penalty. Mallow's C_p the Akaike Information Criterion (AIC), and the Bayesian Information Criterion (BIC) are analogs which impose stiffer penalties for each additional variable, and thus lead to selection of smaller models. These measures are easily obtained in most regression packages; in Stata the `regress` command prints adjusted R^2 by default. The best prediction model is taken to be the one that maximizes adjusted R^2, C_p, AIC, or BIC.

5.2.3 Screening Candidate Models

An efficient way to evaluate candidate prediction models is the *best subsets* procedure, which exhaustively examines models including various numbers of candidate predictors. A hypothetical causal diagram can be used to suggest variables that should be in the initial list, which can be supplemented by alternative predictive measures (eg, waist hip ratio instead of BMI) and transformations. Though not yet available in Stata, implementations in other statistical packages make it possible to select candidate models directly in terms of approximate measures of PE, such as adjusted R^2 and Mallow's C_p. Where an automated best subsets procedure is unavailable, it is possible, though potentially tedious, to implement the procedure by hand, comparing candidate models suggested by the causal diagram in terms of a cross-validation estimate of PE.

Note that estimated PE will be biased low for the selected model, because models have been compared in terms of the cross-validation PE measure. Nonetheless, a model selected by this criterion is likely to predict better than a model that maximizes R^2 or models selected to achieve the other inferential goals described in Sects. 5.3 and 5.4.

5.2.4 Classification and Regression Trees (CART)

Sophisticated methods for selecting optimal prediction models have been developed in the last 20 years. Among the most important are *tree-based* methods (Breiman *et al.*, 1984). Using *recursive partitioning*, a sample is repeatedly subdivided on the basis of predictor variables into groups that are as internally homogeneous as possible, in terms of the outcome. For a continuous outcome, this means minimizing the within-group variance, while maximizing the differences between groups; for a categorical outcome, it means finding groups that are composed of one outcome category to the greatest extent possible. The partitioning is recursive because the two groups formed by the first split are themselves potentially split into two groups each, then the resulting four groups potentially split, and so on until the resulting groups meet criteria for homogeneity or minimum size. Each split is based on an exhaustive search for the partition based on a single predictor that gives the greatest increase in homogeneity. When a tree has been fully grown, cross-validation is used to prune it back in such a way as to minimize PE. This reflects the fact that smaller, slightly biased models often predict better than larger, more variable ones. The resulting tree has the form of a diagnostic flowchart; at least one CART-based decision tree is in widespread use in helping to decide which emergency room patients with chest pain should be admitted for probable heart attack (Goldman *et al.*, 1996).

5.3 Evaluating a Predictor of Primary Interest

In observational data, the main problem in evaluating a predictor of primary interest is to rule out confounding of the association between this predictor and the outcome as persuasively as possible. Potential confounders to be considered include factors identified in previous studies or hypothesized to matter on substantive grounds, as well as variables that behave like confounders by the statistical measures described in Sect. 4.4.

Three classes of covariates would not be considered for inclusion in the model: covariates which are essentially alternative measures of either the outcome or the predictor of interest, and those hypothesized to mediate its effect. A diagram of the proposed causal model can be useful for clarifying hypotheses about these relationships, which can be complex, and selecting variables for consideration. In Fig. 5.2 lipids both mediate and possibly confound the association of HT with CHD events. A possible argument for including lipids in the model is that the resulting estimate for HT would not include the component of the effect mediated through the lipid pathway, implying that the total effect of HT would be larger.

In contrast, mediation of one confounder by another would not affect the estimate for the primary predictor nor its interpretation. In Fig. 5.2 there are many apparent mediating relationships between CHD risk factors. Similarly, high correlation between pairs of adjustment of confounding variables would not necessarily be a compelling reason for removing one of them, if both are seen as necessary on substantive or statistical grounds; the reason is that collinearity between confounding variables will not affect the estimate for the primary predictor or its precision. Covariates which are in some sense alternative measures of the outcome are not always easy to recognize, but should usually be excluded. For example, it would be problematic to include diabetes in a model for glucose, because diabetes is largely defined by elevated glucose. Another example is history of a potentially recurrent outcome like falling in a model for subsequent incidence of the outcome. In both examples, addition of the alternative outcome measure as a predictor to the model tends to attenuate the estimates for other, more interpretable predictors.

5.3.1 Including Predictors for Face Validity

Some variables in the hypothesized causal model may be such well-established causal antecedents of the outcome that it makes sense to include them, essentially to establish the face validity of the model and without regard to the strength or statistical significance of their associations with the primary predictor and outcome in the current data set. The risk factors controlled for in the Nurse's Health Study analysis of the effects of hormone therapy on CHD risk are well-understood and meet this criterion.

5.3.2 Selecting Predictors on Statistical Grounds

In many areas of research, the potential confounders of a predictor of interest may be less well established, so that in the common case where there are many such potential confounders, *a priori* selection of a reasonable subset to adjust for is not a realistic option. However, the inclusion of too many predictors may unacceptably inflate the standard errors of the regression coefficients, especially in smaller samples; in logistic and Cox models bias can also be induced when too many parameters are estimated. We discuss collinearity and the numbers of predictors that can safely be included in Sects. 5.5.1 and 5.5.2. Because of these potential problems, we would like to eliminate variables that are effectively not confounders in the data at hand, because they demonstrate little or no independent association with the outcome after adjustment. Similarly, hypothesized interactions that turn out not to be important on statistical grounds would be eliminated, almost always before either of the interacting main effects are removed.

To do this, we recommend using a backward selection procedure. Our preference for backward selection rather than forward or stepwise procedures is explained in Sect. 5.5.3. However, to rule out confounding more effectively, we recommend a liberal criterion for removal: in particular, only removing variables with P-values ≥ 0.2 (Maldonado and Greenland, 1993). A comparably effective alternative is to retain variables if removing them changes the coefficient for the predictor of interest by more than 10% or 15% (Greenland, 1989; Mickey and Greenland, 1989). These inclusive rules are particularly important in small data sets, where even important confounders may not meet the usual criterion for statistical significance.

5.3.3 Interactions With the Predictor of Primary Interest

A potentially important check on the validity of the selected model is to assess interactions between the primary predictor and important covariates, in particular those that are biologically plausible. Especially for a novel or controversial main finding, it can add credibility to show that the association is similar across subgroups. There is no reason for concern if the association is statistically significant in one subgroup but not in the complementary group, provided the subgroup-specific estimates are similar. However, if a substantial and credible interaction is found, particularly such that the association with the predictor of interest differs qualitatively across subgroups, then the analysis would need to take account of this complexity. For example, Kanaya *et al.* (2004) found an interaction between change in obesity and hormone therapy in predicting CHD and mortality risk which substantively changed the interpretation of the finding. However, since such exploratory analyses are susceptible to false-positive findings, this unexpected and hard-to-explain interaction was cautiously interpreted.

5.3.4 Example: Incontinence as a Risk Factor for Falling

Brown *et al.* (2000) examined urinary incontinence as a risk factor for falling among 6,049 ambulatory, community-dwelling women in the SOF cohort also studied by Orwoll. The hypothesis was that incontinence might cause falling because of hasty trips to the bathroom, especially at night. But it was important to rule out confounding by physical decline, which is strongly associated with both aging and incontinence. The final model included all predictors which were associated with the outcome at $P < 0.2$ in univariable analysis and remained statistically significant at that level after multivariable adjustment. Alternative and more inclusive models with different sets of predictors were also assessed. After adjustment for 12 covariates (age; history of non-spine fracture and falling; living alone; physical activity; use of a cane, walker, or crutch; history of stroke or diabetes; use of two classes of drugs; a physical performance variable; and BMD) weekly or more frequent urge incontinence was independently associated with a 34% increase in risk of falling (95% confidence interval 6–69%, $P = .01$).

In this example, falling was defined as a binary outcome, to be discussed in Chapter 6. In addition, because the outcome was observed over multiple time intervals for each SOF participant, methods presented in Chapter 8 for longitudinal repeated measures were used. A subsequent example in Sect. 5.5.2 uses a Cox proportional hazards model, to be covered in Chapter 7. In using these varied examples, we underscore the fact that predictor selection issues are essentially the same for all the regression models covered in this book.

5.3.5 Randomized Experiments

In clinical trials and other randomized experiments, the intervention is the predictor of primary interest. Other predictors are, in expectation, uncorrelated with the intervention, by virtue of randomization. Thus, in the regression model used to analyze an experiment, covariates do not usually need to be included to rule out confounding of assignment to the intervention. However, there are several other reasons for including covariates in the models used to analyze experiments.

Making valid inferences in stratified designs. Design variables in stratified designs need to be included to obtain correct standard errors, confidence intervals, and *P*-values. At issue is the potential for clustering of outcomes within strata, potentially violating the assumption of independence (Chap. 8). Thus analyses of multicenter clinical trials now commonly take account of clinical center, even though random and equal allocation to treatment within center ensures that treatment is in expectation uncorrelated with this factor. Clustering within center can arise from differences in the populations studied and in the implementation of the intervention.

Increasing precision and power in experiments with continuous outcomes. Adjusting for important baseline predictors of a continuous outcome can increase the precision of the treatment effect estimate by reducing the residual error; because the covariates are in expectation uncorrelated with treatment, the variance inflation factor described in Sect. 4.2.2 is usually negligible. However, Beach and Meier (1989) use simulations to suggest that adjustment may on average increase squared error of the treatment effect estimate in smaller studies or when the selected covariates are not strongly predictive of the outcome. They also explore the difficulties in selecting a reasonable subset of the many baseline covariates typically measured, and conclude that adjusting for covariates which are both imbalanced and strongly predictive of the outcome has the largest expected effect on the statistical significance of the treatment effect estimate. We support adjustment for important prognostic covariates in trials with continuous endpoints, but also endorse the stipulation of Hauck *et al.* (1998) that the adjusted model should be pre-specified in the study protocol, to prevent *post hoc* "shopping" for the set of covariates which gives the smallest treatment effect *P*-value.

"De-attenuating" the treatment effect estimate and increasing power in experiments with binary or failure time outcomes. In contrast to linear models for continuous outcomes, omission of important but balanced predictors, including the stratification variables mentioned previously, from a logistic (Neuhaus and Jewell, 1993; Neuhaus, 1998) or Cox model (Gail *et al.*, 1984; Schmoor and Schumacher, 1997; Henderson and Oman, 1999) used to analyze binary or failure time outcomes attenuates the treatment effect estimate. Hypothesis tests remain valid when the null hypothesis holds (Gail *et al.*, 1988), but power is lost in proportion to the importance of the omitted covariates (Lagakos and Schoenfeld, 1984; Begg and Lagakos, 1993). Note, however, that adjustment for *im*balanced covariates can potentially move the treatment effect estimate *away* from as well as toward the null value, and can decrease both precision and power.

In their review, Hauck *et al.* (1998) recommend adjustment for influential covariates in trials analyzed using logistic and Cox models. Their rationale is not only increased efficiency, but also that the adjusted or de-attenuated treatment effect estimates are more nearly interpretable as *subject-specific* – in contrast to *population-averaged*, a distinction that we explain in Sect. 8.5. We cautiously endorse adjustment for important covariates in trials with binary and failure time endpoints, but only if the adjusted model can be pre-specified and adjustment is likely to make the results more, not less convincing to the intended audience.

Adjusting for baseline imbalances. Adjusted analyses are often conducted when there are apparent imbalances between groups, which can arise by chance, especially in small studies, or because of problems in implementing the randomization. The treatment effect estimate can be badly biased when strongly predictive covariates are imbalanced, even if the imbalance is not statistically significant. It is of course not possible to pre-specify such covariates, but adjustment is commonly undertaken in secondary analyses to demonstrate that the inferences about the treatment effect are not qualitatively affected by any apparent baseline imbalance. Note that the precision and statistical significance of the treatment effect estimate can be eroded by adjustment in this case, whether the endpoint is continuous, binary, or a failure time.

However, a difficult problem can arise when the selection of covariates to adjust for makes a substantive difference in interpretation, as Beach and Meier (1989) show in a re-analysis of time-to-event data from the Chicago Breast Cancer Surgery Study (Meier *et al.*, 1985). In this small trial ($n = 112$), where the unadjusted treatment effect estimate just misses statistical significance ($P = 0.1$), different sets of covariates give qualitatively different results, with some adjusted models showing a statistically significant treatment effect and others weakening and even reversing the direction of the estimate.

5.4 Identifying Multiple Important Predictors

When the focus is on evaluating a predictor of primary interest, covariates are included in order to obtain a minimally confounded estimate of the association of the main predictor with the outcome. A good model rules out confounding of that association as persuasively as possible. However, broadening the focus to multiple important predictors of an outcome can make selecting a single best model considerably more difficult.

For example, inferences about most or all of the predictors retained in the model are now of primary interest, so overfitting and false-positive results are more problematic, particularly for novel associations not strongly motivated *a priori*. Effect modification or interaction will usually be of interest, but systematically assessing the large number of possible interactions can easily lead to false-positive findings, some at least not easily rejected as implausible. It may also be difficult to choose between alternative models that each include one variable from a collinear pair or set. Mediation is also more difficult to handle, to the extent that both the overall effect of a predictor as well as its direct and indirect effects may be of interest. In this case, multiple, nested models may be required, as outlined in Sect. 4.4. Especially in the earlier stages of research, modeling these complex relationships is difficult, prone to error, and likely to be an iterative process.

Fig. 5.2 shows how complex the relationships between multiple predictors can be. In an analysis of a range of CHD risk factors, not focused on the effect of HT, considerable care would have to be taken in dealing with mediation, in particular via pathways involving diabetes. For example, diabetes mediates some of the effects of BMI and age, and its effects on CHD events are in turn partly mediated by lipids and hypertension. In the cohort of participants in the HERS trial of hormone therapy for coronary heart disease (Vittinghoff *et al.*, 2003), BMI predicts CHD events in unadjusted but not adjusted analysis. Because the attenuation of the estimate after adjustment probably results from mediation rather than confounding, and since obesity is a modifiable risk factor, weight control was nonetheless recommended as a means of CHD risk reduction. Diabetes is a powerful independent predictor of CHD events even after adjustment for mediators including lipids and hypertension, suggesting that other causal pathways are involved. It might also be necessary to consider effect modification by other medications not shown in the diagram; for example, if blood pressure differentially predicted CHD events in women according to their use of anti-hypertensive medications. In this complex framework, a series of models, possibly including interactions with treatment, might be necessary to give a full and interpretable picture.

In our view, simplifying the problem by treating each of the candidate predictors in turn as a predictor of primary interest, using the procedures from the previous section, is not a satisfactory solution. This can result in as many different models as there are predictors of interest, especially if covariates are retained because removing them changes the coefficient of the predictor of interest. Such a description of the data is uneconomical and hard to reconcile with an internally consistent causal model. Furthermore, missing values can result in the different models being fit to different subsets of the data.

Given the complexities of assessing multiple independent predictors, the causal diagrams introduced in Sect. 5.1 can be especially useful in sorting out hypotheses about relationships among candidate predictors, which may include interaction and mediation as well as confounding.

5.4.1 Ruling Out Confounding Is Still Central

In exploratory analyses to identify the important predictors of an outcome, confounding remains a primary concern – in this case for any of the independent predictors of interest. Thus some of the same strategies useful when a single predictor is of primary interest are likely to be useful here. In particular, relatively large models that include variables thought necessary for face validity, as well as those that meet a liberal backward selection criterion, are preferable. However, as in the previous section, small sample size and high correlation between predictors may limit the number of variables that can be included. We discuss these issues in more detail in Sects. 5.5.1 and 5.5.2.

5.4.2 Cautious Interpretation Is Also Key

What principally differs in this context is that *any* of the associations in the final model may require substantive interpretation, not just the association with a primary predictor. This may justify a more conservative approach to some minor aspects of the model; for example, poorly motivated and implausible interactions might more readily be excluded. In addition, choices among any set of highly correlated predictors would need to be made.

However, we do not recommend "parsimonious" models that only include predictors that are statistically significant at $P < 0.05$ or even stricter criteria, because the potential for residual confounding in such models is substantial. We also do not recommend explicit correction for multiple comparisons, since in an exploratory analysis it is far from clear how many comparisons to correct for, and by how much.

A better approach is to interpret the results of a larger model cautiously, especially novel, implausible, weak, and borderline statistically significant associations. In some cases it may make sense to focus the presentation of results on a subset of the predictors in the model, leaving the remaining control variables essentially in the background, as in the case of a single primary predictor.

5.4.3 Example: Risk Factors for Coronary Heart Disease

Vittinghoff *et al.* (2003) used multipredictor Cox models to assess the associations between risk factors and coronary heart disease (CHD) events among 2,763 post-menopausal women with established CHD. Risk factors considered included many of those shown in Fig. 5.2. Because of the large number ($n = 361$) of outcome events, it was possible to include all previously identified risk factors that were statistically significant at $P < 0.2$ in unadjusted models and not judged redundant on substantive grounds in the final multipredictor model. Among the 11 risk factors judged to be important on both substantive and statistical grounds were six noted by history (nonwhite ethnicity, lack of exercise, treated diabetes, angina, congestive heart failure, ≥ 2 previous heart attacks) and five that were measured (high blood pressure, lipids including LDL, HDL, and Lp(a), and creatinine clearance).

For face validity and to rule out confounding, the final model also controlled for other known or suspected CHD risk factors, including age, smoking, alcohol use, and obesity, although these were not statistically significant in the adjusted analysis. Mediation of obesity and diabetes, both shown to be associated with risk in single-predictor models, was covered in the discussion section of the paper. The model also controlled for a wide range of CHD-related medications, but because these effects were not of direct interest and hard to interpret, estimates were not presented. However, interactions between risk factors and relevant treatments were examined, on the hypothesis that treatments might modify the association between observed risk factor levels and future CHD risk; the final model included interactions that were statistically significant at $P < 0.2$.

5.4.4 Allen–Cady Modified Backward Selection

Flexible predictor selection procedures, including conventional backward selection, are known to increase the probability of making at least one type-I error. A backward selection procedure (Allen and Cady, 1982) based on a ranking of the candidate variables by importance can be used to help avoid false-positive results, while still reducing the number of covariates in the model. In this procedure, a set of variables may be forced into the model, including predictors of primary interest, as well as confounding variables thought important for face validity. The remaining candidate variables would then be ranked in order of importance. Starting with an initial model including all covariates in these two sets, variables in the second set would be deleted in order of ascending importance until the first variable meeting a criterion for retention is encountered. Then the selection procedure stops.

This procedure is special in that only the remaining variable hypothesized to be least important is eligible for removal at each step, whereas in conventional backward selection, any of the predictors not being forced into the model is eligible. False-positive results are less likely because there is only one pre-specified sequence of models, and selection stops when the first variable not meeting the criterion for removal is encountered. In contrast, conventional stepwise procedures and especially best subsets search over broader classes of models.

5.5 Some Details

5.5.1 Collinearity

In Sect. 4.2 we saw that $s^2_{\beta_j}$, the variance of the regression coefficient estimate for predictor x_j, increases with r_j, the multiple correlation between x_j and the other predictors in the model. When r_j is large, the estimate of β_j can become quite imprecise. Consider the case where two predictors are fairly highly correlated ($r \geq 0.80$). When both are included in the model, the precision of the estimated coefficient for each can be severely degraded, even when both variables are statistically significant predictors in simpler models that include one but not both. In the model including both, an F-test for the joint effect of both variables may be highly statistically significant, while the variable-specific t-tests are not. This pattern indicates that the two variables jointly provide important information for predicting the outcome, but that neither is necessary over and above the other. With modern computers, problems in estimating the independent effects of highly correlated predictors no longer arise from numeric inaccuracy in the computations. Rather, the information is coming from both variables jointly, which makes them both seem unimportant in t-tests evaluating their individual contributions.

> Definition: *Collinearity* denotes correlation between predictors high enough to degrade the precision of the regression coefficient estimates substantially for some or all of the correlated predictors.

How we deal with collinear predictors depends in part on our inferential goals. For a prediction model, inclusion of collinear variables is unlikely to decrease prediction error, which provides a straightforward criterion for choosing one or the other.

Alternatively, suppose that one of two collinear variables is a predictor of primary interest, and the other is a confounder that must be adjusted for on substantive grounds. If the predictor of interest remains statistically significant after adjustment, then the evidence for an independent effect is usually convincing. In small data sets especially it would be necessary to demonstrate that the finding is not the result of a few influential points, and where the data do not precisely meet model assumptions, to show that the inferences are robust, possibly using the bootstrap methods introduced in Sect. 3.6. Alternatively, if the effects of the predictor of interest are clearly confounded by the adjustment variable, we would also have a clearcut result. However, in cases where neither is statistically significant after adjustment, we may need to admit that the data are inadequate to disentangle their effects.

In contrast, where the collinearity is between adjustment variables and does not involve the predictor of primary interest, then inclusion of the collinear variables can sometimes be justified. In this case information about the underlying factor being adjusted for may be increased, but the precision of the estimate for the predictor of interest is unaffected. To see this, consider evaluating the effect of diabetes on HDL, adjusting for BMI. In Sect. 4.7, we found that a quadratic term in BMI added signficantly to the model. However, BMI and its square are clearly collinear ($r = 0.99$). If instead we first "center" BMI (that is, subtract off its sample mean before computing its square), the collinearity disappears ($r = 0.46$). However, the estimate for diabetes and its standard error are unchanged whether or not we center BMI before computing the quadratic term. In short, collinearity between adjustment variables is unlikely to matter.

Finally, when we are attempting to identify multiple independent predictors, an attractive solution is to choose on substantive grounds, such as plausibility as a causal factor. Otherwise, it may make sense to choose the predictor that is measured more accurately or has fewer missing values. As in the case of a predictor of primary interest, the multivariable model may sometimes provide a clear indication of relative importance, in that one of the collinear variables remains statistically significant after adjustment, while the others appear to be unimportant. In this case the usual course would be to include the statistically significant variable and drop the others.

5.5.2 Number of Predictors

The rationale for inclusive predictor selection rules, whether we are assessing a predictor of primary interest or multiple important independent predictors, is to obtain minimally confounded estimates. However, this can make regression coefficient estimates less precise, especially for highly correlated predictors. At the extreme, model performance can be severely degraded by the inclusion of too many predictors.

Rules of thumb have been suggested for number of predictors that can be safely included as a function of sample size or number of events. A commonly used guideline prescribes ten observations for each predictor; with binary or survival outcomes the analogous guideline specifies ten events per predictor (Peduzzi et al., 1995, 1996; Concato et al., 1995). The rationale is to obtain adequately precise estimates, and in the case of the logistic and Cox models, to ensure that the models behave properly.

However, such guidelines are too simple for regular use, although they are useful as flags for potential problems. Their primary limitation is that the precision of coefficient estimates depends on other factors as well as the number of observations or events per predictor. In particular, recall from Sect. 4.2 that the variance of an estimated regression coefficient in a linear model depends on the residual variance of the outcome, which is generally reduced by the inclusion of important covariates. Precision also depends on the multiple correlation between a predictor of interest and other variables in the model. Thus addition of covariates that are at most weakly correlated with the primary predictor but explain substantial outcome variance can actually improve the precision of the estimate for the predictor of interest. In contrast, addition of just one collinear predictor can degrade its precision unacceptably. In addition, the allowable number of predictors depends on effect size, with larger effects being more robust to multiple adjustment than smaller ones.

Rather than applying such rules categorically, we recommend that problems potentially stemming from the number of predictors be assessed by checking for high levels of correlation between a predictor of interest and other covariates, and for large increases in the standard error of its estimated regression coefficient when additional variables are included. For logistic and Cox models, consistency between Wald and likelihood ratio (LR) test results is another useful measure of whether there are enough events to support the number of predictors in the model. Additional validation of a relatively inclusive final model is provided if a more parsimonious model with fewer predictors gives consistent results, in particular for the predictor of interest. If problems do become apparent, a first step would be to make the criterion for retention in backward selection more conservative, possibly $P < 0.15$ or $P < 0.10$. It would also make sense to consider omitting variables included for face validity which do not appear to confound a predictor of primary interest.

Table 5.1. Cox Models for DVT-PE

Predictor variable	RH (95% Confidence Interval) 11-Predictor Model		RH (95% Confidence Interval) 5-Predictor Models		P-values Wald	P-values LR
HT vs. placebo	2.7	(1.4–5.2)	2.7	(1.4–5.1)	0.002	0.001
≥ 53 at LMP	3.6	(2.0–6.4)	3.3	(1.8–5.8)	< 0.001	< 0.001
Inpatient surgery	4.3	(2.1–8.7)	4.7	(2.3–9.5)	< 0.001	< 0.001
Hospitalization	5.6	(2.9–11)	6.7	(3.6–13)	< 0.001	< 0.001
Hip fracture	5.9	(0.8–46)	6.6	(0.9–51)	0.09	0.18
Leg fracture	17.3	(5.1–58)	14.1	(4.2–47)	< 0.001	< 0.001
Cancer	4.1	(1.7–9.7)	3.5	(1.5–8.4)	0.002	0.006
Nonfatal MI	6.0	(2.3–16)	4.4	(1.7–11)	< 0.001	0.002
Stroke/TIA	0.9	(0.1–6.5)	0.9	(0.1–6.4)	0.88	0.88
Aspirin use	0.4	(0.2–0.7)	0.4	(0.2–0.6)	0.003	0.004
Statin use	0.4	(0.2–0.9)	0.4	(0.2–0.7)	0.02	0.02

An analysis of risk factors for deep-vein thrombosis and pulmonary embolism (DVT-PE) among post-menopausal women in the HERS cohort (Grady et al., 2000) is an example of stable results despite violation of the rule of thumb that the number of events per predictor should be at least 10. In this survival analysis of 47 DVT-PE events, 11 predictors were retained in the final model, so that there were only 4.3 events per predictor. However, the largest pairwise correlation between the selected risk factors was only 0.16 and most were below 0.02. As shown in Table 5.1, estimates from the 11-predictor model were consistent with those given by 5-predictor models, in accord with the rule of thumb, which omitted the less important predictors. Although confidence intervals were wide for the strongest and least common risk factors, this was also true for the 5-predictor models. Finally, P-values for the Wald and LR tests based on the larger model were highly consistent.

5.5.3 Alternatives to Backward Selection

Some alternatives to backward selection include best subsets, which was already described; sequential procedures, including forward and stepwise selection; and bivariate screening.

- *Forward selection* begins with the null model with only the intercept, then adds variables sequentially, at each step adding the variable that promises to make the biggest additional contribution to the current model.
- *Stepwise* methods augment the forward procedure by allowing variables to be removed if they no longer meet an inclusion criterion

after other variables have been added. Stata similarly augments backward selection by allowing variables to re-enter after removal. As compared to best subsets, these three sequential procedures are more vulnerable to missing good alternative models that happen not to lie on the sequential path. This implies that plausible alternatives to models selected by stepwise procedures should be examined.

- In *bivariate screening* candidate predictors are evaluated one at a time in single-predictor models. In some cases all predictors that meet the screening criterion are included in the final model; in other cases, screening is used as a first step to reduce the number of predictors then considered in a backward, forward, stepwise, or best subsets selection procedure. Orwoll *et al.* (1996) used a variant of this procedure, including all variables statistically significant at $P < 0.05$ in two-predictor models adjusting for age.

Note that only observations with complete data on all variables under consideration are used in automated selection procedures. The resulting subset can be substantially smaller than the data set used in the final model, and unrepresentative. When implemented by hand, different subsets are commonly used at different steps, for the same reason, and this can also affect results. Findings which depend on the inclusion or exclusion of subsets of observations should be carefully checked.

Why We Prefer Backward Selection

The principal advantage of backward selection is that negatively confounded sets of variables are less likely to be omitted from the model (Sun *et al.*, 1999), since the complete set is included in the initial model. Best subsets shares this advantage. In contrast, forward and stepwise selection procedures will only include such sets if at least one member meets the inclusion criterion in the absence of the others. Univariate screening will only include the complete set if all of them individually meet the screening criterion; moreover, this difficulty is made worse if a relatively conservative criterion is used to reduce the number of false-positive findings in an exploratory analysis.

5.5.4 Model Selection and Checking

Sect. 4.7 focused on methods for checking the linear model which make use of the residuals from a multipredictor model rather than examining bivariate relationships. There we took as a given that the predictors had already been selected. However, transformation of the outcome or of continuous predictors can affect the apparent importance of predictors. For example, in Sect. 4.6.7 we saw that the need for an interaction between treatment with HT and the baseline value of the outcome LDL was eliminated by analyzing treatment effects on percent rather absolute change from baseline. Alternatively, detection of important nonlinearities in the model checking step can uncover associations that were masked by an initial linear specification. As a consequence, predictor selection should be revisited after changes of this kind are made. And then, of course, a modified model would need to be rechecked.

5.5.5 Model Selection Complicates Inference

Underlying the confidence intervals and P-values which play a central role in interpreting regression results is the assumption that the predictors to be included in the model were determined *a priori* without reference to the data at hand. In *confirmatory* analyses in well-developed areas of research, including phase-III clinical trials, prior determination of the model is feasible and important. In contrast, at earlier stages of research, data-driven predictor selection and checking are reasonable, even obligatory, and certainly widely used. However, some of the issues raised for inference include the following–

- The chance of at least one type-I error can greatly exceed the nominal level used to test each term, leading to false-positive results with too-small P-values and too-narrow confidence intervals.
- In small data sets precision and power are often poor, so important predictors may well be omitted from the model, especially if a restrictive inclusion criterion is used. Conversely, in large data sets unimportant predictors are commonly included, reinforcing the need for cautious interpretation of novel, implausible, weak, and borderline statistically significant findings.
- Parameter estimates can be biased away from the null, owing to selection of estimates that are large by chance (Steyerberg *et al.*, 1999).
- Choices between predictors can be poorly motivated, especially between collinear variables. Univariate screening provides no guidance for this problem. Moreover, predictor selection is potentially sensitive to addition or deletion of a few observations, especially when the predictors are highly correlated. Altman and Andersen (1989) propose bootstrap methods for assessing this sensitivity.

Predictor selection driven by P-values is subject to these pitfalls whether it is automated or implemented by hand. How seriously do these problems affect inference for our three inferential goals?

- *Prediction.* In CART, a prediction method introduced in Sect. 5.2.4, candidate cutpoints are exhaustively screened for a potentially large set of candidate predictors, and high-order interactions are routinely included in the model. Breiman (2001) briefly reviews other modern methods which even more aggressively search over candidate models. However, use of GCV measures of prediction error as a criterion for predictor selection effectively protects against both overfitting and invalid inferences. In short, predictor selection does not adversely affect modern procedures for this inferential goal.

- *Evaluating a predictor of primary interest.* Iterative model checking and selection should likewise have relatively small effects on inference about a predictor of primary interest, since it is included by default in all candidate models. In fact, iterative checking and predictor selection should result in better control of confounding, a primary aim for this inferential goal. However, when the primary predictor is of borderline statistical significance, the issue of P-value shopping raised in Sect. 5.3.5 needs to be conscientiously handled, and sensitivity of results to predictor selection reported.

- *Identifying multiple important predictors.* Model selection most clearly complicates inference for this inferential goal, since confidence intervals and P-values for any of the predictors are potentially of direct interest. Note that inclusion of variables for face validity, use of a loose inclusion criterion ($P < 0.2$), and the Allen–Cady procedure all reduce the potential impact of predictor selection on inference. Nonetheless, selection procedures should *only* be used with prior consideration of hypothesized relationships, careful examination of alternative models with other sets of predictors, checks on model fit and robustness, skeptical review of the findings for plausibility, and cautious interpretation of the results, especially novel, borderline statistically significant, and weak associations.

5.6 Summary

We have identified three inferential goals, and recommend predictor selection procedures appropriate to each of them.

For prediction, we recommend using best subsets, where available, to identify a range of candidate models, selecting the model that optimizes a generalized cross-validation measure of prediction error. A causal diagram is useful for identifying the potentially predictive variables to consider for inclusion.

For evaluating a predictor of primary interest, we recommend diagramming the relationships among all potential confounders, effect modifiers, and mediators of the relationship between the primary predictor and outcome. The selected model should include all generally accepted confounders required to ensure its face validity. Other potential confounders that turn out not to be important on statistical grounds can optionally be removed from the model using a backward selection procedure, but with a liberal inclusion criterion to minimize the potential for confounding. Especially in smaller data sets, care must be taken with the inclusion of covariates highly correlated with the predictor of interest, since these can unduly inflate the standard errors of the estimate of its effect. Negative findings for the primary predictor should be carefully interpreted in terms of the point estimate and confidence interval, as described in Sect. 3.7.

For identifying multiple important predictors of an outcome, we recommend a procedure similar to that used for a single predictor of primary interest. A preliminary diagram of the hypothesized relationships between variables can be particularly useful. Strongly motivated covariates may be included by default to ensure the face validity of the model. The Allen–Cady modification of the backward selection procedure is useful for selecting from among the remaining candidate variables while limiting false-positive results. Negative, weak, and/or borderline statistically significant associations retained in the final model as much to control confounding of other associations as for their intrinsic plausibility and importance should be interpreted with particular caution.

5.7 Further Notes and References

Predictor selection may be the most controversial subject covered in this book. Book-length treatments include Miller (1990) and Linhart and Zucchini (1986), while regression texts including Weisberg (1985) and Hosmer and Lemeshow (2000) address predictor selection issues at least briefly. The central place we ascribe to ruling out confounding in the second and third inferential goals owes much to Rothman and Greenland (1998), a standard reference in epidemiology that decribes how substantive considerations can be brought to bear on predictor selection.

Both the theory and application of causal diagrams and models have been advanced substantially in recent years (Pearl, 1995; Greenland *et al.*, 1999) and give additional insights into situations where confounding can be ruled out *a priori*. However, these more advanced methods appear to be most useful in problems where causal pathways are more clearly understood than is our usual experience. Jewell (2004) and Greenland and Brumback (2002) explore the connections between causal diagrams, counterfactuals, and some model selection issues.

Chatfield (1995) reviews work on the influence of predictor selection on inference, while Buckland *et al.* (1997) propose using weighted averages of the results from alternative models as a way of incorporating the extra variability introduced by predictor selection in computing confidence intervals. These would be particularly applicable to the second inferential goal of evaluating a predictor of central interest.

For a sobering view of the difficulty of validly modeling causal pathways using the procedures covered in this book and particularly this chapter, see Breiman (2001). From this point of view, computer-intensive methods validated strictly in terms of prediction error not only give better predictions but may also be more reliable guides to "variable importance" – another term for our third inferential goal of identifying important predictors, and with obvious implications for assessing a predictor of central interest.

Developments in Prediction

Breiman (2001) describes modern methods that do not follow the paradigm, motivated by the bias–variance trade-off, that smaller models are better for prediction. The newer methods tend to keep all the predictors in play, while using various methods to avoid overfitting and control variance; cross-validation retains its central role throughout.

So-called *shrinkage* procedures also play an important role in prediction, especially those made on the basis of small data sets. In this approach overfitting is avoided and prediction improved by shrinking the estimated regression coefficients toward zero, rather than eliminating weak predictors from the model. Analogous shrinkage of predictions for observations in the data at hand is used in the random effects models presented in Chapter 8. A variant of shrinkage is the *LASSO* method, short for *least absolute shrinkage and selection operator* (Tibshirani, 1997).

An alternative to direct shrinkage implements *penalties* in the fitting procedure against coefficient estimates which violate some measure of smoothness. This achieves something like shrinkage of the estimates and thus better predictions; see Le Cessie and Van Houwelingen (1992) and Verweij and Van Houwelingen (1994) for applications to logistic and Cox regression. These methods derive from *ridge regression* (Hoerl and Kennard, 1970), a method for obtaining slightly biased but stabler estimates in linear models with highly correlated predictors.

See Steyerberg *et al.* (2000) and Harrell *et al.* (1996) for guides to implementing prediction procedures in the logistic and survival analysis contexts, respectively.

5.8 Problems

Problem 5.1. Characterize the following contexts for predictor selection as prediction, evaluation of a primary predictor of interest, or identifying the important predictors of an outcome:

- examining the effect of treatment on a secondary endpoint in an RCT
- determining which newborns should be admitted to the neonatal intensive care unit (NICU)
- comparing a measure of treatment success between two surgical procedures for stress incontinence using data from a large longitudinal cohort study
- identifying risk factors for incident hantavirus infection.

Problem 5.2. Consulting Stata documentation, describe how the `sw:` command prefix with options `lockterm1`, `hier`, and `pr()` can be used to implement the Allen–Cady procedure.

Problem 5.3. Think of an outcome under preliminary investigation in the area of your expertise. Following Allen and Cady's prescriptions, try to rank predictors of this outcome in order of importance. Are there any variables that you would include by default?

Problem 5.4. Do any of the variables you have selected in the previous problem potentially mediate the effects of others in your list? If so, how would this affect your decision about what to include in the initial model? What series of models could you use to examine mediation? (See Sect 4.5.)

Problem 5.5. Suppose you included an indicator for diabetes in a multivariable model estimating the independent effect of exercise on glucose. How would you interpret the estimate for exercise? Would you want to consider interactions between exercise and diabetes in this model? How would you deal with use of insulin and oral hypoglycemics?

Problem 5.6. Why are univariate screening and forward selection more likely to miss negatively confounded variables than backward deletion and best subsets?

Problem 5.7. Give an example of a "biologically plausible" relationship that has turned out to be false. Give an example of a biologically *im*plausible relationship that has turned out to be true.

Problem 5.8. Suppose you were using a logistic model to examine the association between a predictor and outcome of interest, and to rule out confounding you needed to include one or two more predictors than would be allowed by the rule of 10 events per variable. In comparing models with and without the two extra predictors, what might signal that you were asking the bigger model to do too much? How would the correlation between the extra variables and the predictor of interest influence your thinking?

5.9 Learning Objectives

1. Diagram hypothetical relationships among confounders, effect modifiers, mediators, and outcome.
2. Describe and implement strategies for predictor selection for
   - prediction
   - evaluation of a primary predictor
   - identifying multiple important predictors.
3. Be familiar with the drawbacks of predictor selection procedures.

6

Logistic Regression

Patients testing positive for a sexually transmitted disease at a clinic are compared to patients with negative tests to investigate the effectiveness of a new barrier contraceptive. One-month mortality following coronary artery bypass graft surgery is compared in groups of patients receiving different dosages of beta blockers. Many clinical and epidemiological studies generate outcomes which take on one of two possible values, reflecting presence/absence of a condition or characteristic at a particular time, or indicating whether a response occurred within a defined period of observation. In addition to evaluating a predictor of primary interest, it is important to investigate the importance of additional variables that may influence the observed association and therefore alter our inferences about the nature of the relationship. In evaluating the effect of contraceptive use in the first example, it would be clearly important to control for age in addition to behaviors potentially linked to infection risk. In the second example, a number of demographic and clinical variables may be related to both the mortality outcome and treatment regime. Both of these examples are characterized by binary outcomes and multiple predictors, some of which are continuous.

Methods for investigating associations involving binary outcomes using contingency table methods were briefly covered in Sect. 3.4. Although these techniques are useful for exploratory investigations, and in situations where the number of predictor variables of interest is limited, they can be cumbersome when multiple predictors are being considered. Further, they are not well suited to situations where predictor variables may take on a large number of possible values (e.g., continuous measurements). Similar to the way linear regression techniques expanded our arsenal of tools to investigate continuous outcomes, the logistic regression model generalizes contingency table methods for binary outcomes. In this chapter, we cover the use of the logistic model to analyze data arising in clinical and epidemiological studies. Because the basic structure of the logistic model mirrors that of the linear regression model, many of the techniques for model construction, interpretation, and assessment will be familiar from Chapters 4 and 5.

6.1 Single Predictor Models

Recall the example in Sect. 3.4 investigating the association between coronary heart disease (CHD) and age for the Western Collaborative Group Study (WCGS). Table 6.1 summarizes the observed proportions (P) of CHD diagnoses for five categories of age, along with the estimated excess risk (ER), relative risk (RR), and odds ratio (OR). The last three measures are computed according to procedures described in Sect. 3.4, using the youngest age group as the baseline category. The estimates show a tendency for increased risk of CHD with increasing age. Although this information provides a useful summary of the relationship between CHD risk and age, the choice of five-year categories for age is arbitrary. A regression representation of the relationship would provide an attractive alternative and obviate the need to choose categories of age.

Table 6.1. CHD for Five Age Categories in the WCGS Sample

Age group	P	$1 - P$	ER	RR	OR
35–40	0.057	0.943	0.000	1.000	1.000
41–45	0.050	0.950	-0.007	0.883	0.877
46–50	0.093	0.907	0.036	1.635	1.700
51–55	0.123	0.877	0.066	2.156	2.319
56–60	0.149	0.851	0.092	2.606	2.886

Recall that in standard linear regression we modeled the average of a continuous outcome variable y as a function of a single continuous predictor x using a linear relationship of the form

$$\mathrm{E}\left[y|x\right] = \beta_0 + \beta_1 x.$$

We might be tempted to use the same model for a binary outcome variable. First, note that if we follow convention and code the values of a binary outcome as one for those experiencing the outcome and zero for everyone else, the observed proportion of outcomes among individuals characterized by a particular value of x is simply the mean (or "expected value") of the binary outcome in this group. In the notation introduced in Sect. 3.4, we symbolize this quantity by $P(x)$. The linear model for our binary outcome might then be expressed as

$$P(x) = \mathrm{E}\left[y|x\right] = \beta_0 + \beta_1 x. \tag{6.1}$$

This has exactly the same form as the linear regression model; the expected value of the outcome is modeled as a linear function of the predictor. Further, changes in the outcome associated with a specified changes in the predictor x have an excess risk interpretation: For example, if x is a binary predictor taking on the values 0 or 1, the effect of increasing x one unit is to add an increment β_1 to the outcome. From equation (6.1),

$$P(1) - P(0) = \beta_1.$$

Referring back to Definition (3.14) in Sect. 3.4, we see that this is the excess risk associated with a unit increase in x. Models with this property are often referred to as *additive risk models* (Clayton and Hills, 1993).

There are several limitations with the linear model (6.1) as a basis for regression analysis of binary outcomes. First, the statistical machinery which allowed us to use this linear model to make inferences about the strength of relationship in Chapter 4 required that the outcome variable follow an approximate normal distribution. For a binary outcome this assumption is clearly incorrect. Second, the outcome in the above model represents a probability or risk. Thus any estimates of the regression coefficients must constrain the estimated probability to lie between zero and one for the model to make sense. The first of these problems is statistical, and addressing it would require generalizing the linear model to accommodate a distribution appropriate for binary outcomes. The second problem is numerical. To ensure sensible estimates, our estimation procedure would have to satisfy the constraints mentioned. Another issue is that in many settings, it seems implausible that outcome risk would change in a strictly linear fashion for the entire range of possible values of a continuous predictor x. Consider a study examining the likelihood of a toxicity response to varying levels of a treatment. We would not expect the relationship between likelihood of toxicity and dose to be strictly linear throughout the range of possible doses. In particular, the likelihood of toxicity should be zero in the absence of treatment and increase to a maximum level, possibly corresponding to the proportion of the sample susceptible to the toxic effect, with increasing dose.

Fig. 6.1 presents four hypothetical models linking the probability $P(x)$ of a binary outcome to a continuous predictor x. In addition to the linear model (A), there is the exponential model (B) that constrains risk to increase exponentially with x, the "step function" model (C) that allows irregular change in risk with increasing values of x, and the smooth S-shaped curve in (D) known as the *logistic* model. The exponential model is also known as *log linear* because it specifies that the logarithm of the outcome risk is linear in x. It presents a problem similar to that noted for the linear model above: Namely, that risk is not obviously constrained to be less than one for large values of $\beta_0 + \beta_1 x$. Model (C) has the desirable properties that risks are clearly constrained to fall in the interval [0, 1], and that the nature of the increase in the interval can be flexibly represented by different "step" heights. However, it lacks smoothness, a property that is biologically plausible in many instances. By contrast, the logistic model allows for a smooth change in risk throughout the range of x, and has the property that risk increases slowly up to a "threshold" range of x, followed by a more rapid increase and

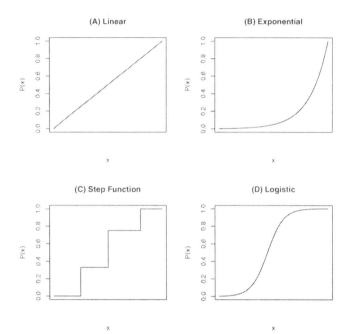

Fig. 6.1. Risk Models for a Binary Outcome and Continuous Predictor

a subsequent leveling off of risk. This shape is consistent with many dose-response relationships (illustrated by the toxicity example from the previous paragraph). As we will see later in this chapter, all of these models represent valid alternatives for assessing how risk of a binary outcome changes with the value of a continuous predictor. However, most of our focus will be on the logistic model.

In addition to a certain degree of biological plausibility, the logistic model does not pose the numerical difficulties associated with the linear and log-linear models, and has a number of other appealing properties that will be described in more detail below. For these reasons, it is by far the most widely used model for binary outcomes in clinical and epidemiological applications, and forms the basis of logistic regression modeling. However, adoption of the logistic model still implies strong assumptions about the relationship between outcome risk and the predictor. In fact, expressed on a transformed scale, the model prescribes a linear relationship between the logarithm of the odds of the outcome and the predictor.

The logistic model plotted in Fig. 6.1(D) is defined by the equation

$$P(x) = \frac{\exp(\beta_0 + \beta_1 x)}{1 + \exp(\beta_0 + \beta_1 x)}. \tag{6.2}$$

In terms of the odds of the outcome associated with the predictor x, the model can also be expressed as

$$\frac{P(x)}{1 - P(x)} = \exp(\beta_0 + \beta_1 x). \tag{6.3}$$

Consider again the simple case where x takes on the values 0 or 1. From the last equation, the ratio of the odds for these two values of x are

$$\frac{P(1)/[1 - P(1)]}{P(0)/[1 - P(0)]} = \exp(\beta_1). \tag{6.4}$$

Expressed in this form, we see that the logistic model specifies that the ratio of the odds associated with these two values of x is given by the factor $\exp(\beta_1)$. Equivalently, the odds for $x = 1$ are obtained by multiplying the odds for $x = 0$ by this factor. Because of this property, the logistic model is an example of a *multiplicative risk model* (Clayton and Hills, 1993). (Note that the log-linear model is also multiplicative in this sense, but is based on the outcome risks rather than the odds.)

Although not easily interpretable in the form given in equations (6.2) and (6.3), expressed as the logarithm of the outcome odds (as given in equation (6.3)), the model becomes linear in the predictor

$$\log \left[\frac{P(x)}{1 - P(x)} \right] = \beta_0 + \beta_1 x. \tag{6.5}$$

This model states that the log odds of the outcome is linearly related to x, with intercept coefficient β_0 and slope coefficient β_1 (i.e., the logistic model is an additive model when expressed on the log odds scale). The logarithm of the outcome odds is also frequently referred to as the *logit* transformation of the outcome probability.

In the language introduced in Chapters 3 and 4, equations (6.2), (6.3), and (6.5) define the systematic part of the logistic regression model, linking the average $P(x)$ of the outcome variable y to the predictor x. The random part of the model specifies the distribution of the outcome variable y_i, conditional on the observed value x_i of the predictor (where the subscript i denotes the value for a particular subject). For binary outcomes, this distribution is called the *binomial* distribution and is completely specified by the mean of y_i conditional on the value x_i. To summarize, the logistic model makes the following assumptions about the outcome y_i:

1. y_i follows a Binomial distribution;
2. the mean $\mathrm{E}\,[y|x] = P(x)$ is given by the logistic function (6.2);
3. values of the outcome are statistically independent.

These assumptions closely parallel those associated with the linear regression (in Sect. 3.3), the primary difference being the use of the binomial distribution for the outcome y. Note that the assumption of constant variance of y across different values of x is not required for the logistic model. Another difference is that the random aspect of the logistic model is not included as an additive term in the regression equation. However, it is still an integral part of estimation and inference regarding model coefficients. (This is discussed further in Sect. 6.6.)

As we will see in the rest of this chapter, both of the alternative expressions (6.2) and (6.5) for the logistic model are useful: the linear logistic form (6.5) is the basis for regression modeling, while the (nonlinear) logistic form (6.2) is useful when we want to express the outcome on its original scale (e.g. to estimate outcome risk associated with a particular value of x).

One of the most significant benefits of the linear logistic formulation (6.5) is that the regression coefficients are interpreted as log odds ratios. These can be expressed as odds ratios via simple exponentiation (as demonstrated above in equation (6.4)), providing a direct generalization of odds ratio methods for frequency tables to the regression setting. This property follows directly from the definition of the model, and is demonstrated in the next section. Finally, we note that there are a number of alternative regression models for binary outcomes that share similar properties to the logistic model. Although none of these comes close to the logistic model in terms of popularity, they offer useful alternatives in some situations. Some of these will be discussed in Sect. 6.5.

6.1.1 Interpretation of Regression Coefficients

Table 6.2 shows the fit of the logistic model (6.5) for the relationship between CHD risk and age in the WCGS study. The coefficient labeled cons in the table is the intercept (β_0), and the coefficient labeled age is the slope (β_1) of the fitted logistic model. Since the outcome for the model is the log odds of CHD risk, and the relationship with age is linear, the slope coefficient β_1 gives the change in the log odds of chd69 associated with a one-year increase in age. We can verify this by using the formula for the model (6.5) and the estimated coefficients to calculate the difference in risk between a 55- and a 56-year-old individual:

$$\log\left[\frac{P(56)}{1 - P(56)}\right] - \log\left[\frac{P(55)}{1 - P(55)}\right]$$
$$= (-5.940 + 0.074 \times 56) - (-5.940 + 0.074 \times 55) = 0.074.$$

This is just the coefficient β_1 as expected; performing the same calculation on an arbitrary one-year age increase would produce the same result (as shown at the end of this section). The corresponding odds ratio for any one-year increase in age can then be computed by simple exponentiation:

$$\exp(0.074) = 1.077.$$

This odds ratio indicates a small (approximately 8%) but statistically significant increase in CHD risk for each one-year age increase. We can estimate

Table 6.2. Logistic Model for the Relationship Between CHD and Age

```
. logistic chd69 age, coef

Logit estimates                              Number of obs   =      3154
                                             LR chi2(1)      =     42.89
                                             Prob > chi2     =    0.0000
Log likelihood = -869.17806                  Pseudo R2       =    0.0241

--------------------------------------------------------------------------
     chd69 |      Coef.   Std. Err.      z     P>|z|    [95% Conf. Interval]
-----------+--------------------------------------------------------------
       age |   .0744226   .0113024     6.58   0.000     .0522703    .0965748
     _cons |  -5.939516    .549322   -10.81   0.000    -7.016167   -4.862865
--------------------------------------------------------------------------
```

the (clinically more relevant) odds ratio associated with a ten-year increase in age the same way, yielding:

$$\exp(0.074 \times 10) = 2.105.$$

Following the same approach we can use equation 6.5 to calculate the log odds ratio and odds ratio for an arbitrary Δ unit increase in a predictor x as follows:

$$\log \left[\frac{\frac{P(x+\Delta)}{1-P(x+\Delta)}}{\frac{P(x)}{1-P(x)}} \right] = \beta_1 \Delta \quad , \quad \frac{\frac{P(x+\Delta)}{1-P(x+\Delta)}}{\frac{P(x)}{1-P(x)}} = \exp(\beta_1 \Delta). \tag{6.6}$$

In addition to computing odds ratios, the estimated coefficients can be used in the logistic function representation of (6.2) to estimate the probability of having CHD during study follow-up for a individual with any specified age. For a 55-year-old individual:

$$P(55) = \frac{\exp(-5.940 + 0.074 \times 55)}{1 + \exp(-5.940 + 0.074 \times 55)}.$$

Of course, such an estimate only makes sense for ages near the values used in fitting the model.

The output in Table 6.2 also gives standard errors and 95% confidence intervals for the model coefficients. The interpretation of these is the same as for the linear regression model. The fact that the lower bound of the interval for the coefficient of age excludes zero indicates statistically significant evidence that the true coefficient is different than zero. Similar to linear regression, the ratio of the coefficients to their standard errors forms the Wald (z) test statistic for the hypothesis that the true coefficients are different than zero. The logarithm of the likelihood for the fitted model along with a likelihood ratio statistic LR chi2(1) and associated P-value (Prob > chi2) are also provided. Maximum likelihood is the standard method of estimating parameters from logistic regression models, and is based on finding the estimates which maximize the joint probability (or *likelihood* – see Sect. 6.6) for the observed data under the chosen model.

The likelihood ratio statistic given in the table compares the likelihood from the fitted model with the corresponding model excluding age, and addresses the hypothesis that there is no (linear) relationship between age and CHD risk. The associated P-value is obtained from the χ^2 distribution with one degree of freedom (corresponding to the single predictor used in the model). (Note that the Pseudo R2 value in the table is intended to provide a measure paralleling that used in linear regression models, and is related to the likelihood ratio statistic. Because the latter measure is more widely used and reported, we will not mention Pseudo R2 further in this book.)

As an additional illustration of the properties of the logistic model, Table 6.3 presents a number of quantities calculated directly from the coefficients in Table 6.2 and equations (6.2) and (6.5). For the ages 40, 50, and 70, the table

Table 6.3. Effects of Age Differences of 1 and 10 Years, by Reference Age

Age (x)	$P(x)$	$P(x+1)$	odds(x)	odds($x+1$)	OR	RR	ER
40	0.049	0.053	0.052	0.056	1.077	1.073	0.004
50	0.098	0.105	0.109	0.117	1.077	1.069	0.007
60	0.186	0.198	0.229	0.247	1.077	1.062	0.012

Age (x)	$P(x)$	$P(x+10)$	odds(x)	odds($x+10$)	OR	RR	ER
40	0.049	0.098	0.052	0.109	2.105	1.996	0.049
50	0.098	0.186	0.109	0.229	2.105	1.899	0.088
60	0.186	0.325	0.229	0.482	2.105	1.746	0.139

gives the estimated response probabilities and odds. These are also calculated for one- and ten-year age increases so that corresponding odds ratios can be computed. As prescribed by the model, the odds ratios associated with a fixed increment change in age remain constant across the age range. Estimates of RR and ER are also computed for one- and ten-year age increments to illustrate that the fitted logistic model can be used to estimate a wide variety of quantities in addition to odds ratios. Note that the estimated values of ER and RR are not constant with increasing age (because the model does not restrict them to be so). Note also that although measures such as ER and RR can be computed from the logistic model, the resulting estimates will not in general correspond to those obtained from a regression model defined on a scale on which ER or RR is assumed constant. We will return to this topic when we consider alternative binary regression approaches in Sect. 6.5.

6.1.2 Categorical Predictors

Similar to the conventional linear regression model, the logistic model (6.5) is equally valid for categorical risk factors. For example, we can use it to look again at the relationship between CHD risk and the binary predictor arcus senilis as shown in Table 6.4. Note that the regression output in Table 6.4 sum-

Table 6.4. Logistic Model for CHD and Arcus Senilis

```
. logistic chd69 arcus

Logistic regression                        Number of obs   =      3152
                                           LR chi2(1)      =     12.98
                                           Prob > chi2     =    0.0003
Log likelihood = -879.10783                Pseudo R2       =    0.0073

------------------------------------------------------------------------
      chd69 | Odds Ratio  Std. Err.     z    P>|z|   [95% Conf. Interval]
------------+-----------------------------------------------------------
      arcus |   1.63528   .2195035    3.66   0.000     1.257    2.127399
------------------------------------------------------------------------
```

marizes the model fit in terms of the odds ratio for the included predictor, and does not include estimates of the regression coefficients. (This is the default option in many statistical packages such as Stata.) Note also that the estimated odds ratio and corresponding 95% confidence interval are virtually the same as the results obtained in Table 3.5. Because **arcus** is a binary predictor (coded as one for individuals with the condition and zero otherwise), entering it directly into the model as if it were a continuous measurement produces the desired result: the coefficient represents the log odds ratio associated with a one-unit increase in the predictor. (In this case, only one, single unit increase is possible by definition.) For two-level categorical variables with levels coded other than zero or one, care must be taken so that they are appropriately treated as categories (and not continuous measurements) by the model-fitting program. Finally, note that if we wish to estimate the probability of CHD, we must re-fit the model requesting the regression coefficients, since the intercept β_0 is not provide by default in Stata.

Categorical risk factors with multiple levels are treated similarly to the procedure introduced in Sect. 4.3 for linear regression. In this way we can repeat the analysis in Table 6.1, dividing study participants into five age groups and taking the youngest group as the reference. In order to estimate odds ratios for each of the four older age groups compared to the youngest group we need to construct four indicator variables. Stata does this automatically, as shown in Table 6.5. Note that the estimated odds ratios correspond very closely with those estimated earlier. In fact, because we are estimating a parameter for each age category except the youngest (reference) group, we are not imposing any restrictions on the parameters (i.e., the logistic assumption does not come into play as it does for continuous predictors). Thus we would expect the estimated odds ratios to be identical to those estimated using the contingency table approach.

The likelihood ratio test for this model compares the likelihood for the model with three indicator variables for age with that from the corresponding

Table 6.5. Logistic Model for CHD and Age as a Categorical Factor

```
. xi: logistic chd69 i.agec
i.agec          _Iagec_0-4          (naturally coded; _Iagec_0 omitted)

Logistic regression                         Number of obs   =       3154
                                            LR chi2(4)      =      44.95
                                            Prob > chi2     =     0.0000
Log likelihood = -868.14866                 Pseudo R2       =     0.0252


-------------------------------------------------------------------------
    chd69 | Odds Ratio   Std. Err.      z    P>|z|     [95% Conf. Interval]
----------+--------------------------------------------------------------
  _Iagec_1 |   .8768215   .2025403    -0.57   0.569     .5575566   1.378902
  _Iagec_2 |   1.70019    .3800503     2.37   0.018     1.097046   2.634935
  _Iagec_3 |   2.318679   .5274959     3.70   0.000     1.484546   3.621493
  _Iagec_4 |   2.886314   .7462191     4.10   0.000     1.738907   4.790829
-------------------------------------------------------------------------
```

model with no predictors. In contrast to the individual Wald tests provided for each level of age, the likelihood ratio test examines the overall effect of age represented as a three-level predictor. The results indicate that inclusion of age affords a statistically significant improvement in the fit of the model.

Estimating regression coefficients for levels of a categorical predictor usually involves selecting a reference category. In the example in Table 6.5, this was chosen automatically by Stata as the age category with the smallest numerical label. (A similar procedure is followed by most major statistical packages.) In cases where a reference group different from the default is of interest, most statistics packages (including Stata and SAS) have methods for changing the default, or the model can be re-fit using a recoded version of the predictor. Note that it is also possible to compute odds ratios comparing arbitrary groups from the coefficients obtained using the default reference group. For example, the odds ratio comparing the fourth age group in Table 6.5 to the third can be shown to be $\frac{2.88}{2.32} = 1.24$. (This calculation is left as an exercise.)

Another important consideration in selecting a reference group for a categorical predictor are the sample sizes in each category. As a general rule, when individuals are unevenly distributed across categories it is desirable to avoid making the smallest group the reference category. This is because standard errors of coefficients for other categories will be inflated due to the small sample size in the reference group.

A final issue that arises in fitting models with categorical predictors formed based on an underlying continuous measurement is the choice of how many categories, and how these should be defined. In the example in Table 6.5, the choice of five-year age groups was somewhat arbitrary. In many cases, categories will correspond to pre-existing hypotheses or be suggested by convention (e.g., ten-year age categories in summaries of cancer rates). In the absence of such information, a good practice is to choose categories of equal size based on quantiles of the distribution of the underlying measure.

How many categories a given model will support depends on the overall sample size as well as the distribution of outcomes in the resulting groups. In the WCGS sample, a logistic model including a coefficient for each unique age (assigning the youngest age as the reference group) yields reasonable estimates and standard errors. There are 266 individuals in the smallest group. (A much simpler model that fits the data adequately can also be constructed using the methods discussed in Sect. 6.4.2.) Care must be taken in defining categories to ensure that there are adequate numbers in the sub-groups (possibly by collapsing categories). In general, avoid categorizations that result in categories that are homogeneous with respect to the outcome or that contain fewer than ten observations. Problems that arise when this is not the case are discussed in Sect. 6.4.4.

6.2 Multipredictor Models

Clinical and epidemiological studies of binary outcomes typically focus on the potential effects of multiple predictors. When these are categorical and few in number, contingency table techniques suffice for data analyses. However, for larger numbers of potential predictors and/or when some are continuous measurements, regression methods have a number of advantages. For example, the WCGS study measured a number of potential predictors of coronary heart disease, including total serum cholesterol, diastolic and systolic blood pressure, smoking, age, body size, and behavior pattern. The investigators recognized that these variables all may contribute to outcome risk in addition to being potentially associated with each other, and that in assessment of the influence of a selected predictor, it might be important to control for the potential confounding influence of others. Because there are a number of candidate predictors, some of which can be viewed as continuous measurements, multiple regression techniques are very appealing in analyzing such data.

The logistic regression model for multiple predictor variables is a direct generalization of the version for a single predictor introduced above (6.5). For a binary outcome y, and p predictors $x_1, x_2, \cdots, x_p$, the systematic part of the model is defined as follows:

$$\log\left[\frac{P(x_1, x_2, \cdots, x_p)}{1 - P(x_1, x_2, \cdots, x_p)}\right] = \beta_0 + \beta_1 x_1 + \beta_2 x_2 + \cdots + \beta_p x_p. \qquad (6.7)$$

This can be re-expressed in terms of the outcome probability as follows:

$$P(x_1, x_2, \cdots, x_p) = \frac{\exp(\beta_0 + \beta_1 x_1 + \beta_2 x_2 + \cdots + \beta_p x_p)}{1 + \exp(\beta_0 + \beta_1 x_1 + \beta_2 x_2 + \cdots + \beta_p x_p)}. \qquad (6.8)$$

As with standard multiple linear regression, the predictors may include continuous and categorical variables. The multiple-predictor version of the logistic model is based on the same assumptions underlying the single predictor version. (These are presented in Sect. 6.3.c.) In addition, it assumes that multiple predictors are related to the outcome in an additive fashion on the log odds scale. The interpretation of the regression coefficients is a direct generalization of that for the simple logistic model:

- For a given predictor x_j, the coefficient β_j gives the change in log odds of the outcome associated with a unit increase in x_j, for arbitrary fixed values for the remaining predictors $x_1, \cdots, x_{j-1}, x_{j+1}, \cdots, x_p$.
- The exponentiated regression coefficient $\exp(\beta_j)$ represents the odds ratio associated with a one unit change in x_j.

Table 6.6 presents the results of fitting a logistic regression model examining the impact on CHD risk of age, cholesterol (mg/dL), systolic blood pressure (mmHg), body mass index (computed as weight in kilograms divided by the square of height in meters) and a binary indicator of whether or not the participant smokes cigarettes, using data from the WCGS sample. This model is of interest because it addresses the question of whether a select group of established risk factors for CHD are independent predictors for the WCGS study.

Table 6.6. Multiple Logistic Model for CHD Risk

```
. logistic chd69 age chol sbp bmi smoke, coef

Logistic regression                             Number of obs   =      3141
                                                LR chi2(5)      =    159.80
                                                Prob > chi2     =    0.0000
Log likelihood = -807.19249                     Pseudo R2       =    0.0901
```

chd69	Coef.	Std. Err.	z	P>|z|	[95% Conf. Interval]
age	.0644476	.0119073	5.41	0.000	.0411097	.0877855
chol	.0107413	.0015172	7.08	0.000	.0077675	.013715
sbp	.0192938	.0040909	4.72	0.000	.0112759	.0273117
bmi	.0574361	.0263540	2.18	0.029	.0057814	.1090907
smoke	.6344778	.1401836	4.53	0.000	.3597231	.9092325
_cons	-12.31099	.977256	-12.60	0.000	-14.22638	-10.3956

Twelve observations were dropped from the analysis in Table 6.6 because of missing cholesterol values. An additional observation was dropped because of an unusually high cholesterol value (645 mg/dL). Note that all predictors are entered as continuous measurements in the model. The coefficient for any one one of these (e.g., chol) gives the log odds ratio (change in the log odds) of CHD for a unit increase in the predictor, adjusted for the presence of the others. The small size of the coefficients for these measures reflects the fact that a unit increase on the measurement scale is a very small change, and does not translate to a substantial change in the log odds.

Log odds ratios associated with larger increases are easily computed as described in Sect. 6.1. Lower bounds of 95% confidence intervals for coefficients of all included predictors exclude zero, indicating that each is a statistically significant independent predictor of outcome risk (as measured by the log odds). Of course, additional assessment of this model would be required before it is adopted as a "final" representation of outcome risk for this study. In particular, we would want to evaluate whether the linearity assumption is met for continuous predictors, evaluate whether additional confounding variables should be adjusted for, and check for possible interactions. These topics are discussed in more detail below.

As an example of an application of the fitted model in Table 6.6, consider calculating the log odds of developing CHD within ten years for a 60-year-old smoker, with 253 mg/dL of total cholesterol, systolic blood pressure of 136 mmHg, and a BMI of 25. Applying equation (6.7) with the estimated coefficients from Table 6.6,

$$\log\left[\frac{P(60, 253, 136, 25, 1)}{1 - P(60, 253, 136, 25, 1)}\right] = -12.311 + .0644 \times 60 + .0107 \times 253$$
$$+ .0193 \times 136 + .0574 \times 25 + .6345 \times 1$$
$$= -1.046.$$

A similar calculation gives the corresponding log odds for a similar individual of age 50:

$$\log\left[\frac{P(50, 253, 136, 25, 1)}{1 - P(50, 253, 136, 25, 1)}\right] = -12.311 + .0644 \times 50 + .0107 \times 253$$
$$+ .0193 \times 136 + .0574 \times 25 + .6345 \times 1$$
$$= -1.690.$$

Finally, the difference between these gives the log odds ratio for CHD associated with a ten year increase in age for individuals with the specified values of all of the included predictors:

$$-1.046 - (-1.690) = 0.644.$$

Closer inspection reveals that this result is just ten times the coefficient for age in Table 6.6. In addition, we see that we could repeat the above calculations for any ten-year increase in age, and for any fixed values of the other predictors and obtain the same result. Thus, the formula (6.6) for computing log odds ratios for arbitrary increases in a single predictor applies here as well. The odds ratio for a ten-year increase in age (adjusted for the other included predictors) is given simply by

$$\exp(0.0644 \times 10) = \exp(.644) = 1.90.$$

Interpretation of regression coefficients for categorical predictors also follow that given for single predictor logistic models. For example, the coefficient (0.633) for the binary predictor variable smoke in Table 6.6 is the log odds ratio comparing smokers to non-smokers for fixed values of `age`, `chol`, `sbp`, and `bmi`. The corresponding odds ratio

$$\exp(0.633) = 1.88$$

measures the proportionate increase in the odds of developing CHD for smokers compared to non-smokers adjusted for age, cholesterol, systolic blood pressure and BMI.

The estimated coefficients for the first four predictors in Table 6.6 are all very close to zero, reflecting the continuous nature of these variables and the fact that a unit change in any one of them does not translate to a large increase in the estimated log odds of CHD. As shown above, we can easily calculate odds ratios associated with clinically more meaningful increases in these predictors. An easier approach is to decide on the degree of change that we would like the estimates to reflect and fit a model based on predictors rescaled to reflect these decisions. For example, if we would like the model to produce odds ratios for ten-year increases in age, we should represent age as the rescaled predictor `age_10 = age/10`. Table 6.7 shows the estimated odds

Table 6.7. Multiple Logistic Model With Rescaled Predictors

```
. logistic chd69 age_10 chol_50 bmi_10 sbp_50 smoke

Logistic regression                          Number of obs   =       3141
                                             LR chi2(5)      =     159.80
                                             Prob > chi2     =     0.0000
Log likelihood = -807.19249                  Pseudo R2       =     0.0901

------------------------------------------------------------------------------
      chd69 | Odds Ratio   Std. Err.      z    P>|z|     [95% Conf. Interval]
------------+-----------------------------------------------------------------
     age_10 |   1.904989    .2268333     5.41   0.000     1.508471    2.405735
    chol_50 |   1.710974    .1297976     7.08   0.000     1.474584    1.985259
     bmi_10 |   1.775995    .4680613     2.18   0.029     1.059518    2.976972
     sbp_50 |   2.623972    .5367141     4.72   0.000     1.757326    3.918016
      smoke |   1.886037    .2643914     4.53   0.000     1.432933    2.482417
------------------------------------------------------------------------------
```

ratios from the model including rescaled versions of the first four predictors in Table 6.6. (The numbers after the underscores in the variable names indicate the magnitude of the scaling.) We also "centered" these predictors before scaling them by subtracting of the mean value for each. (Centering predictors is discussed in Sect. 3.3.1 and Sect. 4.6.) Note that the log-likelihood and Wald test statistics for this model are identical to their counterparts in Table 6.6.

6.2.1 Likelihood Ratio Tests

In Sect. 6.1, we briefly introduced the concept of the likelihood, and the likelihood ratio test for logistic models. The likelihood for a given model is interpreted as the joint probability of the observed outcomes expressed as a function of the chosen regression model. The model coefficients are unknown quantities and are estimated by maximizing this probability (hence the name maximum-likelihood estimation). For numerical reasons, maximum-likelihood estimation in statistical software is usually based on the logarithm of the likelihood. An important property of likelihoods from nested models (i.e., models in which predictors from one are a subset of those contained in the other) is that the maximized value of the likelihood from the larger model will always be at least as large as that for the smaller model.

Although the likelihood (or log-likelihood) for a single model does not have a particularly useful interpretation, the likelihood ratio statistic assessing the difference in likelihoods from two nested models is a valuable tool in model assessment (analogous to the F tests introduced in the Chap. 4). It is especially useful when investigating the contribution of more than one predictor, or for predictors with multiple levels.

For example, consider assessment of the contribution of self-reported behavior pattern to the model summarized in Table 6.7. In the WCGS study, investigators were interested in "type A" behavior as an independent risk factor for coronary heart disease. Behavior was classified as either type A or type B, with each type subdivided into two further levels A_1, A_2, B_1 and B_2 (coded as 1,2,3 and 4, respectively). The expanded model addresses the question of whether behavior pattern contributes to CHD risk when other established risk factors are accounted for.

Table 6.8 displays the results of including the four-level categorical variable **behpat** in the model from Table 6.7. The natural coding of the variable

Table 6.8. Logistic Model for WCGS Behavior Pattern

```
. xi: logistic chd69 age_10 chol_50 sbp_50 bmi_10 smoke i.behpat
i.behpat          _Ibehpat_1-4      (naturally coded; _Ibehpat_1 omitted)

Logistic regression                          Number of obs   =      3141
                                             LR chi2(8)      =    184.57
                                             Prob > chi2     =    0.0000
Log likelihood =    -794.81                  Pseudo R2       =    0.1040
```

chd69	Odds Ratio	Std. Err.	z	P>|z|	[95% Conf. Interval]	
age_10	1.83375	.2198681	5.06	0.000	1.449707	2.319529
chol_50	1.704097	.1301391	6.98	0.000	1.467201	1.979243
sbp_50	2.463505	.5086517	4.37	0.000	1.643621	3.692369
bmi_10	1.739414	.4620339	2.08	0.037	1.033479	2.927551
smoke	1.830672	.2583097	4.29	0.000	1.38837	2.413882
_Ibehpat_2	1.068257	.2363271	0.30	0.765	.6924157	1.648103
_Ibehpat_3	.5141593	.1245592	-2.75	0.006	.3198065	.8266243
_Ibehpat_4	.572071	.1826116	-1.75	0.080	.3060107	1.069457

```
. estimates store mod1
```

results in type A_1 behavior being taken as the reference level. Examination of the coefficients and associated 95% confidence intervals for the remaining indicators reveals that although the second category of type A behavior appears not to differ from the reference level, both categories of type B behavior do display statistically significant differences, and are associated with lower outcome risk.

The likelihood ratio statistic is computed as twice the difference between log-likelihoods from the two models, and can be referred to the χ^2 distribution for significance testing. Because the likelihood for the larger model must be larger than the likelihood for the smaller (nested) model, the difference will always be positive. Twice the difference between the log-likelihood for the model including behpat (Table 6.8) and that for the model excluding this variable (Table 6.6) is

$$2 \times [-794.81 - (-807.19)] = 24.76.$$

This value follows a χ^2 distribution, with degrees of freedom equal to the number of additional variables present in the larger model (three in this case). Statistical packages like Stata can often be used to compute the likelihood ratio test directly by first fitting the larger model (in Table 6.8), and saving the likelihood in the user-defined variable (in this case, in the variable mod1 created in the last line of the table). Next, the reduced model eliminating behpat is fit, followed by a command to evaluate the likelihood ratio test as displayed in the Table 6.9. (See Table 6.6 for the full regression output for this model.) The result agrees with the calculation above, and the associated P-value indicates that collectively, the four-level categorical representation of behavior pattern makes a statistically significant independent contribution to the model.

Table 6.9. Likelihood Ratio Test for Four-Level WCGS Behavior Pattern

```
. lrtest mod1

likelihood-ratio test                         LR chi2(3)  =     24.76
(Assumption: . nested in mod1)                 Prob > chi2 =    0.0000
```

The similarity between the two odds ratios for type A (the reference level and the indicator _Ibehpat_2) and type B (the indicators _Ibehpat_3 and _Ibehpat_4) behavior in Table 6.8 suggests that a single binary indicator distinguishing the A and B patterns might suffice. Note that the logistic model that represents behavior pattern as a two-level indicator (with type B behavior as the reference category) is actually nested within the model in Table 6.8. (The model including the two-level representation is a special case of the four-level version when the coefficients for the two levels of type B and type A behavior, respectively are identical.) Table 6.10 displays the fitted model and likelihood ratio test results for this reduced model including the two-level binary indicator dibpat. The fact that the difference between the likelihoods for the two models is not statistically significant confirms our suspicion that modeling the effect of behavior pattern as a two-level predictor is sufficient to capture the contribution of this variable.

Table 6.10. Likelihood Ratio Test for Two-Level WCGS Behavior Pattern

```
. logistic chd69 age_10 chol_50 sbp_50 bmi_10 smoke dibpat

Logistic regression                          Number of obs   =       3141
                                             LR chi2(6)      =     184.34
                                             Prob > chi2     =     0.0000
Log likelihood = -794.92603                  Pseudo R2       =     0.1039

------------------------------------------------------------------------------
     chd69 | Odds Ratio  Std. Err.      z    P>|z|     [95% Conf. Interval]
-----------+------------------------------------------------------------------
    age_10 |  1.830252   .2190623     5.05   0.000     1.44754      2.314146
   chol_50 |  1.702406   .1299562     6.97   0.000     1.465835     1.977157
    sbp_50 |  2.467919   .5084376     4.38   0.000     1.648039     3.695681
    bmi_10 |  1.732349   .4596114     2.07   0.038     1.029917     2.913859
     smoke |  1.829163   .2580698     4.28   0.000     1.387265     2.411822
    dibpat |  2.006855   .289734      4.82   0.000     1.512259     2.663212
------------------------------------------------------------------------------

. lrtest mod1

likelihood-ratio test                        LR chi2(2)  =        0.23
(Assumption: . nested in mod1)                Prob > chi2 =      0.8904
```

As demonstrated above, the likelihood ratio test is a very useful tool in comparing nested logistic regression models. In moderate to large samples, the results from the likelihood ratio and Wald tests for the effects of single predictors will agree quite closely. However, in smaller samples the results of these two tests may differ substantially. In general, the likelihood ratio test is more reliable than the Wald test, and is preferred when both are available. Finally, note that because the likelihood is computed based on the observations used to fit the model, it is important to ensure that the same observations are included in each candidate model considered in likelihood ratio testing. Likelihoods from models fit on differing sets of observations are not comparable. A more complete discussion of the concepts of likelihood and maximum-likelihood estimation is given in Sect. 6.6.

6.2.2 Confounding

A common goal of multiple logistic regression modeling is to investigate the association between a predictor and the outcome, controlling for the possible influence of additional variables. For example, in evaluating the observed association between behavior pattern (considered in the previous section) and CHD risk, it is important to consider the potential effects of additional variables that might be related to both behavior and CHD occurrence. Recall from Chapter 4 that regression models account for confounding of an association of primary interest by including potential confounding variables in the same model. In this section we briefly review these issues in the logistic regression context.

Consider again the assessment of behavior pattern as a predictor of CHD in the WCGS example considered in the previous section. In the analysis summarized in Table 6.10, we concluded that a two-level indicator (dibpat) distinguishing type A and B behaviors adequately captures the effects of this variable on CHD (in place of a more complex, four-level summary of behavior). In light of the discussion in Sect. 5.3, we should consider the possible causal relationships of the additional variables in the model with both the outcome and behavior pattern before concluding that the association is adequately adjusted for confounding.

Table 6.11. Logistic Model for Type A Behavior Pattern and Selected Predictors

```
. logistic dibpat age_10 chol_50 sbp_50 bmi_10 smoke

Logistic regression                        Number of obs   =      3141
                                           LR chi2(5)      =     53.80
                                           Prob > chi2     =    0.0000
Log likelihood = -2150.1739                Pseudo R2       =    0.0124


-----------------------------------------------------------------------
   dibpat | Odds Ratio   Std. Err.     z     P>|z|    [95% Conf. Interval]
----------+------------------------------------------------------------
   age_10 |  1.324032    .0881552    4.22   0.000     1.16205    1.508594
  chol_50 |  1.084241    .0464136    1.89   0.059    .9969839    1.179135
   sbp_50 |  1.461247    .1876433    2.95   0.003     1.136104   1.879442
   bmi_10 |  1.123846    .1672474    0.78   0.433    .8395252    1.504459
    smoke |   1.26933    .0930786    3.25   0.001     1.099403   1.465522
-----------------------------------------------------------------------
```

Recall that to be a confounder of an association of primary interest, a variable must be associated with both the outcome and the predictor forming the association. From Table 6.10, all of the predictors in addition to dibpat are independently associated with the CHD outcome. Since dibpat is a binary indicator, we can examined its association with these predictors via logistic regression as well. Table 6.11 presents the resulting model. With the exception of BMI (bmi_10), all appear to be associated with behavior pattern. In deciding which variables to adjust for in summarizing the CHD-behavior patter association, it is worth considering the possible causal relationships to help identify distinguish variable with confounding influence from those that could be potential mediators or effect modifiers.

The causal diagram illustrated in Figure 5.2 is partially applicable to the present situation (e.g., if we substitute type 2 behavior for exercise), and illustrates clearly that causal connections are likely to be very complex. For example, cholesterol and SBP (hypertension) could be viewed as mediating variables in the pathway between behavior and CHD. Similarly, smoking and BMI may play either a confounding or mediating role. The unadjusted odds ratio (95% CI) for the association between type A behavior and CHD for this group of individuals is 2.36 (1.79, 3.10). By contrast, the adjusted odds ratio in Table 6.10 is 2.01 (95% CI 1.51, 2.66). Note that dropping any of the adjustment factors from the model singly results in little change to the estimated OR for type A behavior (less than 5%). Thus if any of these variables acts as a mediator, the influence appears to be weak. This suggests that the influence of type A behavior on CHD may act partially through another unmeasured pathway. (Or that this characterization of behavior is itself only a marker for more important behavioral characteristics.) In this case, adjustment for the other variables is appropriate. When we also consider the importance of the adjustment factors as predictors of CHD, it makes sense to include them. See Sect. 5.3 for further discussion of these issues. Finally, before concluding that we have adequately modeled the relationship between behavior pattern and CHD we need to account for possible interactions between included predictors (Sect. 6.2.3), and conduct diagnostic assessments of the model fit (Sect. 6.4).

6.2.3 Interaction

Recall from Chapter 4 that an interaction between two predictors in a regression model means that the degree of association between each predictor and the outcome varies according to levels of the other predictor. The mechanics of fitting logistic regression models including interaction terms is quite similar to standard linear regression (see Sect. 4.6). For example, to fit an interaction between two continuous predictors x_1 and x_2, we include the product $x_1 x_2$ as an additional predictor in a model containing x_1 and x_2 as shown in equation (6.9):

$$\log\left[\frac{P(x_1, x_2, x_1 \times x_2)}{1 - P(x_1, x_2, x_1 x_2)}\right] = \beta_0 + \beta_1 x_1 + \beta_2 x_2 + \beta_3 x_1 \times x_2. \qquad (6.9)$$

Fitting interactions between categorical predictors and between continuous and categorical predictors also follows the procedures outlined in Chapter 4. However, because of the log odds ratio interpretation of regression coefficients in the logistic model, interpreting results of interactions is somewhat different. We review several examples below.

For an illustrative example of a two-way interaction between two binary indicator variables from the WCGS study, consider the regression model presented in Table 6.12. The fitted model includes the indicator arcus for arcus senilis (defined in Sect.3.4), a binary indicator bage_50 for participants over the age of 50, and the product between them, bage_50arcus. The research question addressed is whether the association between arcus and CHD is age-dependent. The statistically significant result of the Wald test for the coefficient associated with the product of the indicators for age and arcus indicates that an interaction is present. This means that we cannot interpret the coefficient for arcus as a log odds ratio without specifying whether or not the

Table 6.12. Logistic Model for Interaction Between Arcus and Age as a Categorical Predictor

```
. logistic chd69 bage_50 arcus bage_50arcus, coef

Logistic regression                          Number of obs   =       3152
                                             LR chi2(3)      =      40.33
                                             Prob > chi2     =     0.0000
Log likelihood = -865.43251                  Pseudo R2       =     0.0228

------------------------------------------------------------------------
     chd69 |     Coef.   Std. Err.      z    P>|z|    [95% Conf. Interval]
-----------+------------------------------------------------------------
   bage_50 |  .8932677   .1721237     5.19   0.000    .5559115    1.230624
     arcus |  .6479628   .1788636     3.62   0.000    .2973966     .9985291
bage_50arcus| -.5920552   .2722265    -2.17   0.030   -1.125609    -.058501
     _cons | -2.882853   .1089259   -26.47   0.000   -3.096344   -2.669362
------------------------------------------------------------------------
```

participant is older than 50. (A similar result holds for the interpretation of bage_50.) The procedure for obtaining the component odds ratios is similar to the methods for obtaining main and interaction effects for linear regression models, and is straightforward using the regression model. If we represent arcus and bage_50 and x_1 as x_2 in equation (6.9), we can compute the log odds for any combination of values of these predictors using coefficients from Table 6.12. For example, the log odds of CHD occurrence for an individual over 50 years old without arcus is given by

$$\log \left[\frac{P(0,1,0)}{1 - P(0,1,0)} \right] = \beta_0 + \beta_2$$
$$= -2.883 + 0.893 = -1.990.$$

Similarly, the log odds for an individual between 39 and 49 years old without arcus is

$$\log \left[\frac{P(0,0,0)}{1 - P(0,0,0)} \right] = \beta_0.$$

With these results, we see that the five expressions below define the component log odds ratios in the example:

$$\log \left[\frac{P(1,0,0)}{1 - P(1,0,0)} \right] - \log \left[\frac{P(0,0,0)}{1 - P(0,0,0)} \right] = \beta_1 = 0.648$$

$$\log \left[\frac{P(1,1,1)}{1 - P(1,1,1)} \right] - \log \left[\frac{P(0,1,0)}{1 - P(0,1,0)} \right] = \beta_1 + \beta_3 = 0.056$$

$$\log \left[\frac{P(0,1,0)}{1 - P(0,1,0)} \right] - \log \left[\frac{P(0,0,0)}{1 - P(0,0,0)} \right] = \beta_2 = 0.893 \qquad (6.10)$$

$$\log \left[\frac{P(1,1,1)}{1 - P(1,1,1)} \right] - \log \left[\frac{P(1,0,0)}{1 - P(1,0,0)} \right] = \beta_2 + \beta_3 = 0.301$$

$$\log \left[\frac{P(1,1,1)}{1 - P(1,1,1)} \right] - \log \left[\frac{P(0,0,0)}{1 - P(0,0,0)} \right] = \beta_1 + \beta_2 + \beta_3 = 0.949.$$

The corresponding odds ratios are then easily calculated by exponentiation, as shown in Table 6.13.

Table 6.13. Component Odds Ratios for Arcus-Age Interaction Model

Odds ratio	Groups compared
$\exp(\beta_1) = 1.91$	arcus vs. no arcus, age 39-49
$\exp(\beta_1 + \beta_3) = 1.06$	arcus vs. no arcus, age 50-59
$\exp(\beta_2) = 2.44$	age 50-59 vs. age 39-49, no arcus
$\exp(\beta_2 + \beta_3) = 1.35$	age 50-59 vs. age 39-49, arcus
$\exp(\beta_1 + \beta_2 + \beta_3) = 2.58$	arcus *and* age 50-59 vs. no arcus and ages 39-49

Referring back to Table 6.12, we see that all of the component odds ratios aren't immediately obvious from standard regression output. However, the log odds ratio and associated 95% confidence intervals for `arcus` among individuals in the younger age group and for older individuals among those without arcus can be read directly. This is because when we set either variable to zero (the reference level), the interaction term evaluates to zero and is eliminated. Estimated log odds ratios corresponding to the non-reference levels of these variables involve the interaction term, and differ from their counterparts by the value of its coefficient (–0.592). Standard errors and 95% confidence intervals for these estimates require additional calculations that cannot be completed without further information about the fitted model. Fortunately, many statistical packages have facilities that greatly simplify these calculations. Table 6.14 illustrates the use of the `lincom` command in Stata to compute the odds ratio comparing the odds of CHD in individuals of age 50 and over with the odds among those under 50, among individuals with arcus. By specifying the

Table 6.14. Example Odds Ratio for Arcus-Age Interaction Model

```
. lincom bage_50 + bage_50arcus

 ( 1)   bage_50 + bage_50arcus = 0

------------------------------------------------------------------------------
      chd69 | Odds Ratio   Std. Err.      z    P>|z|     [95% Conf. Interval]
-------------+----------------------------------------------------------------
        (1) |   1.351497   .2850367     1.43   0.153     .8939077    2.043324
```

correct combination of coefficients (corresponding to those in Table 6.13), the output in the Table 6.14 provides the desired odds ratio estimate along with the 95% confidence interval. Results of the accompanying hypothesis test that the underlying log odds ratio is zero are also provided.

Interactions between a continuous and categorical variable are handled in a similar fashion to those involving binary predictors. In the previous example, the categorization of age was somewhat arbitrary. In fact, because age was represented by two categories, essentially the same results could have been obtained using frequency table techniques (as illustrated in Table 3.9). A more complete assessment of the interaction can be obtained by considering age as a continuous variable (previously considered in Table 6.2). For example, this would allow us to investigate whether increase in CHD risk with increasing age differs in individuals with and without arcus. The logistic model addressing this question is displayed in Table 6.15. With this form of the logistic

Table 6.15. Logistic Model for Interaction Between Arcus and Age as Continuous

```
. xi: logistic chd69 i.arcus*age, coef
i.arcus            _Iarcus_0-1      (naturally coded; _Iarcus_0 omitted)
i.arcus*age        _IarcXage_#      (coded as above)

Logistic regression                          Number of obs   =       3152
                                             LR chi2(3)      =      53.33
                                             Prob > chi2     =     0.0000
Log likelihood = -858.93362                  Pseudo R2       =     0.0301

-------------------------------------------------------------------------
      chd69 |    Coef.   Std. Err.      z    P>|z|    [95% Conf. Interval]
------------+------------------------------------------------------------
   _Iarcus_1 | 2.754185   1.140113    2.42   0.016    .5196041    4.988765
        age |  .089647   .0148903    6.02   0.000    .0604624    .1188315
 _IarcXage_1 | -.0498298   .023343   -2.13   0.033   -.0955812   -.0040784
       _cons | -6.788086  .7179936   -9.45   0.000   -8.195327   -5.380844
-------------------------------------------------------------------------
```

command in Stata, we instruct the program to include an interaction term between the two variables. This is accomplished by inclusion of the product of arcus and age (_IarcXage_1) as well as the individual predictors age and _Iarcus_1. For a fixed age (e.g., 55), the log odds ratio associated with having arcus is calculated as follows, using the estimated coefficients from Table 6.15:

$$
\log\left[\frac{P(1,55,55)}{1-P(1,55,55)}\right] - \log\left[\frac{P(0,55,0)}{1-P(0,55,0)}\right]
$$
$$
= (-6.788 + 2.754 + (0.090 - 0.050) \times 55) - (-6.788 + 0.090 \times 55)
$$
$$
= (2.754 - 0.050 \times 55) = 0.014.
$$

We see that this corresponds to an odds ratio of $\exp(0.014) = 1.01$, which is similar to that calculated for the corresponding age group in Table 6.13. We can obtain this estimate and its 95% confidence interval directly as shown in Table 6.16.

Note that because age is represented as a continuous variable, its value must be specified in interpreting the effect of arcus on the log odds of CHD risk. Similarly, among individuals with arcus, log odds ratios can be computed for any specified increase in age. Fig. 6.2 displays the estimated log odds as a function of age, separately for individuals with and without arcus. The

Table 6.16. Logistic Model for Interaction Between Arcus and Age as a Continuous Predictor

```
. lincom _Iarcus_1 + _IarcXage_1*55

( 1)  _Iarcus_1 + 55 _IarcXage_1 = 0

------------------------------------------------------------------------------
      chd69 | Odds Ratio   Std. Err.     z    P>|z|     [95% Conf. Interval]
------------+-----------------------------------------------------------------
        (1) |   1.013637    .2062327   0.07   0.947     .6802966      1.51031
------------------------------------------------------------------------------
```

equations for these two lines can be obtained directly from the coefficients in Table 6.15 and are printed below for individuals with and without arcus, respectively:

$$\log\left[\frac{P(\text{age})}{1 - P(\text{age})}\right] = (-6.788 + 2.754) + (0.090 - 0.050) \times \text{age}$$
$$= -4.034 + 0.040 \times \text{age}.$$

and

$$\log\left[\frac{P(\text{age})}{1 - P(\text{age})}\right] = -6.788 + 0.0896 \times \text{age}.$$

Fig. 6.2 displays the results obtained above, indicating that CHD risk is higher for younger participants with arcus. However, older participants with arcus seem to be at somewhat lower risk than those without arcus. Of course, further interpretation of these equations should be preceded by thorough checking of the linearity of the relationship between age and the log odds of the outcome, including whether more complicated, higher-order interaction terms are needed.

Recall the discussion in Sect. 6.1 where we motivated the logistic model as an example of a multiplicative risk model (see equation (6.4)). By contrast, the excess risk model (introduced in equation (6.1) and discussed further in Sect. 6.5.2) is an example of an additive risk model. In addition to defining two distinct ways in which a predictor can act to modify outcome risk, this distinction turns out to be very important in the context of interaction: For a specified outcome and predictor pair, it is possible to have interaction under the multiplicative model and not under the additive model, and vice versa.

For example, if we fit the additive risk model to the data from the age/arcus example in Table 6.15, the Wald test P-value for inclusion of the product term (age_50arcus) is 0.15. (The corresponding value from the logistic model was 0.03.) The implications of this are that we should not necessarily regard interaction as mirroring a biological mechanism, but rather as a property of the data and model being fit. In the example, we would want to account for the

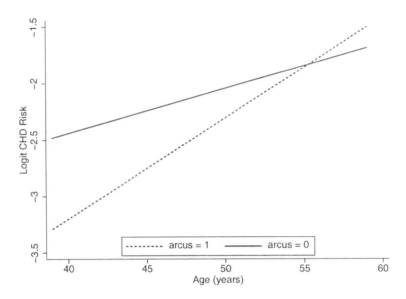

Fig. 6.2. Log Odds of CHD and Age for Individuals With and Without Arcus Senilis

interaction if we were using the logistic model but not necessarily if we were analyzing the WCGS data using the additive model. The additive regression model is described further in Sect. 6.5.2. Also, see Clayton and Hills (1993) and Jewell (2004) for more detailed discussions of the distinction between multiplicative and additive interaction.

6.2.4 Prediction

Frequently, the goal of fitting a logistic model is to predict risk of the binary outcome given a set of risk factors. Recall that in Sect. 6.2.1, we fit a logistic model for the binary coronary heart disease (CHD) outcome in the WCGS sample, using age, cholesterol level, systolic blood pressure, body mass index (BMI), a binary indicator of current cigarette smoking (with non-smokers composing the reference group), and an indicator of type A behavior as predictors. Table 6.10 summarizes the results. Table 6.17 presents an expanded version of this model that includes two additional predictors bmichol and bmisbp for the interactions between BMI and serum cholesterol level and BMI and systolic blood pressure (both centered and scaled as described in Sect. 6.2). These were both found to make statistically significant contributions to the model in further analyses investigating two way interactions between the original predictors in Table 6.10 .

As shown in Sect. 6.2, the estimated coefficients from the model in Table 6.17 can be used directly in the logistic formula (6.8) to compute the log

Table 6.17. Expanded Logistic Model for CHD Events

```
. logistic chd69 age_10 chol_50 sbp_50 bmi_10 smoke dibpat bmichol bmisbp, coef

Logistic regression                           Number of obs   =      3141
                                              LR chi2(8)      =    198.15
                                              Prob > chi2     =    0.0000
Log likelihood = -788.01957                   Pseudo R2       =    0.1117


-------------------------------------------------------------------------
      chd69 |     Coef.   Std. Err.      z    P>|z|    [95% Conf. Interval]
------------+------------------------------------------------------------
     age_10 |  .5949713   .1201092     4.95   0.000    .3595615    .830381
    chol_50 |  .5757131     .07779     7.40   0.000    .4232474   .7281787
     sbp_50 |  1.019647   .2066014     4.94   0.000    .6147159   1.424579
     bmi_10 |  1.048839   .2998176     3.50   0.000    .4612074   1.636471
      smoke |  .6061929   .1410533     4.30   0.000    .3297335   .8826523
     dibpat |  .7234267   .1448996     4.99   0.000    .4394288   1.007425
    bmichol | -.8896932   .2746471    -3.24   0.001   -1.427992  -.3513948
     bmisbp | -1.503455    .631815    -2.38   0.017    -2.74179  -.2651208
      _cons | -3.416061   .1504717   -22.70   0.000    -3.71098  -3.121142
-------------------------------------------------------------------------
```

odds (or the corresponding probability) of CHD for an arbitrary individual by specifying the desired values for the predictors. Table 6.18 displays a few such predictions (labeled prchd) for five individuals in the WCGS sample (obtained using the predict command in Stata).

Table 6.18. Sample Predictions From the Logistic Model in Table 6.17

```
    +----------------------------------------------------------------------+
    | chd69   age   chol   sbp      bmi       smoke   dibpat     prchd |
    |----------------------------------------------------------------------|
 1. |    no    49    225   110   19.78795      smoker   A1,A2   .0433952 |
 2. |    no    42    177   154    22.9551      smoker   A1,A2   .0708145 |
 3. |    no    42    181   110   23.62529   nonsmoker   B3,B4   .0082533 |
 4. |    no    41    132   124     23.109      smoker   B3,B4   .0089318 |
 5. |   yes    59    255   144   21.52041      smoker   B3,B4   .1926046 |
    |----------------------------------------------------------------------|
```

6.2.5 Prediction Accuracy

In some applications, we may be interested in using a logistic regression model as a tool to classify outcomes of newly observed individuals based on values of measured predictors. For the WCGS example just considered, this may involve deciding on treatment strategy based on prognosis as measured by the predicted probability from the logistic model in Table 6.17. Similar to the goals of developing diagnostic tests for detecting diseases, this approach requires us to choose a cut-off or threshold value of the predicted outcome probability above which treatment would be initiated. A fundamental consideration in choosing this threshold is in evaluating the degree of misclassification of outcomes incurred by the choice. For a binary outcome, misclassification can be quantified by calculating the proportion of individuals incorrectly classified as either having the outcome or not. These are known as the *false-positive* and *false-negative* rates, respectively, and are standard measures of prediction error in the logistic regression context. (Recall that prediction error was introduced in Sect. 5.2.) Rather than state prediction performance in terms of misclassification, the following complementary measures are frequently used in assessment of prediction rules for binary outcomes:

Sensitivity The proportion of individuals with the outcome that are correctly classified, calculated as the complement of the false-negative rate.

Specificity The proportion of individuals without the outcome that are correctly classified, calculated as the complement of the false-positive rate.

As the threshold value of a prediction rule varies between zero and one, these quantities can be calculated and compared to evaluate overall performance. A *receiver operating characteristic* (ROC) curve plots the sensitivity against the false-positive rate (i.e., one minus the specificity) for a range of thresholds to help visualize test performance. Figure 6.3 shows the ROC curve for the current example (obtained using the `lroc` command in Stata), along with a diagonal reference line, usually interpreted as representing the ROC curve for a test that is no better than the flip of a coin.

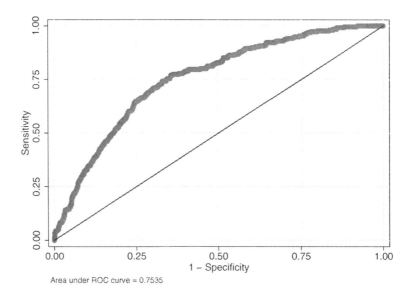

Fig. 6.3. ROC Curve for Logistic Prediction of CHD Events

ROC curves for tests with overall good performance (i.e., low misclassification rates for both positive and negative outcomes) will lie close to the left and topmost margins of the plot. In Figure 6.3, a test with a sensitivity of around 75% is close to optimal in this sense. (The threshold value corresponding to a sensitivity of 0.75 – and a specificity 0f 0.64 – in Figure 6.3 is about 0.07.) Note that in most practical situations, assessment of test performance has a subjective component: The cost of misclassifying an individual as positive may be deemed more serious than the alternative situation, or vice versa. These considerations weigh into evaluation of test results. The area under an ROC curve (also known as the *c-statistic*) provides an overall measure of classification accuracy (representing the overall proportion of individuals correctly classified), with the value of one representing perfect accuracy. In the present case, the value of 0.754 does not indicate very impressive performance.

A clear limitation with the example above is that the individuals used to evaluate the performance are the same as those used to fit the model on which the classification rule is based. Alternative techniques that do not share this limitation include cross-validation and learning set/test set validation (both described in Sect. 5.2). Finally, note that although logistic regression is a valid approach for development of prediction tools, alternative techniques are available. Classification trees (discussed briefly in Sect. 5.2) are very useful in this context, and involve fewer assumptions than the logistic approach. See Goldman *et al.* (1996) for an example of their application in a clinical context.

6.3 Case-Control Studies

In situations where binary outcomes are rare or difficult to observe, it is not always feasible to collect a large enough sample to investigate the relationship between the outcome and predictors of interest. Consider the problem of evaluating dietary risk factors for stomach cancer. Because this disease is relatively rare (accounting for approximately 2% of annual cancer deaths in the U.S.), only a very large cross-sectional or prospective sample would include sufficient numbers of cases to evaluate associations with predictors of interest. Case-control studies address this problem by recruiting a fixed number of individuals with the outcome of interest (the cases) and a number of comparable control individuals free of the outcome. Retrospective histories of predictor variables of interest are then collected via questionnaire after recruitment.

A well-known example of a case-control study is the Ille-et-Vilaine study of cancer conducted in France between 1972 and 1974. It includes 200 cases and 775 comparable controls, and was designed to investigate alcohol, diet, and tobacco consumption as risk factors for esophageal cancer in men. This is known as an *unmatched* study since cases and controls were sampled separately in predetermined numbers. An alternative type of case-control study is based on *matching* a fixed number of controls to each sampled case based on selected characteristics. Methods for matched studies are different and will be covered briefly below in Sect. 6.3.1.

Because the overall proportion of individuals is fixed by design in a case-control study (e.g., 200/995, or approximately five controls per case for Ille-et-Vilaine), it is not meaningful to make direct comparisons of outcome risk (estimated as the proportion of individuals with the outcome) between groups defined by predictor variables, as is conventional in studies where participants are not sampled based on their outcome status. Rather, analyses are based on the distribution of predictors variables compared across case/control status. At first glance, this approach does not seem to address the fundamental question of whether or not the predictor is associated with increased risk of developing the outcome. For example, observing that self-reported alcohol consumption differed between cases and controls in Ille-et-Vilaine does not seemingly translate into a clear statement about esophageal cancer risk associated with alcohol use. Further, application of conventional measures of association to settings where the role of the outcome and predictor are reversed seemingly leads to unintuitive results. For example, observing that individuals with esophageal cancer risk are twice as likely (in terms of the relative risk) as cancer-free individuals to report a specified degree of alcohol consumption does not state the association in a way that makes the possible causal connection clear.

Recall that our definitions of the relative risk, excess risk, and odds ratios in Chapter 3 were stated in terms of the outcome probabilities. This limits their usefulness in retrospective settings such as case-control studies. However, it is a unique property of the odds ratio that it retains its validity as a measure of outcome risk, even for case-control sampling. To demonstrate this for a simple example, Table 6.19 presents odds ratios for the Ille-et-Vilaine study estimated using the `tabodds` procedure in Stata. The first part of the table gives the odds of the binary case-control status indicator `case` compared in two groups defined by the binary indicator `ditob` of moderate to heavy level of smoking (10+ grams/day of tobacco smoked), and the second part gives the corresponding odds ratio comparing moderate-to-heavy level of smoking between cases and controls. The estimated odds ratios are identical. This property does not hold for the excess risk and relative risk.

We can also demonstrate this property directly using the definition of the odds ratio. Table 6.20 presents a hypothetical 2×2 table for a binary outcome and predictor in terms of the frequencies of n individuals in the four possible cross-categorizations (labeled a, b, c and d). We estimate the outcome probability among individuals with and without the predictor with the proportions $a/(a + c)$ and $b/(b + d)$, respectively, and the corresponding odds of the outcome as

$$\frac{a/(a + c)}{c/(a + c)} \quad \text{and} \quad \frac{b/(b + d)}{d/(b + d)}. \tag{6.11}$$

The resulting odds ratio is then ad/bc.

Table 6.19. Odds Ratio for Smoking and Esophageal Cancer

```
. tabodds case ditob, or

-------------------------------------------------------------------------
      ditob |  Odds Ratio        chi2       P>chi2      [95% Conf. Interval]
------------+------------------------------------------------------------
   0-9 g/day |    1.000000          .            .             .          .
   10+ g/day |   10.407051       64.89       0.0000       5.119049  21.157585
-------------------------------------------------------------------------

. tabodds ditob case, or

-------------------------------------------------------------------------
       case |  Odds Ratio        chi2       P>chi2      [95% Conf. Interval]
------------+------------------------------------------------------------
         0 |    1.000000          .            .             .          .
         1 |   10.407051       64.89       0.0000       5.119049  21.157585
-------------------------------------------------------------------------
```

Similarly, we can estimate the exposure probability among individuals with and without the outcome as $a/(a + b)$ and $c/(c + d)$, and the corresponding odds as above. It is easy to verify that the odds ratio based on these is also ad/bc. This property of the odds ratios is central to the wide use of case-control studies, and suggests that logistic regression may be applicable as well. The additional fact that the odds ratio approximates the relative risk for rare outcomes (e.g., many forms of cancer) increases its appeal.

Table 6.20. Outcome by Predictor Status for a Case-Control Study

Predictor	Outcome		Total
Yes	a	b	$a + b$
No	c	d	$c + d$
Total	$a + c$	$b + d$	n

Recall that in the logistic regression model, the intercept coefficient β_0 is interpreted as the "baseline" log odds of outcome risk obtained when no predictors are included in the model (or, equivalently, when all predictors take on the value zero). As we have stated above, this quantity cannot be meaningfully estimated from case-control studies. As a result, the intercept coefficient in logistic regression models for case-control data can not be interpreted as providing an estimate of baseline risk in the population from which the sample was drawn. It is a remarkable fact that the logistic model is nonetheless directly applicable to data from case-control studies, and that estimated regression coefficients for included predictors provide valid estimates of log odds ratios, sharing the interpretation from other study types. Note that the logistic is the only binary regression model with this property.

A primary hypothesis underlying the Ille-et-Vilaine study was that alcohol consumption was related to esophageal cancer. Alcohol consumption was measured in average total daily consumption in grams, estimated directly from questionnaire responses on a number of different types of alcoholic beverages. The investigators recognized that age and smoking were potential confounding influences, and should be accounted for in assessing the association between alcohol consumption and cancer risk. (Dietary factors were also considered, but are not discussed here.)

Table 6.21 presents the results of a logistic regression model fit to these data, including a four-level categorization alcgp of average daily alcohol consumption and controlling for the dichotomous indicator ditob of moderate-to-heavy smoking (introduced above) and age (in years) as a continuous predictor. The lowest level of alcohol consumption (0–39 grams/day) is taken as the reference category, and the three included indicators represent 40–79, 80–119, and 120+ grams/day, respectively. Note that omitting the coef option to the logistic command, Stata returns odds ratio estimates rather than the regression coefficients, and the intercept is not included. The results indicate a clear increase in cancer risk with increasing alcohol consumption, and that this effect is evident when age and smoking are accounted for.

Table 6.21. Logistic Model for Alcohol Consumption and Esophageal Cancer

```
. xi: logistic case i.alcgp ditob age
i.alcgp         _Ialcgp_1-4      (naturally coded; _Ialcgp_1 omitted)

Logit estimates                              Number of obs   =        975
                                             LR chi2(5)      =     280.80
                                             Prob > chi2     =     0.0000
Log likelihood = -354.34556                  Pseudo R2       =     0.2838
```

case	Odds Ratio	Std. Err.	z	P>|z|	[95% Conf. Interval]	
_Ialcgp_2	4.063502	1.024362	5.56	0.000	2.479261	6.660068
_Ialcgp_3	7.526931	2.138601	7.10	0.000	4.312896	13.13611
_Ialcgp_4	32.07349	11.5861	9.60	0.000	15.80016	65.1075
ditob	7.375744	2.732351	5.39	0.000	3.568432	15.24524
age	1.068417	.0087666	8.07	0.000	1.051372	1.085738

Estimated odds ratios in Table 6.21 are larger than 1.0, and lower bounds of the associated 95% confidence intervals exclude 1.0, indicating that each of the predictors is associated with statistically significant increases in risk of esophageal cancer. Further, since esophageal cancer is relatively rare in the general population on which this study was conducted, interpreting the odds ratios as estimated relative risks is approximately correct. A single summary of the contribution of alcohol consumption to a model including age and smoking can be obtained by fitting the same model excluding the indicators for alcohol, and performing a likelihood ratio test, as shown in Table 6.22. This procedure assumes that the full model including alcohol in Table 6.21 is fit first, and the model log-likelihood is stored for future reference as mod1 (in the first line of

Table 6.22. Likelihood Ratio Test for Contribution of `age`

```
. est store mod1
. logistic case ditob age

Logistic regression                          Number of obs   =       975
                                             LR chi2(2)      =    152.11
                                             Prob > chi2     =    0.0000
Log likelihood = -418.68894                  Pseudo R2       =    0.1537

------------------------------------------------------------------------
    case | Odds Ratio   Std. Err.      z    P>|z|    [95% Conf. Interval]
---------+--------------------------------------------------------------
   ditob |  9.463852    3.362354    6.33   0.000     4.716825    18.9883
     age |  1.055568    .0073642    7.75   0.000     1.041232    1.0701
------------------------------------------------------------------------

. lrtest mod1

likelihood-ratio test                        LR chi2(3)   =     128.69
(Assumption: . nested in mod1)                 Prob > chi2 =      0.0000
```

the output in Table 6.22). The results indicate a substantial contribution of the categorical summary `alcgp` of alcohol consumption to the overall fit of the model as summarized by the large log-likelihood ratio statistic (128.7). Further analyses might investigate the relationship between alcohol, smoking, and the log odds of cancer risk in more detail, possibly including these variables as continuous measures. We would naturally want to evaluate the linearity assumption implicit in including the variables (and age) in this form as well.

6.3.1 Matched Case-Control Studies

Consider the issues that would arise in designing a case-control study investigating esophageal cancer in a different population than Ille-et-Vilaine, possibly focusing on exposures other than alcohol as potential risk factors: We certainly would like to take into account known confounding factors such as those considered above as part of our design. If there are many such variables, we may be concerned that they will not be well represented in our chosen sample, and/or that analyses accounting for their influence may be overly complex. If we could recruit study subjects accounting for their profiles for these suspected confounders, we might be able to avoid some of these difficulties. This is the rationale for *matching*. We can build in control for confounding by incorporating knowledge of known confounders into the design of the study. By matching cases with controls that have the same values of these variables, we ensure control for confounding by comparing cases and controls within strata defined by the matching factors. In one of the simplest matched designs, disease cases are paired with controls into *matched sets* having similar values of the matching variables.

Because cases and controls within matched sets are sampled together based on shared values of the matching variables, the structure of the overall sample differs from that of an unmatched study. If we were to try to account for the sampling design via a standard logistic model that accounted for the matched sets with indicator variables, the number of parameters would frequently be too large for reliable estimation. For example, in a matched pair study with 200 matched pairs, as many as 199 parameters would be needed to account for the matching criteria. Clearly another regression approach is called for.

Regression modeling for matched data is based on a modification of the maximum-likelihood estimation approach used for the conventional logistic model (and described in more detail in Sect. 6.6). The *conditional logistic regression model* avoids estimating parameters accounting for the matching via *conditioning*. The parameters for predictors in this model have the log odds ratio interpretation familiar from the standard logistic model. The result is that we can conduct regression analyses exactly as before. However, the variables used in matching are controlled for automatically and not used directly in modeling. The `clogit` command in Stata provides a very convenient way to fit conditional logistic regression models. Most major statistical packages have similar facilities.

Matching is not always a good idea and should never be undertaken lightly. Effective matching (in cases where matching variables are strong confounders) can yield more precise estimates of the disease/exposure relationship. However, in cases where the matching variables do not actually confound the relationship between the exposure of interest and the outcome, the matching can lead to estimates with decreased precision relative to those obtained from an unmatched study. Further, satisfying matching criteria can be difficult and may result in a loss of cases. Good basic references for statistical analysis of data from matched case-control studies include Breslow and Day (1984) and Jewell (2004).

6.4 Checking Model Assumptions and Fit

Chapter 4 (Sect. 4.7) presented a number of techniques for assessing model fit and assumptions for linear regression models. Here we cover many of the same topics for logistic models. Fortunately, many of the issues and techniques are similar and the methods from linear models apply more or less directly. However, the binary nature of the outcomes considered here make construction and interpretation of graphical methods of assessment more complex. We focus here on issues that differ from the approaches discussed in Sect. 4.7.

6.4.1 Outlying and Influential Points

Similar to the definition of residuals for linear regression (in Sect. 4.7), *standardized Pearson residuals* for logistic regression models are based on comparing observed values of the outcome variable with predictions from a fitted model. However, because outcomes in logistic models are binary, the values of these residuals cluster in two groups corresponding to the two values of the outcome. This makes graphical displays of residuals more difficult to interpret than in the linear regression case. An exception occurs when there are relatively few unique covariate patterns in the data (e.g., when predictors are categorical) and residuals and predictions can be grouped.

Figure 6.4 shows standardized Pearson residuals for the model in Table 6.17, plotted against the ordered observation number for the individual subjects. This *index plot* allows observations with unusually large residuals rela-

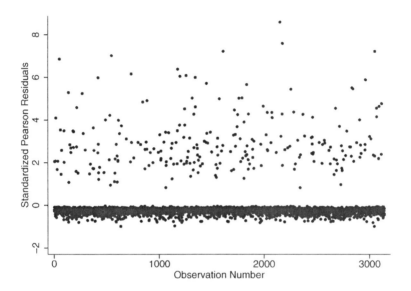

Fig. 6.4. Standardized Pearson Residuals for Logistic Model in Table 6.17

tive to other observations to be identified and investigated as potential outliers. The grouping of residuals based on outcome status is evident from the plot. In this case, although a number of observations have fairly large residuals (i.e., greater than two), none appear to be indicative of outlying observations. A number of other plots based on residuals are possible. In our experience, these are less useful in general than the investigation of influential points discussed in the next paragraph.

Diagnostic techniques for identifying influential observations in logistic regression models are also quite similar in definition and interpretation to their counterparts for linear regression. Most statistical packages that feature logistic regression allow computation of influence statistics that measure how much the estimated coefficients for a fitted model would change if the observation were deleted. Figure 6.5 shows influence statistics (often called DFBETA values) for the model in Table 6.17, plotted against the estimated outcome probabilities.

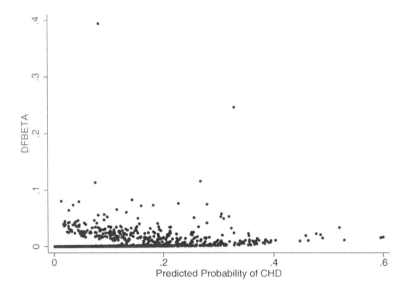

Fig. 6.5. Influence Statistics for Logistic Model in Table 6.17

Two observations appear to have more influence than the rest. The most extreme observation is for an individual who is a non-smoker with CHD, characterized by below average cholesterol (188) and a very high BMI value (39). Deletion of either observation (or both) resulted in no noticeable changes to model coefficients. Since there is no reason to suspect that any of the data are incorrect, both observations were retained.

6.4.2 Linearity

In Table 6.2 we fit a simple logistic regression model relating coronary heart disease (CHD) risk and age for the WCGS data. In addition to providing a simple description of the relationship, the model makes it easy to compute the log odds associated with an arbitrary value of age. However, as in simple linear regression (Sect. 4.7), the uncritical adoption of the assumption that variables are linearly related to the outcome can lead to biased estimates and incorrect inferences. LOWESS scatterplot smoothing methods (introduced in Chap. 2) offer an exploratory approach to assessing the form of the relationship between the log odds of the outcome and age that obviates the need to impose a particular parametric form. In the case of binary outcomes, these average the outcome proportions (or the corresponding log odds) over groups whose size is specified the bandwidth of the selected smoothing method. Fig. 6.6 displays the log odds estimated by LOWESS (obtained using the `lowess` command in Stata with the `logit` option) along with the linear logistic fit. The latter is represented by the dashed line, obtained by simply plotting the log odds

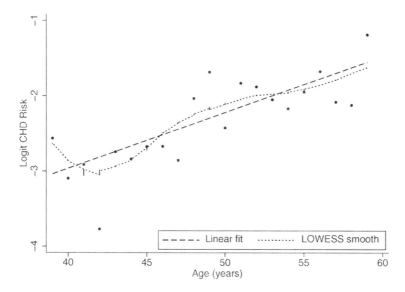

Fig. 6.6. Assessing Linearity in the Relationship Between CHD Risk and Age

estimated by the model for all the (3,154) individuals in the sample. The smoothed estimated is given by the dotted line. The plotted points are the estimated log odds for each of the 20 unique ages in the sample.

Although not conclusive, the results indicate that the linear logistic model fits the data reasonably well. However, the smoothed estimate suggests an initial decrease in the log odds of CHD risk for ages less than 42, followed by a fairly regular increase. The decrease might be due to elevated CHD risk among younger participants. In fact, 7% of the 39-year-olds ($n = 266$) in the study had CHD compared to 4% of the 40-year-old participants. The initial decline in the smoothed estimate is clearly influenced by the observed 2% rate of CHD among the 42-year-olds as well. A reasonable approach to evaluating this further would be to test for particular departures from linearity by adding a polynomial terms in age or using linear splines (similar to the approach described in Sect. 4.9). Table 6.23 displays results from a model including a quadratic term in age (centered to reduce possible collinearity with the linear term). The Wald test statistic clearly indicates that the addition of this term does not afford a statistically significant improvement in the fit over the linear model. We can conclude that the linear model is adequate.

If the role of age in modeling is primarily as an adjustment factor, we would also want to examine whether the assumption of linearity impacts inferences about other predictors. Adoption of the linear form is acceptable if no impacts are seen, but predictions of outcome risk based on the linear model may yield biased results for ages not well represented in the data. Diagnostics for

Table 6.23. Logistic Model Incorporating a Quadratic Effect of Age

```
. logistic chd69 age agesq, coef

Logistic regression                            Number of obs   =      3154
                                               LR chi2(2)      =     42.96
                                               Prob > chi2     =    0.0000
Log likelihood = -869.14333                    Pseudo R2       =    0.0241

-----------------------------------------------------------------------------
     chd69 |     Coef.   Std. Err.      z    P>|z|    [95% Conf. Interval]
-----------+-----------------------------------------------------------------
       age |  .0769963   .0150015     5.13   0.000     .0475938    .1063987
     agesq | -.0005543   .0021066    -0.26   0.792    -.0046831    .0035745
     _cons |  -6.04301    .678737    -8.90   0.000     -7.37331    -4.71271
-----------------------------------------------------------------------------
```

checking linearity in the context of multiple predictor models are somewhat less well developed for logistic models than for linear models. For example, tools like the component plus residual (CPR) plots presented in Sect. 4.7 are not generally available. However, the techniques presented here in combination with likelihood ratio comparisons of models are usually sufficient to diagnose and correct nonlinearity problems. The increased availability of nonparametric regression approaches for binary regression (discussed briefly in Sect. 6.5) is rapidly expanding the arsenal of available tools in this area.

6.4.3 Model Adequacy

The techniques discussed above address potential nonlinearity in the relationship between the log odds of the outcome and the predictor, but implicitly assume that the logistic model is correct. Recall from Sect. 4.7 that transformations of the outcome variable can be used to ensure that the distribution of the errors in a regression model are normally distributed. In a similar way, we can investigate the adequacy of the logistic model.

Specification Tests

A simple (and rather crude) approach to evaluating whether a given logistic model provides an adequate description of the data is through the use of a *specification test*. The linktest procedure in Stata is an example. Table 6.24 presents the results of applying linktest immediately after fitting the model in Table 6.17. This test involves fitting a second model, using the estimated right-hand side (i.e., the linear predictor) from the previously fitted model as a predictor. We would expect that the Wald test result for this predictor (labeled _hat) to be statistically significant if the original model provided a reasonable fit. The model fit by linktest also includes the square of this predictor (labeled _hatsq). The Wald test for inclusion of the latter variable is used to evaluate the hypothesis that the model is adequate; that is, the inclusion of the squared linear predictor should not improve prediction if the

Table 6.24. Link Test for Logistic Model in Table 6.17

```
. linktest

Logit estimates                              Number of obs   =      3141
                                             LR chi2(2)      =    200.40
                                             Prob > chi2     =    0.0000
Log likelihood = -786.89258                  Pseudo R2       =    0.1130

-----------------------------------------------------------------------
   chd69 |    Coef.   Std. Err.     z    P>|z|    [95% Conf. Interval]
---------+-------------------------------------------------------------
    _hat |  .5646788   .306056    1.85   0.065   -.0351799    1.164538
  _hatsq | -.1002356   .0688901  -1.46   0.146   -.2352576    .0347865
   _cons | -.3983753   .3230497  -1.23   0.218   -1.031541    .2347904
-----------------------------------------------------------------------
```

original model was adequate. Rejection indicates that the model is inadequate, and that an alternative binary regression model should be considered. It may also indicate that important predictors have been omitted. The test can not distinguish between these two alternative explanations. It also does not suggest what alternate model form might be preferable.

In the example, the P-value for the Wald test for the predictor _hatsq does not provide strong evidence of inadequacy of the logistic model. However, the fact that the P-value for the predictor _hat in Table 6.24 is also not very small provides some indication that the overall fit may not be very good. (This is consistent with the large residuals noted in Sect 6.4.1.)

Possible alternatives to the logistic model were discussed in Sect. 6.1, and will be covered in more detail in Sect. 6.5. Because these typically involve the use of specialized methods of estimation and result in coefficients with different interpretations, they are rarely used in practice. Fortunately, differences between results from alternative models are often small, and the logistic model applies in a very wide range of problems involving binary outcomes. Problems with fit can frequently be addressed using judicious selection and appropriate transformations of predictors.

Goodness of Fit Tests

Another approach to assessing model adequacy is provided by *goodness of fit* tests. The *Hosmer–Lemeshow* test is an example of this approach applicable to binary regression models such as the logistic. The test works by forming groups of the ordered, estimated outcome probabilities (e.g., ten equal-size groups based on deciles of the distribution of the outcome probabilities) and evaluating the concordance of the expected outcome frequencies in these groups with their empirical counterparts. The underlying hypothesis is that the estimated and observed frequencies agree. Thus, a statistically significant finding (i.e., rejection) indicates lack of fit. A non-significant finding rules out gross lack of fit.

Table 6.25 displays results of the Hosmer–Lemeshow test for the regression model fitted in Table 6.17. The `table` option requests that the observed and expected frequencies of the binary outcome (ones and zeros) for the requested groups be printed as well. The non-significant results do not indicate evidence

Table 6.25. Hosmer–Lemeshow Goodness of Fit Test

```
. lfit, group(10) table

Logistic model for chd69, goodness-of-fit test

(Table collapsed on quantiles of estimated probabilities)
+--------------------------------------------------------------+
| Group |   Prob | Obs_1 | Exp_1 | Obs_0 | Exp_0 | Total |
|-------+--------+-------+-------+-------+-------+-------|
|     1 | 0.0160 |     1 |   3.3 |   314 | 311.7 |   315 |
|     2 | 0.0251 |     6 |   6.5 |   308 | 307.5 |   314 |
|     3 | 0.0344 |    11 |   9.3 |   303 | 304.7 |   314 |
|     4 | 0.0450 |    12 |  12.5 |   302 | 301.5 |   314 |
|     5 | 0.0575 |    18 |  16.0 |   296 | 298.0 |   314 |
|-------+--------+-------+-------+-------+-------+-------|
|     6 | 0.0728 |    10 |  20.4 |   304 | 293.6 |   314 |
|     7 | 0.0963 |    28 |  26.5 |   286 | 287.5 |   314 |
|     8 | 0.1268 |    44 |  34.7 |   270 | 279.3 |   314 |
|     9 | 0.1791 |    50 |  46.7 |   264 | 267.3 |   314 |
|    10 | 0.5996 |    76 |  80.3 |   238 | 233.7 |   314 |
+--------------------------------------------------------------+

          number of observations =      3141
                number of groups =        10
        Hosmer--Lemeshow chi2(8) =       11.36
                    Prob > chi2 =      0.1824
```

for gross lack-of-fit. Increasing the number of groups to 20 yields a larger P-value (0.35), illustrating the sensitivity of the test to the number of groups chosen, and raising the possibility that judicious choice of group size may allow an investigator to choose the number of groups resulting in the most favorable P-value. To avoid this subjectivity, ten groups are generally recommended.

The Hosmer–Lemeshow test has a number of serious limitations. First, it is not sensitive to a number of sources of lack of fit such as misspecification of the model, and lacks power in these situations as a consequence. Further, the results of the test depend on the number of groups specified as well as the distribution of predictor values within these groups. Thus, failure to find a statistically significant result does not necessarily mean that the model fits the data well. This test is most useful as a very crude way to screen for fit problems, and should not be taken as a definitive diagnostic of a "good" fit. Use in conjunction with a specification test (such as the one described above) may provide a bit broader screen to detect problems. However, results of either approach should not be relied on to guarantee model fit in the absence of supplementary investigations, including diagnostic assessment of residuals and influential observations.

6.4.4 Technical Issues in Logistic Model Fitting

In some cases, measures of association for binary outcomes such as odds ratios and relative risks take on the value zero, or are infinite. This happens when subgroups formed by the predictors are homogeneous with respect to outcome status. This translates to estimation problems in regression models, where parameters are typically represented as the logarithm of the underlying association measures.

Table 6.26 presents an example from the WCGS study using a four-level categorization of cholesterol level (0–150, 151–200, 201–250, and 251+) as a predictor of CHD outcome. Note the extremely large estimated odds ratios and

Table 6.26. Logistic Model for CHD and Categorized Cholesterol Level

```
. xi:logistic chd69 i.cholc
i.cholc           _Icholc_0-3         (naturally coded; _Icholc_0 omitted)

Logistic regression                        Number of obs   =       3142
                                           LR chi2(3)      =      68.19
                                           Prob > chi2     =     0.0000
Log likelihood = -855.50635                Pseudo R2       =     0.0383


-----------------------------------------------------------------------
      chd69 | Odds Ratio   Std. Err.     z    P>|z|    [95% Conf. Interval]
------------+----------------------------------------------------------
   _Icholc_1 |    5509191    1108599   77.14  0.000     3713667     8172836
   _Icholc_2 |    8621442          .       .      .           .           .
   _Icholc_3 |    1.91e+07    2719171  117.76  0.000     1.44e+07    2.52e+07
-----------------------------------------------------------------------
note: 89 failures and 0 successes completely determined.
```

the note explaining that "89 failures and 0 successes completely determined." Examination of the data reveals that there are no observed CHD cases among the 89 individuals with cholesterol in the default reference category (0–150 mg/dL). Because the odds of CHD are zero for this group, it is not possible to estimate valid odds ratios for the other categories. Choosing an alternate reference group allows valid estimates to be made. However, the odds ratio of zero for the lowest category still causes a fitting issue: the log odds ratio is infinite, and the parameter can not be estimated.

The problems raised in this example can be easily addressed by choosing a different categorization of cholesterol. However, identifying and resolving problems with fitted models is not always this straightforward. In small samples, frequently no amount of regrouping or re-categorizing will eliminate these issues. In these situations the likelihood ratio test may still be valid and exact contingency tables or logistic regression may be alternatives. However, we recommend that a statistician be consulted to diagnose the exact nature of the problem and suggest solutions.

6.5 Alternative Strategies for Binary Outcomes

A review of current clinical and epidemiological research studies involving binary outcomes will reveal that the overwhelming majority of regression analyses are based on the logistic model. In some instances, specific knowledge about a disease-exposure relationship may suggest a different model. Alternatively, it may be desirable to summarize observed associations using measures such as the relative risk or excess risk in preference to the odds ratio. Because the logistic model yields only the latter, there are situations where alternative regression approaches may be preferred. Finally, diagnostic evaluations may lead to the conclusion that the logistic model is simply not right for a particular data set. In this section we review some examples of alternative approaches to binary regression. We also briefly discuss models for categorical outcomes with more than two levels.

6.5.1 Infectious Disease Transmission Models

Recall the CDC transmission study data discussed in Sect. 3.4 (O'Brien *et al.*, 1994). The goal of this study was to investigate risk factors for sexual transmission of HIV in susceptible female partners of previously infected males. Although the outcomes were restricted to prevalent HIV serostatus measured at enrollment, the infection dates of the male partners were approximately known from transfusion records. In addition, self-reported information on number of unprotected sexual contacts was also collected. These data pertain to contacts that occurred between the time of infection of the male partner and the time of enrollment. (Note that monogamy was an eligibility criterion, to reduce the possibility of infection from other sources.)

Unlike many chronic diseases, the mechanism of acquisition of many infectious diseases is well understood. In these cases, simple probabilistic *transmission models* linking outcomes with exposures are frequently used to quantify infection risk. One of the most basic such models links the cumulative probability of escaping infection following a series of exposed contacts. The model assumes that each contact carries an identical risk λ of infection, and that outcomes of successive contacts are independent. Under these assumptions, the chance of escaping infection following k contacts is

$$(1 - \lambda)^k,$$

with the complementary probability of being infected following k contacts given by

$$P(k) = 1 - (1 - \lambda)^k.$$

This model corresponds well to the observed data from the CDC study: each female partner can be characterized by the binary infection status and the reported number of exposed contacts k (the predictor), with the outcome probability given above. This suggests that a binary regression approach linking these two variables would be ideal for estimating the per-contact transmission probability λ. Unfortunately, the logistic model does not provide a direct estimate. By contrast, an alternative transformation of $P(k)$, known as the complementary log–log, provides a model with a more appealing structure:

$$\log\{-\log[1 - P(k)]\} = \log[-\log(1 - \lambda)] + \log(k). \tag{6.12}$$

This model is similar to the familiar linear model

$$\log\{-\log[1 - P(x)]\} = \beta_0 + \beta_1 x, \tag{6.13}$$

where the intercept coefficient $\beta_0 = \log[-\log(1 - \lambda)]$, but includes the predictor $x = \log(k)$ as a fixed *offset*, with corresponding coefficient $\beta_1 = 1$ as specified by model 6.12. Predictors with fixed coefficients are referred to as *offsets*, and can be easily accommodated by standard statistical software packages. (Part of the model evaluation procedure in this case may include checking whether this is reasonable in terms of fit.) Similar to the logistic model, an inverse transformation allows us to represent this model on the probability scale as follows:

$$P(x) = 1 - \exp[-\exp(\beta_0 + \beta_1 x)], \tag{6.14}$$

Table 6.27 shows the results of fitting model 6.12 using the generalized linear model estimation program `glm` in Stata, which we explain in greater detail in Chapter 9. Note that the logarithm of the number of contacts `logcontacts` appears as an offset, and no coefficient for this predictor was estimated.

An additional calculation inverting the complementary log–log transform of the intercept `_cons` provides the estimate of λ:

$$\lambda = 1 - \exp[-\exp(-7.033)] = 0.0009.$$

The approximate 95% confidence interval (0.0004, 0.0019) can be obtained via a similar calculation applied to confidence limits given in the regression output. Because of the small sample size ($n = 31$), the approximate confidence intervals may not be reliable. For comparison, Table 6.27 also gives bias-corrected 95% bootstrap confidence intervals (calculated using 1000 bootstrap samples) for the same model. The bias-corrected confidence interval (0.0003, 0.0018) for the parameter λ can be obtained from the interval for the intercept coefficient β_0 (represented by `b_cons` in the table) via the calculation used for the approximate interval. The lower bound of this interval is only slightly more conservative than the approximate interval, but otherwise they are remarkably similar. The bootstrap interval should still be considered a better summary of uncertainty about λ.

Clearly, model 6.12 is very simple, and a number of the underlying assumptions are questionable (e.g., that the per-contact risk λ is constant). However, it is a useful "null" model to which more complex alternatives may be compared. Further, the parameter λ is an important ingredient in more complex

Table 6.27. Complementary Log-Log Regression Model for Per-Contact Risk

```
. glm hivp, family(binomial) link(cloglog) offset(logcontacts)

Generalized linear models                    No. of obs      =        31
Optimization     : ML: Newton-Raphson        Residual df     =        30
                                             Scale parameter =         1
Deviance         =    40.8340195             (1/df) Deviance =  1.361134
Pearson          =    84.90572493            (1/df) Pearson  =  2.830191

Variance function: V(u) = u*(1-u)            [Bernoulli]
Link function    : g(u) = ln(-ln(1-u))       [Complementary log-log]
Standard errors  : OIM

Log likelihood   = -20.41700975              AIC             =  1.381743
BIC              = -62.18559663

------------------------------------------------------------------------------
       hivp |      Coef.   Std. Err.      z    P>|z|     [95% Conf. Interval]
------------+-----------------------------------------------------------------
      _cons | -7.033126   .3803284   -18.49   0.000    -7.778556   -6.287696
 logcontacts |  (offset)
------------------------------------------------------------------------------


. bootstrap "glm hivp, family(binomial) link(cloglog) offset(logcontacts)" _b _se,
  reps(1000)

command:      glm hivp, family(binomial) link(cloglog) offset(logcontacts)
statistics:   b_cons   = [hivp]_b[_cons]

Bootstrap statistics                         Number of obs   =        31
                                             Replications    =      1000
------------------------------------------------------------------------------
Variable     | Reps  Observed      Bias  Std. Err. [95% Conf. Interval]
-------------+----------------------------------------------------------------
     b_cons  | 1000 -7.033126 -.0629388  1.163788   -8.216878   -6.296359
------------------------------------------------------------------------------
```

mathematical epidemic models. This model is also interesting because it is an example of a *proportional hazards model*. These arise frequently in studies where controlling for duration of follow-up is an important consideration in data analyses, and are the subject of the next chapter. Finally, model 6.13 and the conventional logistic model are examples of the family of generalized linear models that includes most of the regression models considered in this book.

6.5.2 Regression Models Based on Excess and Relative Risks

A recent study of prevalent human T-cell leukemia/lymphoma virus (HTLV) infection in infants born to mothers in the United Kingdom identified a number of factors associated with infection, including the parent's country of birth and ethnicity of the mother (Ades *et al.*, 2000). The authors found that a regression model based on excess risk provided a better fit to the data than the logistic model, and reported their results accordingly.

Recall the linear regression model defined in equation (6.1) that relates risk for a binary outcome to a single predictor x:

$$P(x) = \beta_0 + \beta_1 x.$$

As noted in Sect. 6.1, the coefficient β_1 measures the excess risk (or risk difference) associated with a unit increase in x. This model is often termed the "additive risk model" because the effect of any unit increase in the predictor x is to add an increment β_1 to the outcome risk. This was the model employed in the HTLV example. Although it provides a valid alternative to logistic regression, it is important to keep in mind the potential problems with fitting and interpretation (raised in Sect. 6.1).

As discussed in Sect. 3.4, the odds ratio is known to approximate the relative risk in the rare outcome setting. Consequently, odds ratios are frequently reported as relative risks in research findings. Unfortunately, this practice is not limited to rare outcomes, and has been the subject of considerable debate in the research literature (Holcomb et al., 2001). This has led many investigators to advocate that regression models based on the relative risk be used in preference to the logistic model (other than in case-control designs where standard regression approaches other than the logistic model do not directly apply). This is possible using the following regression model:

$$\log[P(x)] = \beta_0 + \beta_1 x. \tag{6.15}$$

This is the *log linear* model discussed in Sect. 6.1. The regression coefficient β_1 has the interpretation of the logarithm of the relative risk associated with a unit increase in x. Analogous to the procedure for obtaining odds ratios from logistic models, exponentiated coefficients yield relative risk estimates in this case. Although this model can be fit with many standard software packages, it may present numerical difficulties because of the constraint that the sum of terms on the right-hand side must be no greater than zero for the results to make sense (due to the constraint that the outcome probability $P(x)$ must lie in the interval $[0, 1]$). As a result, convergence of standard fitting algorithms may be unreliable in some cases.

Alternative approaches for obtaining adjusted relative risks from odds ratios estimated using logistic regression have been proposed in the literature (Zhang and Yu, 1998). These are based on simple transformations of the estimated coefficients similar to the illustrative calculations demonstrated in Sect. 6.1.1. Unfortunately, such calculations can produce incorrect estimates for models including multiple predictors and should be avoided in favor of fitting appropriately defined regression models as described above (McNutt et al., 2003).

Table 6.28 presents the results of fitting four alternative generalized linear models for the relationship between coronary heart disease and age using the WCGS data. (Results were obtained with the Stata generalized linear models procedure glm, also applied in Table 6.27.) These correspond to the alternative formulations considered in this section (i.e., equations (6.1), (6.2), (6.13), and (6.15)). Results for the intercept parameter β_0 are similar. Note that the estimated regression coefficients cannot be directly compared because the models are based on different representations of the outcome. However, since all of them are based on the same number of parameters, comparison of the likelihoods provides a cursory look at how well they describe the data in relative terms. Although the likelihood for the logistic model is slightly larger, there is very little overall difference between the models. Similarly, the estimated coefficients for the log, complementary log–log, and logit models are remarkably similar. (The coefficients for the excess risk model differ because the outcome is modeled without transformation.) Finally, the estimated probabilities for a 55-year-old individual $(P(55))$ are also quite similar. Based on these results, there would be no particular reason to prefer any alternatives over the logistic model.

Table 6.28. Generalized Linear Models for CHD Risk (P) and Age (x)

Model	β_1 (95% CI)	Log-likelihood	$P(55)$
$P(x)$	$0.005(0.004, 0.007)$	-869.96	0.130
$\log[P(x)]$	$0.067(0.048, 0.087)$	-869.24	0.136
$\log\{-\log[1-P(x)]\}$	$0.072(0.050, 0.092)$	-869.21	0.136
$\log\{P(x)/[1-P(x)]\}$	$0.074(0.052, 0.097)$	-869.18	0.136

The results in Table 6.28 illustrate that a variety of models other than the logistic may be appropriate for a given problem. However, given the ease of interpretation, wide use, and software availability of the logistic model, it is by far the most common choice in practice. In general, we advocate fitting the logistic model unless another model is preferable on scientific grounds. Lack of fit can often be dealt with via the techniques discussed in Sect. 4.7, obviating the need to investigate alternative model formulations. Finally, note that the approaches discussed here are not directly applicable to data from case-control studies (Scott and Wild, 1997).

6.5.3 Nonparametric Binary Regression

The examples of alternative techniques for binary regression considered above represent only a small subset of the available possibilities for estimating the relationship between a binary outcome and a predictor variable. The goal of *nonparametric regression* methods is to provide estimates of this relationship based on minimal assumptions about its form.

Recall the assessment of linearity for the logistic model for the relationship between coronary heart disease and age in the WCGS data in Sect. 6.4.2. The smoothed LOWESS estimate displayed in Figure 6.6 is an example of a nonparametric logistic regression model for this relationship. Although the assumption that the predictor is related to the disease outcome in an additive fashion via the log odds is retained, this technique allowed us to relax the assumption that the relationship is linear by assuming only that the change in CHD risk with age has a certain degree of smoothness. This can prove very useful in exploring the form of the relationship between outcome and predictor, but does not yield readily interpretable parameter estimates or generalize easily to models including more than one predictor. The class of *generalized additive models* provide an extension to the LOWESS technique, allowing multiple predictors to be fit simultaneously, each of which can be represented as a smooth function (Hastie and Tibshirani, 1990). Although very useful in evaluating outcome-predictor relationships, these models are frequently difficult to fit and interpret.

Methods for significance testing, confidence intervals, and model evaluation are less well developed for nonparametric alternatives than for conventional logistic regression. In addition, decisions about degree of smoothness and interpretation of resulting estimates is often very complex. Finally, practical implementations of nonparametric binary regression that handle multiple predictors are not widely available in standard statistical packages. For these reasons, we recommend that flexible parametric approaches be used in accounting for nonlinearities in the relationship between predictor and outcome, and that nonparametric alternatives be used primarily for exploratory purposes.

Classification trees (Breiman *et al.*, 1984) are another popular approach to nonparametric binary regression. As discussed in Sect. 5.2 and Sect. 6.2.4, these lack the linear and additive structure shared by other approaches, and are extremely useful in developing flexible prediction tools for using measured characteristics to correctly distinguish binary outcomes. However, classification trees can also be used to explore complex relationships between multiple predictors and a binary response. Because they do not yield estimates of association parameters, interpretation of the contribution of individual predictors to the outcome risk is complex. However, like the nonparametric regression approaches discussed above, they are very useful tools in exploratory analyses and can be very helpful in discovering and interpreting interaction.

6.5.4 More Than Two Outcome Levels

Research studies frequently yield outcomes that have multiple categories. (See Chap. 2 for definitions of categorical variable types.) Consider the back pain example introduced in Sect. 1.1, where pain intensity was measured on an ordered, ten-point scale. In addition to the *ordinal* categorical outcome just considered, *nominal* categorical outcome measures are also commonplace in clinical research. For example, the outcome in a study of cancer outcomes by cell type is a nominal categorical variable. Both type of outcomes can be investigated using contingency table methods. The limitations of these when multiple predictors are involved are clear. For certain questions, considering a binary representation might also be reasonable. For example, to investigate factors that distinguish patients suffering from severe pain from all others in the pain example. In this case, logistic regression is an appropriate tool to consider. However, there is clearly information lost in reducing ten levels down to two. In the remainder of this section we briefly review regression methods for nominal and ordinal categorical outcomes.

Ordinal Categorical Outcomes

The *proportional odds model* is a commonly used generalization of the logistic model that accommodates a multilevel categorical response with ordered categories. Rather that modeling the probability of response in a particular category, this model is based on the cumulative probability that the response is not greater than a chosen category. The dependence of this response on predictors is identical to the form of the logistic model. For the back pain example, (assuming a ten-level response and a single predictor x), the form of this model for a response probability of severity no greater than 5 is given by

$$\log\left[\frac{\Pr(y \le 5)}{\Pr(y > 5)}\right] = \alpha_5 - \beta x.$$

A similar expression applies to all ten levels of the response. (We assume that the levels of the response are coded $1, 2, \cdots, 10$.)

Note that the intercept parameter α_5 is unique to this response level, and represents the probability of a response of no more than 5 among individuals with $x = 0$. Because the response is expressed as a cumulative probability, the intercept coefficients are constrained as $\alpha_1 \le \alpha_2 \le \cdots \le \alpha_{10}$. The coefficient β is interpreted as the log odds ratio associated with a unit increase in x, assumed to be constant across response levels. (i.e., response levels are parallel, each with slope β.) This assumption amounts to a strong restriction on the effect of the predictor on the response, and needs to be validated.

Note that there are many alternatives to the proportional odds model, including the *continuation ratio model*. We refer the reader to the references provided below for additional information on these.

Nominal Categorical Outcomes

When there is no natural ordering implicit in a categorical response, or when the assumptions implicit in the models above do not apply to an ordinal outcome, the *polytomous logistic* model can be for regression analyses. For a single predictor x, the model specifies that each response level follows a logistic regression model for x, with a selected level specified as the reference. The regression coefficients for each level are unique; so for the pain example the model would include nine intercept and slope coefficients. For level 5, and specifying the first level as the reference category, the model would take the form

$$\log \left[\frac{\Pr(y = 5)}{\Pr(y = 1)} \right] = \alpha_5 + \beta_5 x.$$

Thus, the log odds ratio for a unit increase in x is given by β_5. Because this model does not involve the restrictions implicit in the proportional odds model, it is an attractive alternative when the proportional odds assumption is not satisfied. However, because of the potentially large number of parameters and the flexibility of choice for the reference group, the polytomous logistic model can be challenging to interpret.

The models outlined here represent a few of those available for analyzing categorical responses. For further information on these and other models, including examples and a description of available software resources, see Ananth and Kleinbaum (1997) and Greenland (1994).

6.6 Likelihood

One of the common themes uniting methods presented in this book is the principle of using observed data to estimate unknown quantities of interest. The majority of the methods presented are regression models relating outcome and predictor variables measured on a sample of individuals. The principal unknown quantities in the models are the regression parameters. Once these are estimated, inferences can be made about the true values of these parameters and related quantities of interest such as predicted outcomes. All available information about the parameters is contained in the observed data. A standard approach to estimating parameters in models like the ones covered here is known as *maximum-likelihood estimation*. Although not required for applications, a basic understanding of this topic helps in unifying the concepts underlying estimation and inference in most of the regression models covered in this book. Here we provide a brief discussion of some of the key ideas in the binary regression context.

The *likelihood* associated with a set of independent observations of an outcome is just the product of their respective probabilities of occurrence under the assumed model relating outcomes to predictors. Because this represents the joint probability of observing all of the outcomes in the sample, the likelihood can then be interpreted as a measure of support provided for the model by the data. The maximum-likelihood estimate of the parameter(s) is the most likely value for the parameter(s) given the observed data (i.e., the value that yields the maximum value of the likelihood).

To take a very simple example from the binary outcome context, consider the problem of estimating the prevalence of HIV for the sample of 31 female partners of previously infected males from the CDC transmission study considered in the examples presented above and in Sect. 3.4. The assumed model is that the actual prevalence in the target population is represented by a constant that we can symbolize by P (similar to the definition introduced earlier in this chapter). We can think of P as the probability that a randomly sampled individual will test positive. The corresponding probability of observing a negative is $1 - P$. However, P is unknown. The observed data consist of the 31 indicators of HIV status, and the likelihood, as defined above, is just the product of the individual outcome probabilities:

$$P^7 \times (1 - P)^{23}.$$

The likelihood is formed as the product of the individual outcome probabilities because these are independent events. It is a function of the unknown constant P, with the observed infection indicators providing the number of positive and negative individuals. Fig. 6.7 presents a plot of this function for a range of values for P. The maximum-likelihood estimator of P is just the

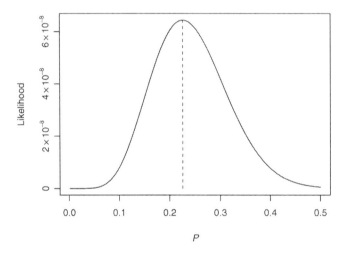

Fig. 6.7. Likelihood Function for HIV Prevalence

value of P that maximizes the likelihood function. This value is indicated in the figure. The maximum can be found easily in this example using calculus. Not surprisingly, it corresponds exactly to the intuitive estimate of the actual prevalence of HIV-positive individuals in the sample of 31: Because there are seven such individuals in the sample, the estimated prevalence is 0.226. For more complicated models (e.g., regression models with multiple predictors) computing the maximum typically involves iterative calculations on a computer.

Likelihood functions for binary regression models are defined following the procedure used above, but the outcome probability P for each individual is replaced with the form defined by the logistic model (equation 6.2). To take another example from the CDC study, consider a regression model relating HIV status of the female partners to a binary indicator of presence of an AIDS diagnosis in the male. (This example was already considered in Sect. 3.4.) Following our conventional notation, we will represent the outcome as Y and the predictor as x. The observed data now include both Y and the binary predictor x for each individual in the sample. The likelihood takes exactly the same form as in the last example, except the constant P is replaced with the expression for the logistic model, substituting in each individual's value of the predictor (i.e. x_i for the i_{th} individual):

$$\prod_{i=1}^{31} \left[\frac{\exp(\beta_0 + \beta_1 x_i)}{1 + \exp(\beta_0 + \beta_1 x_i)} \right]^{Y_i} \times \left[1 - \frac{\exp(\beta_0 + \beta_1 x_i)}{1 + \exp(\beta_0 + \beta_1 x_i)} \right]^{1-Y_i}.$$

Since both Y and x (the indicator of AIDS status) are observed, the only unknown quantities are the regression parameters β_0 and β_1. These are generally estimated using an iterative maximization algorithm. Fig. 6.8 presents a plot of the logarithm of this function for a range of values for P. Because the

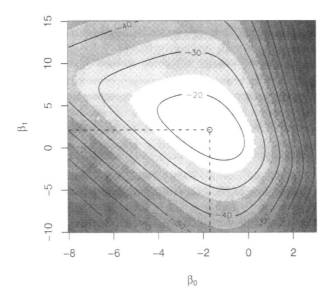

Fig. 6.8. Likelihood Function for a Two-Parameter Logistic Model

likelihood function depends on two unknown parameters, it has the form of a "surface" when plotted in three dimensions. The two-dimensional figure represents the contours of this surface as seen from above. The maximum value is indicated, and the corresponding maximum-likelihood estimates for β_0 and β_1 are -1.705 and 2.110, respectively.

Because likelihoods are formed from the product of outcome probabilities for all individuals in a sample, the numerical value of a given likelihood depends on the sample size and is not particularly interpretable by itself. However, comparing likelihoods from nested models is a direct way to evaluate improvements in fit. This is the basis of the likelihood ratio test.

Finally, we note that although the discussion here is limited to the binary outcome context, estimation methods for most of the regression models presented in this book are likelihood-based. For example, least squares estimation and F-testing for comparing nested models in linear regression and analysis of variance models are examples of likelihood methods. Further, likelihood methods are fundamental to the family of generalized linear models discussed in Chapter 9.

6.7 Summary

The logistic regression model extends frequency table techniques for investigating the association between a binary outcome and categorical predictor to include continuous predictors and allow simultaneous consideration of multiple (continuous and categorical) predictors.

Modeling techniques for logistic regression mirror those for linear regression, allowing many of the concepts and methods learned in Chapters 4 and 5 to be applied directly to studies involving binary outcomes. However, interpretation of logistic regression models is slightly more complex due to the model's nonlinear relationship between outcome risk and predictors. In particular, regression coefficients need to be transformed to be interpretable as odds ratios.

Although a powerful and useful tool, there are a number of situations where logistic regression is not the best method for analyzing binary outcome data. As we have seen in several examples, when attention is restricted to one or a few categorical predictors, regression techniques are not needed. Another example arises in studies yielding binary outcomes that are duration-dependent. In such studies, additional information about the time to development of an outcome is often available in addition to whether or not the outcome occurs. Further, duration of follow-up between individuals may vary and is informative about the amount of time each was observed to be "at risk" of outcome development. Although the WCGS data arose from such a study, complete follow-up of participants allows a comparable assessment of whether or not the outcome occurred for each. Thus, use of logistic regression is warranted. However, this will not allow us to answer questions regarding differences in how quickly participants experienced outcomes after study onset without consideration of additional data. The methods covered in Chapter 7 are more appropriate for answering these types of questions.

6.8 Further Notes and References

There are a number of excellent text books on logistic regression, including Breslow and Day (1984), Hosmer and Lemeshow (2000), Kleinbaum (2002), and Collett (2003). All of these provide more details and cover a broader range of topics than provided here. Although we have focused on Stata in our example analyses, most modern statistical software packages provide extensive facilities for fitting and interpretation of logistic models, including R, SAS, S-PLUS, and SPSS. Exact logistic regression and contingency table methods are available in the programs StatXact and LogXact.

Throughout this chapter we have concentrated on analysis of data where the outcomes and predictors were measured without substantial error and missing observations were not considered a major problem. In many studies we cannot assume that this is the case. There is an extensive literature on the impacts of misclassified outcomes and measurement error in predictors in the context of logistic regression (Carroll *et al.*, 1995; Magder and Hughes, 1997).

Missing data are an issue in most studies involving binary outcomes, and arise through a variety of mechanisms. When relatively few observations are involved, the problem can be handled via the default procedure in most available software programs (i.e., to eliminate any observations with one or more missing values among the predictors). The validity of this approach rests on the assumption that the individuals dropped from the analysis are "missing completely at random." However, when a substantial fraction of observations involve missing values, more care is required. In addition to the obvious problem of the reduction in power incurred by dropping observations there are substantial concerns that the results based on the remaining complete data may be biased. There are a number of approaches to handling missing observations, including sensitivity analyses, imputation and modified maximum-likelihood estimation methods. (See Jewell (2004) for a more complete discussion.) These tend to be complex to apply and are not generally well represented in standard software.

6.9 Problems

Problem 6.1. Verify that the numerical average (mean) of the following sample of 25 binary outcomes equals the proportion of positive outcomes (ones) in the sample:

$$(1, 0, 0, 0, 1, 0, 0, 0, 1, 1, 0, 0, 0, 1, 1, 1, 0, 0, 1, 1, 1, 0, 1, 0, 0)$$

Problem 6.2. Use the regression coefficients from the logistic model presented in Table 6.2 in the logistic formula (6.2) to estimate the quantities in Table 6.3 for a 65-year-old individual. Use additional calculations to add a new section to Table 6.3 for an age increment of five years.

Problem 6.3. Perform the basic algebra necessary to verify the properties of the logistic regression coefficient β_1 stated in equation (6.6).

Problem 6.4. The output in the Table 6.29 provides the regression coefficients corresponding to the model fitted in Table 6.5. Use the coefficients and

Table 6.29. Logistic Model for CHD and Age

```
. xi: logistic chd69 i.agec, coef
i.agec          _Iagec_0-4              (naturally coded; _Iagec_0 omitted)

Logistic regression                         Number of obs   =       3154
                                            LR chi2(4)      =      44.95
                                            Prob > chi2     =     0.0000
Log likelihood = -868.14866                 Pseudo R2       =     0.0252
```

chd69	Coef.	Std. Err.	z	P>|z|	[95% Conf. Interval]	
_Iagec_1	-.1314518	.2309937	-0.57	0.569	-.5841912	.3212876
_Iagec_2	.5307399	.223534	2.37	0.018	.0926212	.9688585
_Iagec_3	.8409976	.2274985	3.70	0.000	.3951088	1.286886
_Iagec_4	1.05998	.2585371	4.10	0.000	.5532569	1.566704
_cons	-2.804337	.1849626	-15.16	0.000	-3.166857	-2.441817

calculations similar to those illustrated in Sect. 6.1.1 to compute the log odds ratio comparing CHD risk in the fourth age category (_Iagec_4) with the third (_Iagec_3). Also, compute the odds ratio for this comparison. Comment on how we might obtain a estimated standard error and 95% confidence interval for this quantity.

Problem 6.5. For the fitted logistic regression model in Table 6.6, calculate the log odds for a 60-year-old smoker with cholesterol, SBP, and BMI values of 250 mg/dL , 150 mmHg, and 20, respectively. Now calculate the log odds for an individual with a cholesterol level of 200 mg/dL, holding the values of the other predictors fixed. Use these two calculations to estimate an odds ratio associated with a 50 mg/dL increase in cholesterol. Repeat the above calculations for a 70-year-old individual with identical values of the other predictors. Comment on any differences between the two estimated odds ratios.

Problem 6.6. Use the regression output in Table 6.15 and a calculation similar to that presented in equation (6.11) to compute the odds ratio comparing the odds of CHD in a 55-year-old individual with arcus to the corresponding odds for a 40-year-old who also has arcus.

Problem 6.7. Use the WCGS data set to fit the regression model presented in Table 6.17. Perform the Hosmer–Lemeshow goodness of fit test for the following number of groups: 10, 15, 20, and 25. Comment on the differences. The data set is available at `http://www.biostat.ucsf.edu/vgsm`.

Problem 6.8. Verify that the odds ratio formed from the two odds presented in equation (6.11) is given by ad/bc. Verify that the same odds ratio is obtained if the two component odds are computed based on the probability of exposure conditional on outcome status.

Problem 6.9. Compute the approximate 95% confidence interval for the following per-contact infection risk based on the intercept coefficient and associated standard errors given in Table 6.27:

$$1 - \exp\left[-\exp(-7.033)\right].$$

6.10 Learning Objectives

1. Describe situations in which logistic regression analysis is needed.
2. Translate research questions appropriate for a logistic regression model into specific questions about model parameters.
3. Use logistic regression models to test hypotheses about relationships between a binary outcome variable and a continuous or categorical predictor.
4. Describe the logistic regression model, its key assumptions, and their implications.
5. State the relationships between–
   - odds ratios and logistic regression coefficients
   - a two×two table analysis of the association between a binary outcome and single categorical predictor and a logistic regression model for the same variables.
6. Know how a statistical package is used to fit a logistic regression model to continuous and categorical predictors
7. Interpret logistic regression model output, including–
   - regression parameter estimates, hypothesis tests, confidence intervals
   - statistics which quantify the fit of the model.

7

Survival Analysis

Children receiving a kidney transplant may be followed to identify predictors of mortality. Specifically, is mortality risk lower in recipients of kidneys obtained from a living donor? If so, is this effect explained by the time the transplanted kidney is in transport or how well the donor and recipient match on characteristics that affect immune response? Similarly, HIV-infected subjects may be followed to assess the effects of a new form of therapy on incidence of opportunistic infections. Or patients with liver cirrhosis may be followed to assess whether liver biopsy results predict mortality.

The common interest in these studies is to examine predictors of time to an event. The special feature of the survival analysis methods presented in this chapter is that they take time directly into account: in our examples, time to transplant rejection, incidence of opportunistic infections, or death from liver failure. Basic tools for the analysis of such *time-to-event* data were reviewed in Sect. 3.5. This chapter covers multipredictor regression techniques for the analysis of outcomes of this kind.

7.1 Survival Data

7.1.1 Why Linear and Logistic Regression Won't Work

In Sect. 3.5 we saw that a defining characteristic of survival data is *right-censoring*:

> Definition: A survival time is said to be *right-censored* at time t if it is only known to be greater than t.

Because of right-censoring, survival times cannot simply be analyzed as continuous outcomes. But survival data also involves an outcome *event*, so why isn't logistic regression applicable? The reason is unequal length of follow-up. In Chapter 6 the logistic model was used to study coronary heart disease events among men in the Western Collaborative Group Study (Rosenman *et al.*,

1964). But in that study, the investigators were able to determine whether each one of the study participants experienced the outcome event at any time in the well-defined ten-year follow-up period; follow-up was constant across participants.

In contrast, follow-up times were quite variable in ACTG 019 (Volberding *et al.*, 1990), a randomized double-blind placebo-controlled clinical trial of zidovudine (ZDV) for prevention of AIDS and death among patients with HIV infection. Between April 1987 and July 1989, 453 patients were randomized to ZDV and 428 to placebo. When the data were analyzed in July 1989, some had been in the study for less than a month, while others had been observed for more than two years. These data could only be forced into the logistic framework by restricting attention to the events that occur within the shortest observed follow-up time – a huge waste of information.

7.1.2 Hazard Function

In Sect. 3.5 we introduced the survival function and its complement, the cumulative incidence function, as useful summaries of the distribution of a survival time.

> Definition: The *survival function* at time t, denoted $S(t)$, is the probability of being event-free at t. The *cumulative incidence function* at time t, denoted $F(t) = 1 - S(t)$, is the complementary probability that the event has occurred time by t.

Another useful summary is the hazard function $h(t)$.

> Definition: The *hazard function* $h(t)$ is the short-term event rate for subjects who have not yet experienced the outcome event.

The hazard function is systematically related to both the survival and cumulative incidence functions.

Table 7.1 shows mortality rates for children who have recently undergone kidney transplantation, on each of the first ten days after surgery, using data from the United Network for Organ Sharing (UNOS). At the beginning of fifth day after surgery, for example, 9,653 children remained alive and in the study, and of these, 3 died during the next 24 hours, yielding an estimated death rate of 0.31 deaths per 1,000 subjects per day. From the rightmost column of the table, it appears that the mortality rate declines over the first ten days, although the estimates spike on days 8 and 10.

In Fig. 7.1, daily death rates, smoothed by LOWESS, are used to estimate the mortality hazard for a much longer time period, the first 12 years after transplantation. The mortality hazard declines rapidly over the course of the first two years, reaching a plateau approximately three years after transplantation.

Table 7.1. Mortality Among Pediatric Kidney Transplant Recipients

Days since transplant	No. in follow-up	No. died	No. censored	Death rate per 1,000 subject-days
1	9,752	7	14	$7/9,752 \times 1,000 = 0.72$
2	9,731	5	8	$5/9,731 \times 1,000 = 0.51$
3	9,718	5	12	$5/9,718 \times 1,000 = 0.51$
4	9,701	7	41	$7/9,701 \times 1,000 = 0.72$
5	9,653	3	54	$3/9,653 \times 1,000 = 0.31$
6	9,596	2	57	$2/9,596 \times 1,000 = 0.21$
7	9,537	0	50	$0/9,537 \times 1,000 = 0.00$
8	9,487	4	49	$4/9,487 \times 1,000 = 0.42$
9	9,434	1	49	$1/9,434 \times 1,000 = 0.11$
10	9,384	3	28	$3/9,384 \times 1,000 = 0.32$

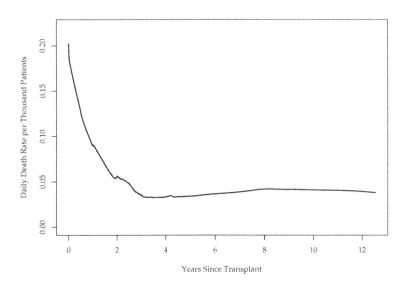

Fig. 7.1. Mortality Rate for Pediatric Kidney Transplant Recipients

7.1.3 Hazard Ratio

We now compare the hazard functions for children whose transplanted kidney was provided by a living donor, commonly a family member, and those for whom the source was recently deceased. Fig. 7.2 shows LOWESS-smoothed death rates for the recipients of kidneys from living and recently deceased donors. The mortality rate is considerably lower among the recipients of kidneys from living donors at all time points, but the curves are similar in shape.

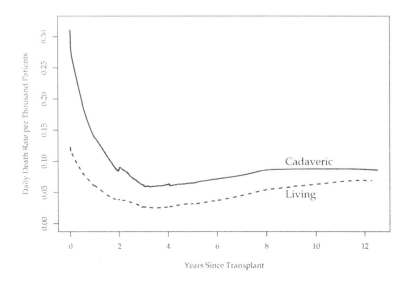

Fig. 7.2. Smoothed Mortality Rates for Recipients by Kidney Donor Type

Table 7.2. Smoothed Death Rates (per 1,000 Days) by Donor Type

Years since transplantation	Smoothed rates Cadaveric	Living	Death rate ratio
0.25	0.235	0.098	2.40
0.50	0.193	0.082	2.36
1.00	0.138	0.061	2.27
2.00	0.088	0.038	2.30
3.00	0.061	0.027	2.25
4.00	0.063	0.026	2.37
5.00	0.065	0.032	2.03

Table 7.2 gives the values of the LOWESS-smoothed death rates shown in Fig. 7.2 for selected time points, which estimate the hazard functions in each group, as well as the death rate ratio, an estimate of the *hazard ratio*. We could write the hazard ratio as

$$\mathrm{HR}(t) = h_c(t)/h_l(t), \tag{7.1}$$

where $h_c(t)$ is the hazard function in the recipients of kidneys from recently deceased donors, and $h_l(t)$ is the corresponding hazard function in the reference group, the recipients of kidneys from living donors.

7.1.4 Proportional Hazards Assumption

The results in Table 7.2 show that while the mortality hazards decline over time in both groups of pediatric kidney transplant recipients, the hazard ratio is almost constant. In other words the hazard in the comparison group is a constant proportion of the hazard in the reference group.

> Definition: Under the *proportional hazards assumption* the hazard ra-tio does not vary with time. That is, $\mathrm{HR}(t) \equiv \mathrm{HR}$.

Provided the hazards are proportional in this sense, the effect of donor source on post-transplant mortality risk can be summarized by a single number. This simplification is important but not necessary for the Cox proportional hazards model described in the next section. A rough analog is multiple linear regression without interaction terms. In Sect. 7.4.2 below we show how the proportional hazards assumption can be checked and violations addressed by including interactions between time and the predictors causing trouble. This is implemented using *time-dependent covariates*, an extension of the basic Cox model introduced in Sect. 7.3.1.

7.2 Cox Proportional Hazards Model

The Cox proportional hazards regression model is a flexible tool for assessing the relationship of multiple predictors to a right-censored, time-to-event outcome, and has much in common with linear and logistic models. To understand how the Cox model works, we first consider the broader class of proportional hazards models.

7.2.1 Proportional Hazards Models

In the linear model for continuous outcomes, covered in Chapters 3 and 4, the linear predictor $\beta_1 x_1 + \ldots + \beta_p x_p$, which captures the effects of predictors, is linked directly to the conditional mean of the outcome, $E[y|\mathbf{x}]$:

$$E[y|\mathbf{x}] = \beta_0 + \beta_1 x_1 + \ldots + \beta_p x_p. \tag{7.2}$$

In the logistic model for binary outcomes, covered in Chapter 6, the linear predictor is linked to the conditional mean through the logit transformation:

$$\log \frac{p(\mathbf{x})}{1 - p(\mathbf{x})} = \beta_0 + \beta_1 x_1 + \ldots + \beta_p x_p. \tag{7.3}$$

In (7.3), $p(\mathbf{x}) = E[y|\mathbf{x}]$ is the probability of the outcome event for a observations with predictor values $\mathbf{x} = (x_1, \ldots, x_p)$.

In proportional hazards regression models, the linear predictor is linked through the log transformation to the hazard ratio introduced in Sect. 7.1.3.

If the hazard ratio obeys the proportional hazards assumption, and thus does not depend on time, we can write

$$\log\left[\text{HR}(\mathbf{x})\right] = \log\frac{h(t|\mathbf{x})}{h_0(t)} = \beta_1 x_1 + \ldots + \beta_p x_p. \tag{7.4}$$

In (7.4), $h(t|\mathbf{x})$ is the hazard at time t for an observation with covariate value $\mathbf{x}$, and $h_0(t)$ is the *baseline hazard function*, defined as the hazard at time t for observations with all predictors equal to zero. As with the intercept in linear and logistic regression, this may mean that the baseline hazard does not apply to any possible observation, and argues for centering continuous predictors.

Solving (7.4) for $h(t|\mathbf{x})$ gives

$$\begin{aligned} h(t|\mathbf{x}) &= h_0(t)\exp(\beta_1 x_1 + \ldots + \beta_p x_p) \\ &= h_0(t)\text{HR}(\mathbf{x}). \end{aligned} \tag{7.5}$$

Note that exponentiating the linear predictor ensures that $\text{HR}(\mathbf{x})$ cannot be negative, as required. Furthermore, taking the log of both sides of (7.5), we obtain

$$\log[h(t|\mathbf{x})] = \log[h_0(t)] + \beta_1 x_1 + \ldots + \beta_p x_p. \tag{7.6}$$

This shows that the log baseline hazard plays the role of the intercept in other regression models, though in this case it can change over time. Furthermore, (7.6) defines a *log-linear* model, which implies that the log of the hazard is assumed to change linearly with any continuous predictors.

Note also that (7.5) defines a *multiplicative* model, in the sense that the predictor effects act to multiply the baseline hazard. This is like the logistic model, where the linear predictor acts multiplicatively on the baseline odds. In contrast, (7.2) shows that in the linear model the predictor effects are *additive* with respect to the intercept β_0.

7.2.2 Parametric vs. Semi-Parametric Models

We have two options in dealing with the baseline hazard $h_0(t)$. One is to model it with a parametric function, as in the Weibull or exponential survival models. In this case the baseline hazard $h_0(t)$ is specified by a small number of additional parameters, which are estimated along with $\beta_1, \beta_2, \ldots, \beta_p$. If the baseline hazard is specified correctly, this approach is efficient, handles right-censoring as well as more complicated censoring schemes with ease, and makes it simple (though still risky) to extrapolate beyond the data. Of course the adequacy of the model for the baseline hazard has to be checked.

In contrast to parametric models, the Cox model does not require us to specify a parametric form for the baseline hazard, $h_0(t)$. Because we still specify (7.4) as the model for the log hazard ratio, the Cox model is considered semi-parametric. Nonetheless, estimation of the regression parameters $\beta_1, \beta_2, \ldots, \beta_p$ is done without having to estimate the baseline hazard function. The nonparametric Breslow estimate of the hazard function (Kalbfleisch and Prentice, 1980) available from Stata is after-the-fact and based on the coefficient estimates. The Cox model is more robust than parametric proportional hazards models because it is not vulnerable to misspecification of the baseline hazard. Furthermore, the robustness is commonly achieved with little loss of precision in the estimated predictor effects.

Proportionality and Multiplicativity

Fig. 7.2 and the summary statistics in Table 7.2 showed that the two mortality hazards for pediatric recipients of kidney transplants from living and recently deceased donors were very nearly proportional over time, in the sense that the ratio of the LOWESS-smoothed death rates was virtually constant. So the Cox model appears appropriate for these data, because the proportional hazards assumption appears to be met for this important predictor. Table 7.3 shows the unadjusted Cox model hazard ratio estimate for `txtype`, a binary indicator identifying the group receiving transplants from recently deceased donors. The estimated hazard ratio of 2.1 (95% CI 1.6–2.6, $P < 0.0005$) is

Table 7.3. Cox Model for Type of Donor

```
stcox txtype

No. of subjects =          9752             Number of obs    =     9752
No. of failures =           461
Time at risk    =   15621.88767             LR chi2(1)       =    34.07
Log likelihood  =     -2452.7587            Prob > chi2      =   0.0000
```

_t	Haz. Ratio	Std. Err.	z	P>|z|	[95% Conf. Interval]	
txtype	2.056687	.2606945	5.69	0.000	1.604259	2.636707

quite consistent with the estimates shown in Table 7.2, and suggests that receiving a transplant from a recently deceased donor roughly doubles the mortality risk at every point over the 12 years of follow-up.

Another important determinant of mortality after kidney transplant is the age of the recipient. Using results from a Cox model with age as continuous (results not shown), Fig. 7.3 shows fitted hazards for 6-, 11-, and 21-year-olds. The hazards for the three groups differ proportionally. However, it is important to point out that the perfect proportionality of the hazard functions plotted in Fig. 7.3 is imposed under the fitted model, like the perfectly parallel regression lines for the additive linear model without interaction terms shown in Fig. 4.2. This is in contrast to the apparently proportional relationship between the independently smoothed death rates in Fig. 7.2, which are based only on the data.

While the hazard ratio is assumed to be constant over time in the basic Cox model, under this multiplicative model the between-group *differences* in

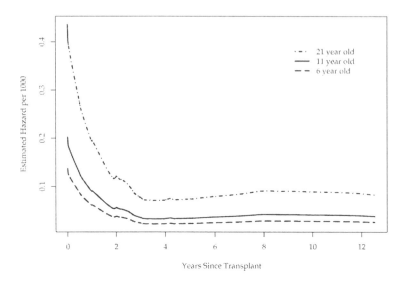

Fig. 7.3. Hazard Functions for 6-, 11-, and 21-Year-Old Transplant Recipients

the hazard can easily be shown to depend on $h_0(t)$ and thus on time. This is reflected in the fact that the hazard functions in Fig. 7.3 are considerably farther apart immediately after transplant when the baseline hazard (for 11-year-olds in this case) is higher.

DPCA Study of Primary Biliary Cirrhosis (PBC)

To illustrate interpretation of Cox model results, we consider a cohort of 312 participants in a placebo-controlled clinical trial of D-penicillamine (DPCA) for primary biliary cirrhosis (PBC) (Dickson *et al.*, 1989). PBC destroys bile ducts in the liver, causing bile to accumulate. Tissue damage is progressive and ultimately leads to liver failure. Time from diagnosis to end-stage liver disease ranges from a few months to 20 years. During the approximate ten-year follow-up period, 125 study participants died.

Predicting survival in PBC patients is important for clinical decision making. The investigators collected data on age as well as baseline laboratory values and clinical signs including serum bilirubin levels, enlargement of the liver (hepatomegaly), accumulation of water in the legs (edema), and visible veins in the chest and shoulders (spiders) – all signs of liver damage.

7.2.3 Hazard Ratios, Risk, and Survival Times

Table 7.4 displays a Cox model for the effects of treatment with DPCA (rx) and bilirubin (bilirubin) on mortality risk in the PBC cohort. The haz-

Table 7.4. Cox Model for Treatment and Bilirubin

```
stcox rx bilirubin
                                           LR chi2(2)        =      85.79
Log likelihood  =   -597.08411            Prob > chi2       =     0.0000


------------------------------------------------------------------------
        _t | Haz. Ratio   Std. Err.      z    P>|z|    [95% Conf. Interval]
-----------+------------------------------------------------------------
        rx |   .8181612   .1500579    -1.09   0.274    .5711117   1.172078
  bilirubin |  1.163459   .0154566    11.40   0.000    1.133556   1.194151
------------------------------------------------------------------------
```

ard ratio for treatment, 0.82, means that estimated short-term mortality risk among patients assigned to DPCA was 82% of the risk in the placebo group. This ratio is assumed to be constant over the ten years of follow-up. Likewise, the hazard ratio for bilirubin levels means that for each mg/dL increase in bilirubin, short term risk is increased by a factor of 1.16.

More broadly, (7.6) implies that in a model with predictors $x_1, x_2, \ldots, x_p$, coefficient β_j is the increase in the log hazard ratio for a one-unit increase in predictor x_j, holding the values of the other predictors constant. It follows that $\exp \beta_j$ is the hazard ratio for a one-unit increase in x_j. Below we show how this applies to continuous as well as binary and categorical predictors. Furthermore, for predictors with hazard ratios less than 1 ($\beta < 0$), increasing values of the predictors are associated with lower risk and longer survival times. Conversely, when hazard ratios are greater than 1 ($\beta > 0$), increasing values of the predictor are associated with increased risk and shorter survival times. In using the term *risk* in this context, it is important to keep in mind the definition of the hazard as a short-term rate and distinguish risk in this sense from cumulative risk over a defined follow-up period.

7.2.4 Hypothesis Tests and Confidence Intervals

In the Cox model, as in the logistic model, the estimated coefficients have an approximate normal distribution when there are adequate numbers of events in the sample. The normal approximation is better for the coefficient estimates than for the hazard ratios, so hypothesis tests and confidence intervals are based on calculations involving the coefficients and their standard errors. If there are fewer than 15–25 events, the normal approximation is suspect and bootstrap confidence intervals may work better; see Sect. 7.5.1 below. Table 7.5 displays the Cox model for the effects of DPCA and bilirubin on mortality risk with results on the coefficient rather than the hazard ratio scale.

Table 7.5. Cox Model for Treatment and Bilirubin Showing Coefficients

```
stcox rx bilirubin, nohr
                                        LR chi2(2)      =     85.79
Log likelihood  =   -597.08411         Prob > chi2     =     0.0000


------------------------------------------------------------------------
      _t |     Coef.    Std. Err.     z      P>|z|    [95% Conf. Interval]
---------+--------------------------------------------------------------
      rx | -.2006959   .1834088    -1.09    0.274    -.5601705    .1587787
bilirubin |  .1513976   .0132851    11.40    0.000     .1253594    .1774358
------------------------------------------------------------------------
```

For each predictor in the model, Wald Z-tests are the default used by
Stata to test the null hypothesis H_0: $\beta = 0$, or equivalently that the hazard
ratio equals 1. Under the null, the ratio of the coefficient estimate to its
standard error tends to a standard normal, or Z, distribution with mean 0
and standard deviation 1. In Table 7.5 the Z-statistics and associated P-
values for rx and bilirubin appear in the columns headed |z| and P >
|z| respectively. The evidence for the efficacy of DPCA is not persuasive
($P = 0.27$), but there is strong evidence that bilirubin levels are associated
with mortality risk ($P < 0.0005$). You can verify that the test results in Table
7.4 are identical to those in Table 7.5 and refer to the Z-test involving the
actual coefficients and their standard errors, and not to a Z-test involving the
ratio of the hazard ratio to its standard error (Problem 7.1).

Since Cox regression is a likelihood-based method, tests for predictors can
also be obtained using the likelihood ratio (LR) tests introduced in Sect. 6.2.1
for the logistic regression model. The procedure is the same in this setting,
comparing twice the difference in log-likelihoods for nested models to a χ^2
distribution with degrees of freedom equal to the between-model difference
in the number of parameters. For instance, to obtain an LR test of the null
hypothesis that the hazard ratio for treatment is 1, we would compare the
log-likelihood for the model in Table 7.4 to the log-likelihood for a model with
bilirubin as the only predictor. These log-likelihoods are –597.1 and –597.7,
yielding a LR test statistic of $2[(-597.1)-(-597.7)] = 1.2$, with an associated
P-value of 0.27.

In this case the Wald and LR results are essentially identical. In most
situations these tests give results which are similar but not exactly the same.
The results be will closest when the sample size is large or the estimated
hazard ratio is near 1. However, in data sets with few events, the LR test
gives more accurate P-values, and so is recommended in that context. As
noted in Sect. 5.5.2, qualitative discrepancies between the two test results
may indicate that the model includes too many predictors for the number of
events.

A 95% confidence interval (CI) for each β is obtained by computing $\hat{\beta}\pm$
1.96 SE($\hat{\beta}$). Stata and other packages usually make it possible to compute

confidence intervals with other significance, or α, levels; this just involves replacing 1.96 with the upper $1 - \alpha/2$ percentile of the Z distribution.

In turn, confidence intervals for the hazard ratios are obtained by exponentiating the upper and lower limits of the CIs for the coefficients, again because the normal approximation is better on the coefficient scale. From Table 7.4, the confidence interval for rx, the indicator for treatment with DPCA, shows that the data are consistent with risk reductions as large as 43%, but also with risk increases of 17%. It is also clear that the increase in risk associated with each mg/dL increase in bilirubin is rather precisely estimated (95% CI for the hazard ratio 1.13–1.19).

You can also verify that the confidence intervals in Table 7.4 are *not* equal to the estimated hazard ratio plus or minus 1.96 times *its* standard error (Problem 7.1). For rx, that calculation would yield (0.52–1.11) rather than (0.57–1.17). In reasonably large samples like this one, the two intervals are usually very similar. However, since the intervals based on exponentiating the confidence limits for the coefficients are more accurate in small samples, they are the ones used in Stata.

7.2.5 Binary Predictors

Interpretation of the binary predictor rx is simplified by coding assignment to the DPCA arm as 1 and to placebo as 0. Then the exponentiated coefficient gives the hazard ratio for treatment versus placebo (and retains its literal interpretation as the hazard ratio for a one-unit increase in the predictor). Some alternative codings, (e.g., placebo $= 1$ and treatment $= 2$) would give the same results in this instance, but would complicate interpretation in the presence of an interaction involving the binary predictor. This would also make the baseline hazard harder to interpret; in the DPCA example, the baseline hazard would not refer to either the placebo or the treatment group. Thus, we recommend the 0/1 coding for all binary predictors in this context as well (Problem 7.2).

7.2.6 Multilevel Categorical Predictors

Patients in the PBC study underwent a liver biopsy to determine their level of tissue damage. The scores ranged from 1 to 4, with increasing values reflecting greater damage. As in the linear and logistic models, the Stata command prefix xi: and variable prefix i. ensures that variable histology is treated as categorical in the Cox model. By default, the group with the lowest score is used as the reference category. Results are shown in Table 7.6. Estimated hazard ratios with respect to the reference group are 5.0, 8.6, and 21.4 for the groups with ratings of 2, 3, and 4, respectively, suggesting a steady increase in the hazard with higher ratings.

In addition to the default comparisons with the selected reference group, pairwise comparisons between any two categories can be obtained using the

Table 7.6. Categorical Fit for Histology

```
xi: stcox i.histology
                                           LR chi2(3)       =      52.72
Log likelihood  =   -613.62114             Prob > chi2      =     0.0000

-----------------------------------------------------------------------------
        _t | Haz. Ratio   Std. Err.      z    P>|z|     [95% Conf. Interval]
-----------+-----------------------------------------------------------------
 _Ihistol_2 |   4.987976    5.143153    1.56   0.119     .6610611    37.63631
 _Ihistol_3 |   8.580321    8.685371    2.12   0.034     1.179996    62.39165
 _Ihistol_4 |   21.38031    21.57046    3.04   0.002     2.959663    154.4493
-----------------------------------------------------------------------------


testparm _Ihistol*

 ( 1)   _Ihistol_2 = 0
 ( 2)   _Ihistol_3 = 0
 ( 3)   _Ihistol_4 = 0

         chi2(  3) =    43.90
       Prob > chi2 =    0.0000

lincom _Ihistol_4 - _Ihistol_3, hr

 ( 1) - _Ihistol_3 + _Ihistol_4 = 0

-----------------------------------------------------------------------------
        _t | Haz. Ratio   Std. Err.      z    P>|z|     [95% Conf. Interval]
-----------+-----------------------------------------------------------------
       (1) |   2.491785    .4923268    4.62   0.000     1.691727     3.67021
-----------------------------------------------------------------------------
```

lincom command, as shown in Table 7.6 for groups 3 and 4. The hazard in group 4 is 2.5 times higher than in group 3 (95% CI 1.7–3.7, $P < 0.0001$).

Categories With No Events

In our example, the default reference category is sensible and does not cause problems. However, categories may sometimes include no events, because the group is small or cumulative risk is low. Hazard ratios with respect to a reference category with no events are infinite, and the accompanying hypothesis tests and confidence intervals are hard to interpret. In this case, selecting an alternative reference group can correct the problem, although the hazard ratio, Wald test, and confidence interval for the category without events, with respect to the new reference category, will remain difficult to interpret.

Global Hypothesis Tests

As in logistic models, global hypothesis tests for the overall effect of a multilevel categorical predictor can be conducted using Wald or likelihood ratio (LR) χ^2 tests, with degrees of freedom equal to the number of categories minus 1. The Wald test result ($\chi^2 = 43.9, P < 0.00005$), obtained using the **testparm** command, is displayed in Table 7.6. The LR test result

$(\chi^2 = 52.7, P < 0.00005)$ also appears in the upper right corner of the table. Note that if covariates were included in the model, this default Stata output would refer to a test of the overall effect of *all* covariates in the model, not just `histology`; thus a LR test focused on the overall effect of `histology` would require combining the results of models with and without this predictor. Finally, a logrank test, as in Sect. 3.5.6, is available; this yields a χ^2 of 53.8 $(P < 0.0001)$. The tests agree closely and all show that the groups with different histology scores do not have equal survival.

The statistical significance of pairwise comparisons should be interpreted with caution, especially if the global hypothesis test is not statistically significant, as discussed in Sect. 4.3.4. With a large number of categories, multiple comparisons can lead to inflation of the type-I error rate; in addition, some comparisons may lack power due to small numbers in either of the categories being compared.

Ordinal Predictors and Tests for Trend

The histology score is ordinal, suggesting a more specific question: does the log mortality hazard increase linearly with higher `histology` ratings? This question can be addressed using tests for trend across categories like those introduced in Sect. 4.3.5. Note that these tests, like other hypothesis tests for the Cox model, are conducted using the coefficients and their standard errors, rather than the relative hazards. Thus for the Cox model these linear trend tests assess log-linearity of the hazard ratios. From Table 4.5, the trend test for a four-category variable such as `histology` is

$$-\beta_2 + \beta_3 + 3\beta_4 = 0. \tag{7.7}$$

Results presented in Table 7.7 $(\chi^2 = 10.23, P = 0.0014)$ confirm an increasing linear trend across `histology` categories.

Table 7.7. Linear Trend Test for Histology

```
. test -1* _Ihistol_2 + _Ihistol_3 +3* _Ihistol_4=0

 ( 1) - _Ihistol_2 + _Ihistol_3 + 3 _Ihistol_4 = 0

           chi2(  1) =    10.23
         Prob > chi2 =    0.0014
```

It is also possible to check whether the linear trend adequately captures the pattern of the coefficients across categories, or whether there are also important departures from this trend. To do this, we use a model with both categorical and log-linear terms for `histology`, as shown in Table 7.8. Then a Wald test for the joint effect of the categorical terms, obtained using the `testparm` command, can be used to assess the departure from log-linearity.

Table 7.8. Test of Departure From Linear Trend

```
xi: stcox histology i.histology
                                          LR chi2(3)     =     52.72
Log likelihood  =   -613.62114            Prob > chi2    =     0.0000


-----------------------------------------------------------------------
      _t | Haz. Ratio   Std. Err.     z    P>|z|     [95% Conf. Interval]
---------+-------------------------------------------------------------
  histol |   2.77548    .9333879    3.04   0.002     1.435756    5.365316
_Ihistol_2 |   1.797158   1.282043    0.82   0.411     .4439793    7.274612
_Ihistol_3 |   1.113852    .4188115    0.29   0.774     .5330565    2.327457
-----------------------------------------------------------------------


testparm _Ih*

 ( 1)  _Ihistol_2 = 0
 ( 2)  _Ihistol_3 = 0

          chi2(  2) =     1.24
        Prob > chi2 =    0.5385
```

The result ($\chi^2 = 1.24$, $P = 0.54$) suggests that a linear trend across categories is an adequate description of the association between histology score and mortality risk. However, it is not uncommon for both trend and departure from trend to be statistically significant, signaling a more complex pattern in risk.

7.2.7 Continuous Predictors

Age at enrollment of participants in the PBC study was recorded in years. The Cox model shown in Table 7.9 shows that the hazard ratio for a one-year increase in age is 1.04 (95% CI 1.02–1.06, $P < 0.0005$).

Table 7.9. Cox Model for Age in One-Year Units

```
stcox age
                                          LR chi2(1)     =     20.51
Log likelihood  =   -629.72592            Prob > chi2    =     0.0000


-----------------------------------------------------------------------
      _t | Haz. Ratio   Std. Err.     z    P>|z|     [95% Conf. Interval]
---------+-------------------------------------------------------------
     age |   1.04081    .0091713    4.54   0.000     1.022989    1.058941
-----------------------------------------------------------------------
```

The hazard ratio for continuous predictors is affected by the scale of measurement, and a one-unit increase may not have a meaningful interpretation. In the PBC study ages range from 26 to 78; thus, a one-year difference is age is small compared to the range of values. A five-year increase in age might provide a more clinically interpretable result (Problem 7.4).

Using (7.5) we can write down the ratio of the hazards for any two patients who differ in age by k years – that is, for a patient at age $x + k$ compared with another at age x:

$$\frac{h_0(t)e^{\beta(x+k)}}{h_0(t)e^{\beta x}} = \frac{e^{\beta(x+k)}}{e^{\beta x}}$$

$$= e^{\beta(x+k)-\beta x}$$

$$= e^{\beta k}. \tag{7.8}$$

Thus a k-unit change in a predictor multiplies the hazard by $\exp(\beta k)$, no matter what reference value x is considered. Obviously, $\exp(\beta)$ is the hazard ratio for a one-unit increase in the predictor.

Applying (7.8), with $\hat{\beta} = \log(1.04081)$ being the log of the hazard ratio for age from Table 7.9, the hazard ratio for an increase in age of five years is $\exp(\hat{\beta}5) = 1.22$. The same transformation can be applied to the confidence limits for age giving a 95% CI for a five-year increase in age of 1.12–1.33. Equivalently, we could raise the hazard ratio estimate for an increase of one unit to the fifth power, that is, $[\exp(\beta)]^k$, and apply the same operation to the confidence limits (Problem 7.5).

The hazard ratio for a five-unit change can also be obtained by defining a new variable age5 equal to age in years divided by 5. The Cox model for age5 appears in Table 7.10. Note that the Wald and LR test results are identical

Table 7.10. Cox Model for Age in Five-Year Units

```
stcox age5
                                          LR chi2(1)     =      20.51
Log likelihood  =   -629.72592           Prob > chi2    =     0.0000
```

_t	Haz. Ratio	Std. Err.	z	P>|z|	[95% Conf. Interval]	
age5	1.221397	.0538127	4.54	0.000	1.120352	1.331556

in Tables 7.9 and 7.10; changes in the scale of a continuous variable do not affect these tests.

Hazard ratios can be interpreted in terms of percent changes in risk. It is easy to see from Table 7.9 that estimated mortality risk among PBC patients increases about 4% for every year increase in age. We could also compute the percent increase risk associated with larger increases in age. A k-unit increase in the predictor implies a $100(\exp \hat{\beta} k - 1)\%$ change in risk. Note that this is the back transformation presented in Sect. 4.7.5 for linear regression models with log-transformed outcomes. Using the log of the hazard ratio estimate from Table 7.9 in place of $\hat{\beta}$, this calculation gives 22% for the increase in mortality risk associated with a five-year increase in age, a result we could get more directly from Table 7.10.

7.2.8 Confounding

The definition of confounding in Sect. 4.4 is not specific to the linear regression model. The conceptual issues and statistical framework for dealing with confounding are similar across all regression models covered in this book. To illustrate these concepts for the Cox model, we examined the association between bilirubin levels and survival among patients in the DPCA trial. We first fit the simple Cox model which appears in Table 7.11. For each one-point increase in baseline bilirubin, the hazard is increased by 16% – the same result as shown in Table 7.4 where the estimate is adjusted for treatment assignment (why?). However, patients with higher bilirubin may also be more likely to

Table 7.11. Unadjusted Cox Model for Bilirubin

```
stcox bilirubin
                                       LR chi2(1)      =      84.59
Log likelihood  =    -597.6845         Prob > chi2     =     0.0000
```

_t	Haz. Ratio	Std. Err.	z	P>|z|	[95% Conf. Interval]
bilirubin	1.160509	.0151044	11.44	0.000	1.131279	1.190494

have hepatomegaly, edema, or spiders – other signs of liver damage which are correlated with elevated bilirubin levels but not mediators of its effects, and all associated with higher mortality risk. Table 7.12 shows the estimated effect of bilirubin on mortality risk adjusted for hepatomegaly, edema, and spiders.

Table 7.12. Adjusted Cox Model for Bilirubin

```
stcox bilirubin edema hepatom spiders
                                       LR chi2(4)      =     118.82
Log likelihood  =    -580.56805        Prob > chi2     =     0.0000
```

_t	Haz. Ratio	Std. Err.	z	P>|z|	[95% Conf. Interval]
bilirubin	1.118276	.0166316	7.52	0.000	1.086149	1.151353
edema	2.126428	.4724983	3.40	0.001	1.375661	3.286927
hepatom	2.050617	.434457	3.39	0.001	1.353765	3.106173
spiders	1.474788	.28727	1.99	0.046	1.00676	2.160393

The adjusted hazard ratio for a one-point increase in bilirubin is 1.12 (95% CI 1.09–1.15, $P < 0.0005$). This coefficient represents the effect of a one-unit change in bilirubin while holding edema, hepatomegaly, and spiders constant. The other predictors, which may reflect other aspects of PBC-associated damage to the liver, account for about 25% of the unadjusted effect of bilirubin, and clearly contribute independent information about mortality risk. The attenuation of the unadjusted hazard ratio for bilirubin in the adjusted model is typical of confounding.

7.2.9 Mediation

Mediation can also be addressed using the Cox model, using the strategies outlined in Sect. 4.5. The key element is comparing the estimated effects of the predictor of interest before and after adjustment for the hypothesized mediators. Lin *et al.* (1997) give a complete statistical framework for assessing mediation using the Cox model, including tests and confidence intervals for PTE, the proportion of the treatment effect explained (Sect 4.5).

7.2.10 Interaction

The concept of interaction presented in Sect. 4.6 is also common to other multipredictor models. To illustrate its application to the Cox model, we checked for interaction between two binary variables in the PBC data, treatment with DPCA (rx), and the presence of liver enlargement or hepatomegaly (hepatom). This analysis examines the hypothesis that treatment is differentially effective according to this baseline covariate. As in linear and logistic models, interaction is handled by including product terms in the model. Defining rxhepa as the product of rx and hepatom, the resulting interaction model is shown in Table 7.13.

Table 7.13. Cox Model With Interaction

```
stcox rx hepatom rxhepa


                                        LR chi2(3)      =     40.54
Log likelihood  =    -619.7079          Prob > chi2     =    0.0000


--------------------------------------------------------------------------
     _t | Haz. Ratio   Std. Err.      z    P>|z|    [95% Conf. Interval]
--------+-----------------------------------------------------------------
     rx |   .8365301   .2778607    -0.54   0.591    .4362622    1.604041
hepatom |   2.865553   1.735658     1.74   0.082    .8742547    9.392452
 rxhepa |   1.099791   .4343044     0.24   0.810    .5071929    2.384775
--------------------------------------------------------------------------


. lincom rx + rxhepa, hr
 ( 1)   rx + rxhepa = 0


--------------------------------------------------------------------------
     _t | Haz. Ratio   Std. Err.      z    P>|z|    [95% Conf. Interval]
--------+-----------------------------------------------------------------
    (1) |   .9200085   .1963396    -0.39   0.696    .6055309    1.397807
--------------------------------------------------------------------------
```

Table 7.14. Cox Model With Interaction

group	rx	hepatom	rxhepa	$h(t\vert\mathbf{x})$
1	0	0	0	$h_0(t)$
2	1	0	0	$h_0(t)\exp(\beta_1)$
3	0	1	0	$h_0(t)\exp(\beta_2)$
4	1	1	1	$h_0(t)\exp(\beta_1 + \beta_2 + \beta_3)$
				$= h_0(t)\exp(\beta_1)\exp(\beta_2)\exp(\beta_3)$

Table 7.14 shows the hazard functions for the four groups defined by treatment and hepatomegaly (Problem 7.6). The coefficients β_1, β_2, and β_3 correspond to the predictors rx, hepatom and rxhepa, respectively. We obtain the hazard ratios of interest by taking ratios of the hazard functions for the different rows. Specifically, the ratio of the hazard for group 2 to the hazard for group 1, or $\exp(\beta_1)$ gives the effect of DPCA in the absence of hepatomegaly. In Table 7.13, the estimated hazard ratio for rx is 0.84 (95% CI 0.44–1.60, $P = 0.6$).

Similarly, the ratio of the hazard for group 4 to the hazard for group 3, or $\exp(\beta_1)\exp(\beta_3)$, gives the effect of DPCA in the presence of hepatomegaly. From Table 7.13, the estimate can be calculated as the product of the hazard ratios for rx and rxhepa, or $0.8365301 \times 1.099791 = 0.9$. This estimate, along with a 95% confidence interval (0.61–1.40) and P-value (0.7), can also be obtained using the lincom command shown in Table 7.13.

It follows that the interaction hazard ratio $\exp(\beta_3)$ gives the ratio of the DPCA treatment effects among patients with and without hepatomegaly. In Table 7.13, the estimated hazard ratio for rxhepa is 1.1 (95% CI 0.5–2.4, $P = 0.81$). The Z-test of H_0: $\beta_3 = 0$ assesses the equality of the effects of DPCA in the two groups.

To interpret these negative findings fully, as discussed in Sect. 3.7, both the point estimates and confidence intervals need to be considered. Both stratum-specific treatment effect estimates as well as the interaction are weakly negative, in the sense that the point estimates represent almost no effect or interaction, but the confidence limits are consistent with fairly large effects. In view of the weak evidence for interaction, the overall – also negative – finding for treatment with DPCA is the more sensible summary.

Similar methods can be used to obtain estimates of the effect of hepatomegaly stratified by treatment assignment: that is, by comparing groups 3 and 1, then 4 and 2. However, unlike the DPCA effect estimates, these estimates are potentially confounded (why?) and so are of less interest.

Interactions involving continuous or multilevel categorical predictors can also be modeled using product terms, but as Sect. 4.6 explains, care must be taken with these more complex cases.

7.2.11 Adjusted Survival Curves for Comparing Groups

Suppose we would like to examine the survival experience of pediatric recipients of kidney from living as compared to recently deceased donors, using the UNOS data. Kaplan–Meier curves, introduced in Sect. 3.5.2, would be a good place to start and are shown in Fig. 7.4.

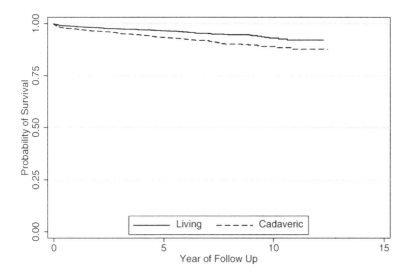

Fig. 7.4. Kaplan–Meier Curves for Transplant Recipients by Donor Type

In accord with the hazard ratio of 2.1 estimated by the unadjusted Cox model shown in Table 7.3, the curves show superior survival in the group with living donors. However, there are two potentially important confounders of this effect. First, living donors are more likely to be related and thus are closer tissue matches, as reflected in the number of matching human leukocyte antigen (HLA) loci (range 0–6). Second, cold ischemia time (essentially the time spent in transport) is shorter for kidneys obtained from living donors. After adjustment for these two factors, the hazard ratio for donor type is reduced to 1.3 (95% CI 0.9–1.9, $P = 0.19$). On the scale of the coefficients, almost two-thirds of the association of donor type with mortality risk is explained by cold ischemia time and number of HLA matches.

To see how adjusted survival curves might be constructed, first recall that adjustment for these covariates implies that adjusted curves for the two groups should differ only by donor type, with the other covariates being held constant. Curves meeting these criteria can be obtained using the coefficient estimates from the Cox model and an estimate of the baseline survival function, $\hat{S}_0(t)$,

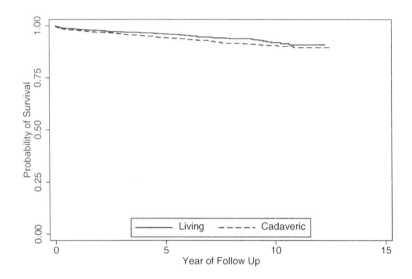

Fig. 7.5. Adjusted Survival Curves for Transplant Recipients by Donor Type

based on the Breslow baseline hazard estimate described earlier. Like the baseline hazard, the baseline survival function refers to observations with all predictor values equal to zero. Then, for an observation with predictor values $(x_1, \ldots, x_p)$, the estimated survival function follows:

$$\{\hat{S}_0(t)\}^{\exp(\hat{\beta}_1 x_1 + \ldots + \hat{\beta}_p x_p)}. \tag{7.9}$$

That is, we raise the baseline survival to the $\exp(\hat{\beta}_1 x_1 + \ldots + \hat{\beta}_p x_p)$ power. To evaluate (7.9), we need to specify a value for each of the predictors. In our example with three predictors, we would need to choose and hold constant values for x_2 (cold ischemia time) and x_3 (number of matching HLA loci), then generate the two curves by varying the predictor x_1 (recently deceased versus living donor).

It is conventional to use values for the adjustment variables which are close to the "center" of the data. Thus we centered cold ischemia time at its mean value of 10.8 hours and number of matching variable HLA loci at its median, three. With this centering, the baseline hazard and survival functions now refer to observations with cold ischemia time of 10.8 hours, three matching HLA loci, and a living donor. Then our adjusted estimate of the survival function for the group with living donors, holding the covariates constant at the chosen values, is $\hat{S}_0(t)$, while the corresponding estimate for the group with recently deceased donors is $\{\hat{S}_0(t)\}^{\exp(\hat{\beta}_1)}$. These adjusted curves, obtained in Stata using the `stcurve` command, are shown in Fig. 7.5. The differences between the survival curves are, as expected, narrower after adjustment. Note that the adjusted survival curves could also have be estimated using a stratified Cox model, as discussed in Sect. 7.3.2.

7.2.12 Predicted Survival for Specific Covariate Patterns

The estimated survival function (7.9) is also useful for making predictions for specific covariate patterns (Problem 7.7). For example, consider predicting survival for a PBC patient based on hepatomegaly status and bilirubin level, the two strongest predictors in the model shown in Table 7.12. Fig. 7.6 displays

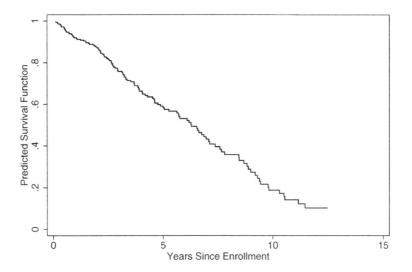

Fig. 7.6. Predicted Survival Curve for PBC Covariate Pattern

the predicted survival curve for a PBC patient with hepatomegaly and a bilirubin level of 4.5 mg/dL. From the curve, the median survival function for this covariate pattern is 6.3 years. Survival probabilities at key time points can likewise be read from the plot: at five years, predicted survival for this covariate pattern is below 60%, and by ten years, it has dropped to less than 20%. However, mean survival cannot be estimated in this case, because the longest follow-up time in the PBCA data is censored (Sect. 3.5).

7.3 Extensions to the Cox Model

7.3.1 Time-Dependent Covariates

So far we have only considered fixed predictors measured at study baseline, such as bilirubin in the DPCA study. However, multiple bilirubin measurements were made over the ten years of follow-up, and these could provide extra prognostic information. A special feature of the Cox model is that these valuable predictors can be included as *time-dependent covariates* (TDCs).

> Definition: A *time-dependent covariate* in a Cox model is a predictor whose values may vary with time.

In some cases, use of TDCs is critical to obtaining reasonable effect estimates. For example, Aurora *et al.* (1999) followed 124 patients to study the effect of lung transplantation on survival in children with cystic fibrosis. The natural time origin in this study is the time of *listing* for transplantation, not transplantation itself, because the children are most comparable at that point. However, waiting times for a suitable transplant can be long, and there is considerable mortality among children on the waiting list.

In this context, lung transplantation has to be treated as a TDC. To see this, consider the alternative in which transplantation is modeled as a fixed binary covariate, in effect comparing mortality risk in the group of children who undergo transplantation during the study to risk among those who do not. This method can make transplantation look more protective than it really is. Here is how the artefact comes about:

- Because transplanted patients must survive long enough to undergo transplantation, and waiting times can be long, the survival times measured from listing forward will on average be longer in the transplanted group even if transplantation has no protective effect.
- Because of this, children in the transplanted group are selected for better prognosis. So the randomization assumption discussed in Sect. 4.4.4 does not hold.
- Children are counted as having received a transplant from the time of listing forward, in many cases well before transplantation occurs. As a result they appear to be protected by a procedure that has not yet taken place. This illustrates the general principle that we can get into trouble by using information from the future to estimate current risk.

Treating transplantation as a TDC avoids this artefact. For each child we define an indicator of transplantation $X(t)$, which takes on value 0 before transplantation and 1 subsequently. For children who are not observed to undergo transplantation, $X(t)$ retains its original value of 0. Thus in an unadjusted model, the hazard at time t can be written as

$$h(t|x) = h_0(t)\exp\{\beta X(t)\}$$
$$= \begin{cases} h_0(t) & \text{before transplantation} \\ h_0(t)\exp(\beta) & \text{at or after transplantation.} \end{cases} \quad (7.10)$$

So now, all children are properly classified at t as having undergone transplantation or not, and we avoid the artefact that comes from treating transplantation as a fixed covariate. Note that Kalbfleisch and Prentice (1980) cite additional conditions concerning the allocation of transplants that must be met for the randomization assumption to hold and an unbiased estimate of the effect of transplantation to be obtained.

The transplantation TDC is relatively simple, because it is binary and cannot change back in value from 1 to 0. In practice, however, use of TDCs in Cox models is complicated. Some of the potential difficulties include the following:

- In most prospective studies, predictors like bilirubin will only be measured occasionally, but we need a value at each event time. A commonly used approach is to evaluate $X(t)$ using the most recent measurement before t. More difficult is a two-stage approach in which we first model the mean trajectory of the TDC for each subject. Then in the second stage we can set $X(t)$ equal to its expected value at t, based on the first-stage model. However, fitting and inference are both complicated in this procedure (DeGruttola and Tu, 1994; Wulfsohn and Tsiatis, 1997; Self and Pawitan, 1992).
- While $X(t)$ cannot legitimately be evaluated using information from the future, it oftentimes should be evaluated using all available information up until t. Consider two PBC patients, one with bilirubin values of 0.8 and 3.5 at baseline and year two, and the other with values of 2.5 and 3.5 at those times. In evaluating a TDC for bilirubin at year two, it might not be adequate to account only for the most recent values. A commonly used approach is to include the baseline value as a fixed covariate along with the change since baseline as a TDC. But other combinations of baseline and time-dependent covariates summarizing history up to t may be more appropriate.
- Mediation can be evaluated using TDCs, but must be handled carefully. For example, we could assess mediation of the effects of ZDV via its effects on CD4 counts in the ACTG 019 trial by comparing the unadjusted coefficient estimate for treatment to an estimate adjusted for a TDC defined using post-randomization CD4 values. However, in observational studies where the inferential goal is to identify multiple important predictors of an outcome event, mediating relationships can severely complicate the use and interpretation of TDCs. Moreover, note that if randomization had failed to balance the groups properly, then only baseline CD4 should be adjusted for; adjusting for the post-randomization values would result in an attenuated estimate for ZDV that captures only its effects via other pathways.

- One TDC may both confound and mediate the effects of another TDC. Suppose we wanted to evaluate the effect of highly active anti-retroviral therapy (HAART) on progression to diagnosis of AIDS, and that follow-up data were available from an observational cohort. Then, as in the lung transplantation example, treatment with HAART would need to be modeled as a TDC. However, an unadjusted estimate would almost always be *confounded by indication*, since in this setting patients with more advanced disease are more likely to be treated. Suppose that we try to control for prognosis at the time of initiation of HAART by including CD4 count and HIV viral load, both powerful prognostic variables, as TDCs. Now consider what happens if we continue to update the TDCs for these adjustment variables after HAART is begun. It is well known that the protective effects of HAART are mediated via its effects on CD4 count and viral load. Thus we would only obtain an estimate of the effects of HAART via other pathways. See Hernan *et al.* (2001) for a solution to this problem using *marginal structural models*.
- The TDCs most likely to cause trouble are *internal* covariates which reflect subject-level causal processes. In contrast, *external* covariates like calendar time, season of the year, or air pollution pose fewer difficulties (Kalbfleisch and Prentice, 1980).
- Ideally TDCs are measured at regularly scheduled visits, so acertainment does not depend on prognosis. Missing visits can induce bias if the missingness is related to the value of the TDC that would have been obtained. Likewise, ascertainment of TDCs by clinical chart review can be fraught with pitfalls.
- Fitting a model with TDCs in Stata involves making a *long data set* that reflects changes in the TDCs. For the lung transplantation example, this would be straightforward, requiring only a second record for children who undergo transplantation during follow-up. But in more complicated situations, many records may be required for each observation if the value of the TDC potentially changes continuously. The PHREG procedure in SAS is an exception in making it easy to use TDCs without first making a long data set.

In view of these difficulties, we recommend working closely with an experienced biostatistician to implement Cox models with TDCs.

7.3.2 Stratified Cox Model

Suppose we want to model the effect of edema among patients with PBC in the DPCA cohort. We could do this using the binary predictor edema, coded 1 for patients with edema and 0 for others. Then in an unadjusted model the hazard for patients with edema is $h(t|x) = h_0(t)\exp(\beta)$, while for other patients it is just $h_0(t)$. So the hazard for patients with edema is modeled as a constant proportion $\exp(\beta)$ of the baseline hazard $h_0(t)$.

However, we will show in Sect. 7.4.2 below that the proportional hazards assumption does not hold for edema. We can accommodate the violation by fitting a stratified Cox model in which a separate baseline hazard is used for patients with and without edema. Specifically, we let

$$h(t|\texttt{edema} = 1) = h_{01}(t) \tag{7.11}$$

for patients with edema, and

$$h(t|\texttt{edema} = 0) = h_{00}(t) \tag{7.12}$$

for other patients. Now the hazards for the two groups can differ arbitrarily.

Generalizing from edema to a stratification variable with two or more levels, and to a model with covariates $(x_1, \ldots, x_p)$, the hazard for an observation in stratum j would have the form

$$h_{0j}(t) \exp(\beta_1 x_1 + \ldots + \beta_p x_p). \tag{7.13}$$

Note that in this model we assume that the effect of each of the covariates is the same across strata; below we examine methods for relaxing this assumption. It is also important to point out that while the stratified, adjusted survival curves presented in Sect. 7.2.11 above can give a clear visual impression of the effect of the stratification variable after adjustment, current methods for the stratified Cox model do not allow us to estimate or test the statistical significance of its effect. Thus stratification could be used in our example to adjust for edema, but might be less useful if edema were a predictor of primary interest. In Sect. 7.4.2 below we show how time-dependent covariates can be used to obtain valid estimates of the effects of a predictor which violates the proportional hazards assumtion.

Stratification is also useful in the analysis of stratified randomized trials. We pointed out in Sect. 5.3.5 that we need to take account of the stratification to make valid inferences. But we also need to avoid making an unwarranted assumption of proportional hazards for the stratification variable that could potentially bias the treatment effect estimate.

The stratified Cox model is easy to implement in Stata as well as other statistical packages. In ACTG 019 participants were randomized within two strata defined by baseline CD4 count. To conduct the stratified analysis, we defined `strcd4` as an indicator coded 1 for the stratum with baseline CD4 count of 200–499 cells/mm^3 and 0 for the stratum with baseline CD4 of less than 200. The stratified model for the effect of ZDV treatment (`rx`) is shown in Table 7.15. In this instance, the estimated 54% reduction in risk for treatment with ZDV is the same as an estimate reported below in Sect. 7.5.3, which was adjusted for rather than stratified on CD4.

Table 7.15. Cox Model for Treatment With ZDV, Stratified by Baseline CD4

```
stcox rx, strata(strcd4)
                                    LR chi2(1)    =     7.36
Log likelihood =  -276.45001        Prob > chi2   =    0.0067

-------------------------------------------------------------------
        _t | Haz. Ratio   Std. Err.      z    P>|z|    [95% Conf. Interval]
-----------+-------------------------------------------------------
        rx |  .4646665    .1362697    -2.61   0.009    .2615261    .8255963
-------------------------------------------------------------------
                                            Stratified by strcd4
```

Number of Strata

Stratification is a flexible approach to adjustment for a nominal categorical variable with a large number of levels. An example is in a multicenter randomized trial with many centers. For stratification to work well, there do need to be a reasonable number of events in each stratum. When the number of strata gets large, there can be some loss of efficiency in estimation of the treatment or other covariate effects, since the stratified model does not "borrow strength" across strata. Nonetheless, Glidden and Vittinghoff (2004) showed that in this situation the stratified Cox model performs better than an unstratified model in which the covariate is treated as a nominal categorical predictor.

Interaction Between Stratum and a Predictor of Interest

In Table 7.15, the model assumes that the ZDV effect is the same in both strata. It is possible, however, that patients with less severe HIV disease, as reflected in higher CD4 counts, may respond better to ZDV. Such an interaction between stratum and treatment can be examined by including a product term between the treatment and stratum indicators. Note that in the stratified model only the product term **inter** and the treatment indicator **rx** term are entered as predictors, while **strcd4** is still incorporated as a stratification factor. In Table 7.16 we see only weak evidence for a protective effect of ZDV in the stratum with lower baseline CD4 (hazard ratio 0.71, 95% CI 0.32–1.65, $P = 0.43$). From the **lincom** result there is more persuasive evidence for protection in the stratum with higher CD4 (hazard ratio 0.32, 95% CI 0.14–0.74, $P = 0.008$). There is weak but not convincing evidence for interaction (hazard ratio 0.45, 95% CI 0.14–1.48, $P = 0.19$), so the overall estimate shown above in Table 7.15 may be the preferable summary estimate of the effect of ZDV.

Stratified and Adjusted Survival Curves

In Sect. 7.2.11 we presented adjusted survival curves for pediatric kidney transplant recipients according to donor type, based on an adjusted model in which the effect of donor type was modeled as proportional. We can also obtain

Table 7.16. Stratified Fit With Interaction Term

```
. stcox rx inter, strata(strcd4)
                                              LR chi2(2)      =       9.14
Log likelihood  =   -275.56324               Prob > chi2     =     0.0104


-----------------------------------------------------------------------------
     _t | Haz. Ratio   Std. Err.      z    P>|z|     [95% Conf. Interval]
--------+--------------------------------------------------------------------
     rx |   .7124052    .305808    -0.79   0.430      .307142    1.652399
  inter |   .4508514   .2728073    -1.32   0.188     .1377136    1.476012
-----------------------------------------------------------------------------
                                              Stratified by strcd4
. lincom rx + inter, hr
 ( 1)   rx + inter = 0


-----------------------------------------------------------------------------
     _t | Haz. Ratio   Std. Err.      z    P>|z|     [95% Conf.Interval]
--------+--------------------------------------------------------------------
    (1) |   .3211889    .136976    -2.66   0.008     .1392362     .7409156
-----------------------------------------------------------------------------
```

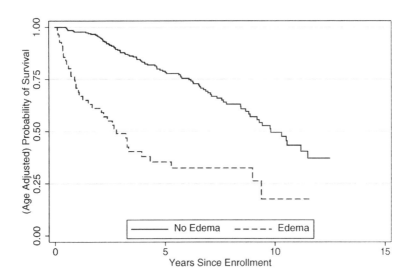

Fig. 7.7. Stratified Survival Curves for Edema Adjusted for Age

adjusted survival curves according to the levels of a stratification factor. We will show in Sect. 7.4.2 that the effects of baseline edema on mortality risk among PBC patients in the DPCA cohort were not proportional. Suppose we would like to compare the survival curves according to edema, adjusting for age. As in the earlier example, we need to specify a value for age in order to estimate the survival curves, and make a similar choice in centering age on its mean of 50. Under the stratified Cox model, the survivor function for a PBC subject with centered agec is given by

$$[S_{0j}(t)]^{\exp(\beta \text{agec})}. \tag{7.14}$$

The adjusted survival curves for the edema ($j = 1$) and no edema ($j = 0$) strata, adjusted to age 50 (i.e., `agec` $= 0$), are therefore $S_{01}(t)$ and $S_{00}(t)$ respectively. Fig. 7.7 shows shorter survival in patients with edema at baseline. However, these stratum-specific survival functions also suggest that the multiplicative effect of `edema` on the mortality hazard is not constant over time. We examine this more carefully in Sect. 7.4.2.

7.4 Checking Model Assumptions and Fit

Two basic assumptions of the Cox model are *log-linearity* and *proportional hazards*. Just as with other regression models, these assumptions can be examined, and extensions of the model can be used to deal with violations and model more complex effects.

7.4.1 Log-Linearity

In Sect. 7.2.1, we saw that equation (7.6) defines a log-linear model in which each unit change in a continuous predictor is assumed to have the same effect on the log of the hazard. This implies that the hazard ratio is log-linear in the continuous predictors.

  Unlike the linear model, but like the logistic, diagnostics for violations of log-linearity using plots of residuals do not work very well for the Cox model. However, violations of this assumption are easy to accommodate, using the same tools covered in Sect. 4.7.1 for the linear model. Thus a workable method for assessing violations of log-linearity is to assess more complicated models for improvements in fit. For example, we can add polynomial terms in the predictor in question to the model and then check effect sizes and P-values to determine whether the higher order terms are important; or the predictor can be log-transformed and the log-likelihoods informally compared (Problem 7.3). Alternatively, the continuous predictor can be categorized using well-chosen cutpoints; then log-linearity is checked using the methods outlined above in Sect. 7.2.2 for assessing both trend and departures from trend in ordinal predictors. Linear splines (Sect. 4.9) are another alternative implemented in Stata.

7.4.2 Proportional Hazards

The adjusted Cox model shown in Table 7.12 shows that mortality risk is increased about twofold in PBC patients with edema at baseline. However, Fig. 7.7 suggests that edema may violate the proportional hazards assumption: specifically, the increase in risk is greatest in the first few years and then diminishes. Thus the effect of edema on the hazard is time-dependent. A transformed version of Fig. 7.7 turns out to be more useful for examining violations of the proportional hazards assumption.

Log-Minus-Log Survival Plots

To illustrate the use of transformed survival plots for assessing proportionality for binary or categorical predictors, we consider the treatment indicator (rx) in the DPCA trial. This method exploits the relationship between the survival and hazard functions. If proportional hazards hold for rx, then by (7.9)

$$S_1(t) = [S_0(t)]^{\exp(\beta)}, \tag{7.15}$$

where $S_0(t)$ is the survival function for placebo patients and $S_1(t)$ is the corresponding survival function for the DPCA-treated patients. Then, the *log-minus-log* transformation of (7.15) gives

$$\log\{-\log[S_1(t)]\} = \beta + \log\{-\log[S_0(t)]\}. \tag{7.16}$$

Thus when proportional hazards holds, the two transformed survival functions will be a constant distance β apart, where β is the log of the hazard ratio for treatment with DPCA.

This result enables us to use a simple graphical method for examining the proportional hazards assumption. Specifically, log-minus-log transformed Kaplan–Meier estimates of the survival functions for the placebo and DPCA groups are plotted against follow-up time. In Stata, this plot is implemented in the stphplot command. The log-minus-log survival plot for DPCA is shown in Fig. 7.8.

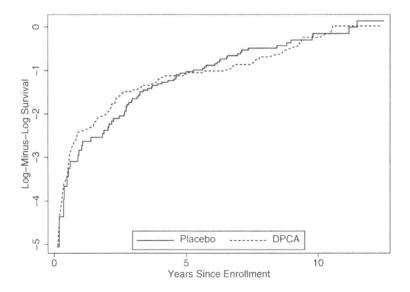

Fig. 7.8. Log-Minus-Log Survival Plot for DPCA Treatment

In assessing the log-minus-log survival plot for evidence of non-proportionality, the patterns to look for are convergence, divergence, or crossing of the curves followed by divergence. Convergent curves suggest that the difference between the groups decreases with time, and vice versa. If the curves converge, cross, and then diverge, then the non-proportionality may be more important; for example, this might indicate that treatment is harmful early on but protective later. In Fig. 7.8, however, the curves for DPCA and placebo remain close over the entire follow-up period and do not suggest non-proportionality.

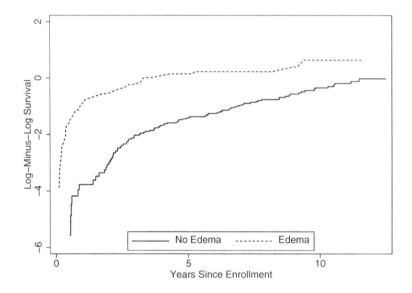

Fig. 7.9. Log-Minus-Log Survival Plot for Edema

In contrast, the log-minus-log survival plot for edema in Fig. 7.9 shows rather clear evidence of a violation of proportionality. While there is a pronounced difference between the groups at all time points, showing that patients with edema have poorer survival, the difference between the groups diminishes with follow-up. Specifically, the distances between the curves – that is, the implied log hazard ratios – are 4.7, 1.8, 1.1, and 1.0 at years 1, 4, 7, and 10, respectively.

Smoothing the Hazard Ratio

Log-minus-log survival plots are good diagnostic tools for violations of the proportional hazards assumption. To address such a violation, however, we may need more information about how the log-hazard ratio changes with follow-up time. We can do this using a nonparametric, smoothed estimate of the hazard ratio against time, analogous to the LOWESS estimates of the regression function used in diagnosing problems in linear models in Sect 4.7. If the smoothed estimate of the hazard ratio is nearly constant, then the assumption of proportional hazards is approximately satisfied. Conversely, when curvature is pronounced, the shape of the smooth helps us determine how to model the hazard ratio as a function of time.

The method works as follows. As in checking the linear model, the Cox model with all the important predictors is first estimated. Then we obtain *scaled Schoenfeld residuals*; in Stata this is done using `scaledsch` option for the `stcox` command, which generates a residual for each observation and predictor. Then the Schoenfeld residuals for each predictor are smoothed against time using LOWESS, providing a nonparametric estimate of the log hazard ratio for that predictor as it changes over time. In Stata the plot can be generated using the `stphtest` command with the `plot` option. Fig. 7.10 shows the

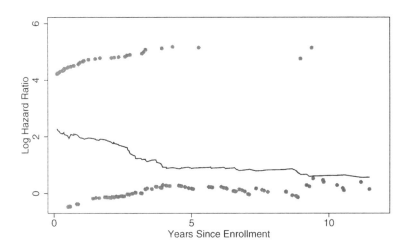

Fig. 7.10. Smoothed Estimate of Log Hazard Ratio for Edema

smoothed Schoenfeld residual plot for **edema**. A non-constant trend is readily apparent: the log-hazard ratio decreases steadily over the first four years and then remains constant.

A final note: relatively influential points are identifiable from the plots of the Schoenfeld residuals. DFBETA statistics, the influence measure we recommend for the linear and logistic models, are defined for the Cox model, but not directly available in Stata. See Sect. 4.7.4 for approaches to dealing with this problem.

Schoenfeld Test

Schoenfeld (1980) provides a test for violation of proportional hazards which is closely related to the diagnostic plot using LOWESS smooths of scaled Schoenfeld residuals just described. The test assesses the correlation between the scaled Schoenfeld residuals and time. This is equivalent to fitting a simple linear regression model with time as the predictor and the residuals as the outcome, and the parametric analog of smoothing the residuals against time using LOWESS. If the hazard ratio is constant, the correlation should be zero.

Table 7.17. Schoenfeld Tests of Proportional Hazards Assumption

```
stphtest, detail

        Test of proportional hazards assumption

        Time:  Time
        ----------------------------------------------------------------
                     |    rho          chi2       df      Prob>chi2
        -------------+--------------------------------------------------
        rx           |  -0.06393       0.51        1       0.4766
        -------------+--------------------------------------------------
        global test  |                 0.51        1       0.4766
        ----------------------------------------------------------------


        Test of proportional hazards assumption

        Time:  Time
        ----------------------------------------------------------------
                     |    rho          chi2       df      Prob>chi2
        -------------+--------------------------------------------------
        edema        |  -0.35779      14.40        1       0.0001
        -------------+--------------------------------------------------
        global test  |                14.40        1       0.0001
        ----------------------------------------------------------------
```

The Schoenfeld tests for rx and edema are shown in Table 7.17. Positive values of the correlation rho suggest that the log hazard ratio increases with time and vice versa. In accord with the graphical results, the Schoenfeld test finds strong evidence for a declining log hazard ratio for edema (rho = -0.36, $P = 0.0001$), but does not suggest problems with rx (rho = -0.07, $P = 0.5$).

The Schoenfeld test is most sensitive in cases where the log hazard ratio is linearly increasing or decreasing with time. However, because the test is based on a linear regression model, it is sensitive to a few large residual values. Such values should be evident on the scatterplot of the scaled Schoenfeld residuals against time. Useful examples and discussion of the application of the Schoenfeld test appear in Sect. 6.5 of Therneau and Grambsch (2000).

Graphical Diagnostics Versus Testing

We have described both graphical and hypothesis testing methods for examining the proportional hazards assumption. The Schoenfeld test is widely used and gives two easily interpretable numbers that quantify the violation of the proportional hazards assumption. However, as pointed out in Sect. 4.7, such tests may lack power to detect important violations in small samples, while in large samples they may find statistically significant evidence of model violations which do not meaningfully change the conclusions. While also lacking sensitivity in small samples, graphical methods give extra information about the magnitude and nature of model violation, and should be the first-line approach in examining the fit of the model.

Stratification

The stratified Cox model introduced in Sect. 7.3.2 is an attractive option for handling binary or categorical predictors which violate the proportional hazards assumption. We explained there that no assumption is made about the relationships between the stratified hazard functions specific to the different levels of the predictor. Because the resulting fit to the stratification variable is unrestricted, this is a particularly good way to rule out confounding of a predictor of interest by a covariate that violates the proportional hazards assumption. However, because no estimates, confidence intervals, or P-values are obtained for the stratification variable, this approach is less useful for any predictor of direct interest.

Note that we can apply this approach to a continuous variable by first categorizing it. How many categories to use involves a trade-off (Problem 7.8). Using more strata more effectively controls confounding, but as we suggested in Sect. 7.3.2, precision and power can suffer if the confounder is stratified too finely, because strength is not borrowed across strata. Five or six strata generally suffice, but there should be at least 5–7 events per stratum.

Modeling Interactions With Time

In this section we briefly outline a widely used approach to addressing violations of the proportional hazards assumption using interactions with time, and implemented using time-dependent covariates (TDCs), as described above in Sect. 7.3.1. We return to the edema example and show how the declining hazard ratio can be modeled. To begin, let $h_1(t)$ and $h_0(t)$ denote the hazard functions for PBC patients with and without edema. Because proportional hazards does not hold, the hazard ratio

$$\mathrm{HR}(t) = \frac{h_1(t)}{h_0(t)} \tag{7.17}$$

is a function of t. To address this, we define $\beta(t) = \log\{\mathrm{HR}(t)\}$ as a coefficient for edema which changes with time. This is equivalent to a hazard function of the form

$$h(t|\mathsf{edema}) = h_0(t)\exp\{\beta(t)\mathsf{edema}\} \tag{7.18}$$

where as before edema is a 0/1 indicator of the presence of edema. This can be modeled in one of two ways.

- We can model the log hazard ratio for edema as a linear function of time. This is implemented using a main effect, edema, plus an interaction term, edemat, defined as a TDC, the product of edema and t. That is, we set

$$\beta(t)\text{edema} = (\beta_0 + \beta_1 t)\text{edema}$$
$$= \beta_0\text{edema} + \beta_1 t\text{edema}$$
$$= \beta_0\text{edema} + \beta_1\text{edemat}. \qquad (7.19)$$

Alternatively, we could model the log hazard ratio as linear in log time, defining the product term with $\log(t)$ in place of t; this might be preferable in the edema example, since the decline in the log hazard ratio shown in Fig. 7.10 grows less steep with follow-up (Sect. 4.7.1).

- We can split follow-up time into sequential periods and model the log hazard ratio for edema as a step function with a different value in each period. For example, we could estimate one log hazard ratio for edema in years 0–4, and another in years 5–10, again motivated by Fig. 7.10. We could do this by defining two TDCs:
  - edema04, equal to 1 during the first four years for patients with edema, and 0 otherwise.
  - edema5on, equal to 1 during subsequent follow-up for patients with edema, and 0 otherwise.

  Then we set

$$\beta(t)\text{edema} = \beta_1\text{edema04} + \beta_2\text{edema5on}. \qquad (7.20)$$

  This approach is analogous to categorizing a continuous predictor to model nonlinear effects (Sect. 4.7.1).

The first alternative is more realistic because it models the hazard ratio for edema as a smooth function of time. But it is harder to implement because the TDC edemat changes continuously for patients with edema from randomization forward; up to one record for every distinct time at which an outcome event occurs would be required for these patients in the "long" data set used for the analysis in Stata. (Implementation is considerably easier in SAS.) In contrast, the second alternative is less realistic but easier to implement, only requiring two records for patients with edema and more than four years of follow-up, and one record per patient otherwise.

7.5 Some Details

7.5.1 Bootstrap Confidence Intervals

The ACTG 019 data set includes 880 observations but only 55 failures. We can check the validity of the standard confidence intervals in the Cox model for ZDV treatment (**rx**) and baseline CD4 cell count (**cd4**) using the bootstrap (Sect. 3.6). The results are reported on the coefficient scale in Table 7.18. The standard and bias-corrected bootstrap confidence intervals, based

Table 7.18. Cox Model for ZDV and CD4 With Bootstrap Confidence Intervals

```
stcox rx cd4, nohr

No. of subjects =          880            Number of obs   =          880
No. of failures =           55
Time at risk    =       354872
                                          LR chi2(2)      =        34.46
Log likelihood  =   -314.17559            Prob > chi2     =       0.0000

------------------------------------------------------------------------
       _t |     Coef.   Std. Err.      z    P>|z|    [95% Conf. Interval]
----------+-------------------------------------------------------------
       rx |  -.7851153   .2930517   -2.68   0.007   -1.359486   -.2107446
      cd4 |  -.0065752   .0012376   -5.31   0.000   -.0090009   -.0041495
------------------------------------------------------------------------

. bootstrap '"stcox rx cd4"' _b, reps(1000)

command:      stcox rx cd4
statistics:   b_rx     = _b[rx]
              b_cd4    = _b[cd4]

Bootstrap statistics                      Number of obs   =          880
                                          Replications    =         1000

------------------------------------------------------------------------
Variable    | Reps  Observed     Bias  Std. Err. [95% Conf. Interval]
------------+-----------------------------------------------------------
       b_rx | 1000 -.7851154 -.0221124  .2900748 -1.354341  -.2158895  (N)
            |                                     -1.388345  -.2682095  (P)
            |                                     -1.372091  -.2503596  (BC)
      b_cd4 | 1000 -.0065752 -.0001226  .0014062 -.0093347  -.0038157  (N)
            |                                     -.0095451  -.0039595  (P)
            |                                      -.009454  -.0038239  (BC)
------------------------------------------------------------------------

Note: N  = normal
      P  = percentile
      BC = bias-corrected
```

on 1,000 resampled data sets, yield very similar results, confirming that the semi-parametric model works well in this case, even though there are only moderate numbers of events.

7.5.2 Prediction

Evaluating prediction error using some form of cross-validation, as described in Sect. 5.2, is more complicated with time-to-event outcomes. Comparing observed to expected survival times is ruled out for censored observations in the test set; moreover, as we explained above in Sect. 7.2.12, expected – that is, mean – survival times are usually undefined under the Cox model. Comparing the *occurrence* of events in the test set with predictions based on the learning set, as with binary outcomes analyzed using a logistic model, is relatively tractable, but complicated by variations in follow-up time, in particular extrapolations for any follow-up times in the test set that exceed the longest times in the learning set.

Dickson *et al.* (1989) showed one way in which predictions based on a Cox model can be cross-validated using a test data set. To see how this was done, note that the survival curves for different covariate patterns, like the one estimated in the previous section using (7.9), differ only in the value of the linear predictor $x_1\beta_1 + \ldots + x_p\beta_p$, termed the *risk score* in this context. You can verify that the higher the risk score, the poorer the predicted survival. The risk score was used to group patients into four predicted survival categories, choosing cut points which gave approximately equal numbers of events in the four groups. The investigators then validated the model by calculating the risk score for a new set of PBC patients, *using the coefficient estimates from the original model*, and grouping the new observations based on the cutpoints they had developed. Finally, they showed that the predicted curves were similar to Kaplan–Meier survival curves based only on data for the four groups in the test set.

Begg *et al.* (2000) compare three methods for assessing predictiveness for survival of tumor staging and grading systems among cancer patients.

7.5.3 Adjusting for Non-Confounding Covariates

If a covariate is strongly predictive of survival but uncorrelated with a predictor of interest, omitting it from a Cox model will nonetheless attenuate the estimated hazard ratio for the predictor of interest, as discussed in Sect. 5.3.5 (Gail *et al.*, 1984; Schmoor and Schumacher, 1997; Henderson and Oman, 1999). Omitting important covariates from logistic models also induces such attenuation. Although the gain in precision is usually modest at best, it can be advantageous to include such a prognostic factor in order to avoid the attenuation.

A compelling example is provided by ACTG 019, the randomized clinical trial of ZDV for prevention of AIDS and death in HIV infection discussed earlier in this chapter. As expected in a clinical trial, there was no between-group difference in mean baseline CD4 count, known to be an important prognostic variable. Thus by definition, baseline CD4 count could not have confounded the effect of ZDV. However, when CD4 count is added to the model, the estimated reduction in risk of progression to AIDS or death afforded by ZDV goes from 49% to 54%, an increase of about 12%. More discussion of whether to adjust for covariates in a clinical trial is given in Sect. 5.3.5.

7.5.4 Independent Censoring

To deal with right-censoring, we have to make the assumption of *independent censoring*. The essence of this assumption is that after adjustment for covariates, future event risk for a censored subject does not differ from the risk among other subjects who remain in follow-up and have the same covariate values. Under this assumption, subjects are censored independent of their future risk.

To see how this assumption may be violated, consider a study of mortality risk among patients followed from admission to the intensive care unit until hospital discharge. Suppose no survival information is available after discharge, so subjects have to be censored at that time. In general subjects are likely to be discharged because they have recovered and are thus at lower risk than patients who remain hospitalized. Unless we can completely capture the differences in risk using baseline and time-dependent covariates, the assumption of independent censoring would be violated.

Dependent censoring can also arise from informative loss to follow-up. In prospective cohorts it is not unlikely that prognosis for dropouts differs from that for participants remaining in follow-up in ways that can be difficult to capture with variables routinely ascertained.

It can also be difficult to diagnose dependent censoring definitively, because that would require precisely the information that is missing – for example, mortality data after discharge from the ICU. But that is a case where an experienced investigator might recognize on substantive grounds that censoring is likely to be dependent. Furthermore, the problem could be addressed in that study by ascertaining mortality for a reasonable period after discharge. Similarly, losses to follow-up are best addressed by methods to maximize study retention; but it also helps to collect as much information about censored subjects as possible. Valid modeling in the presence of dependent censoring requires statistical methods that are still under development and well beyond the scope of this book.

7.5.5 Interval Censoring

We also assume that the time of events occurring during the study is known more or less exactly. This is almost always the case for well-documented events like death, hospitalization, or diagnosis of AIDS. But the timing of many events is not observed with this level of precision. For example, in prospective cohort studies of people at risk for HIV infection, it is common to test participants for infection at semi-annual visits (Buchbinder *et al.*, 1996). Thus the actual time of an incident infection is only known up to an interval of possible values; in technical terms, it is *interval-censored* between the last visit at which the participant tested negative and the first at which the result was positive. Another example is development of abnormal cellular changes in the cervix, which must be assessed by clinical exam. These exams may be performed periodically, perhaps months or even years apart. As with HIV infection, newly observed changes may have occurred at any time since the last exam. Interval-censored data are common and require specialized methods of analysis, again beyond the scope of this book.

7.5.6 Left Truncation

This chapter deals solely with right-censored survival data in which observation begins at the predefined *origin* of the survival times (i.e., $t = 0$). In ACTG 019 and as well as the leukemia trial described in Sect. 3.5, the natural origin is time of randomization. In studies of risk factors for progression of a disease, the natural origin is time of disease onset. For example, in studies of HIV disease, it is time of HIV infection.

However, if study participants are not followed prospectively from the time origin forward, the survival times are said to be *left-truncated*. In the early years of the AIDS epidemic, for example, participants in cohort studies of HIV infection were for the most part already infected at recruitment in the mid-1980s. Simply using time of recruitment as the time origin in such *prevalent cohorts* can induce bias under a range of circumstances (Brookmeyer and Gail, 1987; Brookmeyer *et al.*, 1987).

In contrast, for many participants in the San Francisco City Clinic Cohort (SFCCC), time of infection was known from analysis of blood samples stored in the course of an earlier study of a candidate hepatitis-B vaccine conducted in the late 1970s, at about the time the HIV epidemic began in San Francisco. To be included in an analysis of risk factors associated with progression to AIDS (Vittinghoff *et al.*, 2001), SFCCC participants had to survive and remain AIDS-free during the period from their infection, mostly before 1981, until recruitment into the SFCCC, which began in 1984. The effect of this delay was selectively to exclude from the analysis men whose disease progressed most rapidly; this subset had either died or progressed to AIDS before information on risk factors could be collected at the baseline SFCCC visit. The survival analysis of risk factors was conducted on the natural time-scale with origin at HIV infection. However, special methods beyond the scope of this chapter were required to account for the left truncation of these survival times. Survival analyses that ignore the fact that observation begins at some $t > 0$ can give badly biased parameter estimates. Methods for handling left truncation are easily implemented in many statistical packages including Stata.

7.6 Summary

This chapter has shown how right-censored survival data can be analyzed using the Cox model. This model has much in common with other regression models: in particular, issues of confounding, mediation, and interaction are dealt with in similar ways. The Cox model summarizes predictor effects in terms of their multiplicative effect on the hazard rate. A feature of note is the ability of the Cox model to accommodate time-dependent covariates.

7.7 Further Notes and References

The Cox model has proven popular because it is computationally feasible, does not require us to specify the baseline hazard, and is flexible. Alternatives include the *accelerated failure time model* and the *proportional odds model*. These models are less popular and statistical techniques for these models are less well developed. By contrast, there are extensively developed techniques for parametric survival regression. Parametric models require us to make assumptions about the form of the baseline hazard function and have proved less popular because the parametric assumptions sacrifice robustness without substantial efficiency gains.

Some more complex survival data settings were not discussed in this chapter. For instance, there may be more than a single event per subject, yielding clustered or hierarchical survival data. While data like these can be analyzed using the standard methods by ignoring events after the first, important information may be discarded. See Wei and Glidden (1997) for an overview of possible approaches, including analogs of the *marginal* and *random effects* models described for repeated continuous and binary outcomes in Chapter 8. In addition, left truncation and interval censoring were discussed only in passing. The Cox model can be extended to accommodate both, and there is an extensive literature in this area.

Sometimes time-to-event data can be more effectively handled using an alternative framework. In particular, consider cohort studies in which interval-censored outcomes are ascertained at each follow-up visit. One alternative is to use the continuation ratio model, referenced in Chapter 6, for time to the first such event. This can be seen as a discrete-time survival model, where the time scale is measured in visits (or intervals). Where appropriate, another, often more powerful, alterative is to use a logistic model for repeated binary measures, covered in Chapter 8. Finally, some time-to-event data has no censored values. In that situation, techniques covered in Chapter 9 can provide a useful regression framework for dealing with the skewness and heteroscedasticity such data are likely to exhibit.

Applied book-length treatments on survival analysis are available by Miller *et al.* (1981) and Marubini and Valsecchi (1995). These two texts strike a nice balance in their completeness and orientation toward biomedical applications.

The texts by Klein and Moeschberger (1997) and Therneau and Grambsch (2000) are very complete in their coverage of tools for survival analysis in general and the Cox model in particular, but are geared toward statisticians.

Stata provides extensive capabilities for fitting and assessing Cox models. A complete suite of parametric survival analysis methods are also provided. The flexible stset command handles complex patterns of censoring and truncation. However, the PHREG procedure in SAS makes TDCs easier to handle.

7.8 Problems

Problem 7.1. Divide the hazard ratio for bilirubin by its standard error in Table 7.4 and compare the result to the listed value of z. Also compute a confidence interval for this hazard ratio by adding and subtracting 1.96 times its standard error from the hazard ratio estimate. Are the results very different from the confidence interval listed in the output, which is based on computations on the coefficient scale?

Problem 7.2. In the ACTG 019 data, treatment rx is coded $ZDV = 1$ and placebo $= 0$. Define a new variable rx2 which is coded $ZDV = 12$ and placebo $= 11$; this can be done using the Stata command generate rx2=rx+10. Fit a Cox model with rx2 as the only predictor, then fit a second Cox model with rx as the only predictor. How do the two results compare? Now define rx3 coded placebo $= 1$, $ZDV = 0$ (Stata command generate rx3 = 1 - rx). How does this fit compare with the one for rx? Why? The ACTG 019 data set is available at http://www.biostat.ucsf.edu/vgsm.

Problem 7.3. Using the PBC data set, calculate the hazard ratio for values of albumin $= 2.5$, 3.5, and 4.0, using albumin = 3 as the reference level, under the assumption of log-linearity. The PBC data set is available at http://www.biostat.ucsf.edu/vgsm.

Problem 7.4. For the PBC data set, fit a model with cholesterol and bilirubin. Interpret the results, as you would in a paper, reporting the hazard ratios for a 100 mg/dL increase in cholesterol and a 10 mg/dL increase in bilirubin. Is the relationship between cholesterol and survival confounded by bilirubin?

Problem 7.5. Calculate a hazard ratio and confidence interval for a five-year increase in age by computing the fifth power of the estimated hazard ratio and its confidence limits, using the results for a one-year increase in Table 7.9.

Problem 7.6. For the ACTG 019 data set, write out the Cox model allowing for an interaction between ZDV treatment rx and the baseline CD4 cell count cd4.

- Express the test of the null hypothesis of no interaction between CD4 and treatment in terms of the parameters of the model.

- Again using the parameters of the model, what is the hazard ratio for a ZDV-treated subject with x CD4 cells compared with a placebo-treated subject with x CD4 cells?
- Fit the model. Does there appear to be an interaction between treatment and CD4 stratum? If so, what is the interpretation?
- What are the hazard ratios for ZDV as compared to placebo for patients with 500, 109, and 50 CD4 cells, respectively?

Problem 7.7. Using equation (7.9), you can calculate the probability of survival at time t, the odds, and the odds ratio (OR) for survival in exposed and reference groups. Calculate the OR for times t_1, t_2 and t_3 when $S_0(t_1) = 0.95$, $S_0(t_2) = 0.90$ and $S_0(t_3) = 0.80$. Try hazard ratio (i.e., e^β) values of both 2 and 10. Comment on the relationship between the hazard ratio and the OR for survival. How is it affected by the magnitude of the hazard, baseline survival probability, and time?

Problem 7.8. We can also control for the effect of bilirubin in the PBC mortality data using stratification rather than adjustment. One way to categorize bilirubin is by quantile. In Stata, for example, you can create a categorical variable for quintile of bilirubin using the command `xtile cat5=bilirubin, nq(5)`. Try fitting a Cox model for `cholesterol` stratified by `bilirubin`, stratified at 2, 3, 10, and 50 levels. What is the trade-off in increasing the number of levels? What number of levels works best? (Hint: Balance adjustment against the size of the standard error).

Problem 7.9. Using the PBC data set, apply the methods of Sect. 7.4.2 for examining proportional hazards to the variable `hepatomegaly` and interpret the results.

7.9 Learning Objectives

1. Define right-censoring, hazard function, proportional hazards, and time-dependent covariates.
2. Be able to
   - convert a predictor to a new unit scale
   - derive the hazard ratio between two groups defined by their predictor values
   - interpret hazard ratio estimates, Wald test P-values, and confidence intervals
   - calculate and interpret the likelihood-ratio test comparing two nested Cox models
   - detect and model interaction using the Cox model
   - detect non-proportional hazards using log-minus-log and smoothed hazard ratio plots, and the Schoenfeld test

- use stratification to control for a covariate with non-proportional effects.

3. Understand
    - when to use survival techniques
    - why the semi-parametric form of the Cox model is desirable
    - why the Cox model is "multiplicative"
    - how the stratified Cox model relaxes the proportional hazard assumption
    - recognize settings which are beyond the scope of this chapter, including left truncation, interval and dependent censoring, and repeated-events data.

8

Repeated Measures and Longitudinal Data Analysis

Knee radiographs are taken yearly in order to understand the onset of osteoarthritis. Echocardiograms are measured 1, 3, and 6 days after admission to the hospital for a brain hemorrhage. Groups of patients in a urinary incontinence trial are assembled from different treatment centers. Susceptibility to tuberculosis is measured in family members. All of these are examples of what is called repeated measures data or hierarchical or clustered data. Such data structures are quite common in medical research and a multitude of other fields.

Two features of this type of data are noteworthy and significantly impact the modes of statistical analysis. First, the outcomes are correlated across observations. Yearly radiographs on a person are more similar to one another than to radiographs on other people. Echocardiograms on the same person over time are more similar to one another than to those on other people. And groups of patients from a single center may yield similar responses because of treatment protocol variations from center to center, the persons or machines providing the measurements, or the similarity of individuals that choose to participate in a study at that center.

A second important feature of this type of data is that predictor variables can be associated with different levels of a hierarchy. Consider a study of the choice of type of surgery to treat a brain aneurysm either by clipping the base of the aneurysm or implanting a small coil. The study is conducted by measuring the type of surgery a patient receives from a number of surgeons at a number of different institutions. This is thus a hierarchical data set with multiple patients clustered within a surgeon and multiple surgeons clustered within a hospital. Predictor variables can be specific to any level of this hierarchy. We might be interested in the volume of operations at the hospital, or whether it is a for-profit or not-for-profit hospital. We might be interested in the years of experience of the surgeon or where she was trained. Or we might be interested in how the choice of surgery type depends on the age and gender of the patient.

Accommodation of these two features of the data, predictors specific to different levels in the data structure, and correlated data, are the topics of the chapter. We begin by illustrating the basic ideas in a simple example and then describe hierarchical models through two examples. In Sect. 8.4 we introduce the first of the methods of dealing with correlation structures, namely, generalized estimating equations. Sect. 8.5 introduces an example that we use throughout the rest of the chapter to illustrate the use of the models. Sect. 8.6 considers an alternative to generalized estimating equations, called random effects modeling, and the remaining sections contrast these approaches.

8.1 A Simple Repeated Measures Example: Fecal Fat

Lack of digestive enzymes in the intestine can cause bowel absorption problems. This will be indicated by excess fat in the feces. Pancreatic enzyme supplements can be given to ameliorate the problem. The data in Table 8.1 come from a study to determine if the form of the supplement makes a difference (Graham, 1977).

Table 8.1. Fecal Fat (g/day) for Six Subjects

Subject number	None	Tablet	Capsule	Coated	Subject average
1	44.5	7.3	3.4	12.4	16.9
2	33.0	21.0	23.1	25.4	25.6
3	19.1	5.0	11.8	22.0	14.5
4	9.4	4.6	4.6	5.8	6.1
5	71.3	23.3	25.6	68.2	47.1
6	51.2	38.0	36.0	52.6	44.5
Pill type average	38.1	16.5	17.4	31.1	25.8

We can think of this as either a repeated measures data set, since there are four measurements on each patient or, alternatively, as a hierarchical data set, where observations are clustered by patient. This simple example has as its only predictor pill type, which is specific to both the person and the period of time during which the measurement was taken. We do not have predictors at the patient level, though it is easy to envision predictors like age or a history of irritable bowel syndrome.

We identify a continuous outcome variable, fecal fat, and a single categorical predictor of interest, pill type. If we were to handle this analysis using the tools of Chapter 3, the appropriate technique would be a one-way ANOVA, with an overall F-test, or, perhaps better, a pre-planned set of linear contrasts. Table 8.2 gives the one-way ANOVA for the fecal fat example.

Table 8.2. One-Way ANOVA for the Fecal Fat Example

```
anova fecfat pilltype
```

Number of obs =	24	R-squared	=	0.2183	
Root MSE	= 18.9649	Adj R-squared =	0.1010		

Source	Partial SS	df	MS	F	Prob > F
Model	2008.6017	3	669.533901	1.86	0.1687
pilltype	2008.6017	3	669.533901	1.86	0.1687
Residual	7193.36328	20	359.668164		
Total	9201.96498	23	400.085434		

Following the prescription in Chapter 3, the F-test indicates ($P = 0.1687$) that there are not statistically significant differences between the pill types. But this analysis is incorrect. The assumptions of the one-way ANOVA require that all observations be independent, whereas we have repeated measures on the same six subjects, which are undoubtedly correlated. The one-way ANOVA would be appropriate if we had collected data on six *different* subjects for each pill type.

Should we have conducted the experiment with different subjects for each pill type? Almost certainly not. We gain precision by comparing the pill types within a subject rather than between subjects. We just need to accommodate this fact when we conduct the analysis. This is analogous to the gain in using a paired t-test.

In this situation, the remedy is simple: we conduct a two-way ANOVA, additionally removing the variability between subjects. Table 8.3 gives the two-way ANOVA.

The results are now dramatically different, with pill type being highly statistically significant. In comparing Tables 8.2 and 8.3 we can see that a large portion (about 5,588 out of 7,193 or almost 78%) of what was residual variation in Table 8.2 has been attributed to subject-to-subject variation in Table 8.3, thus sharpening the comparison of the pill types.

This is an illustration of a very common occurrence: failure to take into account the correlated nature of the data can have a huge impact on both the analysis strategy and the results.

Table 8.3. Two-Way ANOVA for the Fecal Fat Example

```
anova fecfat subject pilltype

              Number of obs =      24    R-squared     =  0.8256
              Root MSE      =  10.344    Adj R-squared =  0.7326

    Source |  Partial SS    df      MS            F      Prob > F
-----------+----------------------------------------------------
     Model |  7596.98166     8  949.622708        8.88    0.0002
           |
   subject |  5588.37996     5  1117.67599       10.45    0.0002
  pilltype |  2008.6017      3  669.533901        6.26    0.0057
           |
  Residual |  1604.98332    15  106.998888
-----------+----------------------------------------------------
     Total |  9201.96498    23  400.085434
```

8.1.1 Model Equations for the Fecal Fat Example

We next write down model equations appropriate for the fecal fat example to represent more precisely the differences between the two analyses from the previous section. The analysis in Table 8.2 follows the one-way ANOVA model from Chapter 3.

$$\text{FECFAT}_{ij} = \text{fecal fat measurement for person } i \text{ with pill type } j$$
$$= \mu + \text{PILLTYPE}_j + \varepsilon_{ij}, \tag{8.1}$$

where, as usual, we would assume $\varepsilon_{ij} \sim$ i.i.d $\mathcal{N}(0, \sigma_\varepsilon^2)$, meaning that it is independently and identically distributed with mean zero and variance σ_ε^2 (Sect. 3.3.2).

As noted above, there is no account taken of the effect of each subject. We would expect some subjects to generally have higher values and others to generally have lower values. To accommodate this we include a subject effect in the model, which simultaneously raises or lowers all the measurements on that subject:

$$\text{FECFAT}_{ij} = \text{fecal fat measurement for person } i \text{ with pill type } j$$
$$= \mu + \text{SUBJECT}_i + \text{PILLTYPE}_j + \varepsilon_{ij}, \tag{8.2}$$

with

$$\varepsilon_{ij} \sim \text{i.i.d } \mathcal{N}(0, \sigma_\varepsilon^2).$$

To this we add one more piece. We assume that the subject effects are also selected from a distribution of possible subject effects: $\text{SUBJECT}_i \sim$ i.i.d $\mathcal{N}(0, \sigma_{subj}^2)$, independently of ε_{ij}.

This additional piece serves two purposes. First, it captures the idea that the subjects in our experiment are assumed to be a random sample from a larger population of subjects to which we wish to draw inferences. Otherwise, the conclusions from our experiment would be scientifically uninteresting, as they would apply only to a select group of six subjects. Second, the inclusion of a subject effect (along with an assigned distribution) models a correlation in the outcomes. Once we have added this subject effect to our model, we must accordingly modify the analysis to become the two-way ANOVA shown in Table 8.3.

8.1.2 Correlations Within Subjects

The main reason the results in Tables 8.2 and 8.3 differ so dramatically is the failure of the analysis in 8.2 to accommodate the repeated measures or correlated nature of the data. How highly correlated are measurements within the same person? The model given in (8.2) gives us a way to calculate this. The observations on the same subject are modeled as correlated through their shared random subject effect. The larger the subject effects in relation to the error term, the larger the correlation (relatively large subject effects means the observations on one subject are quite different than those on another subject, but, conversely, that observations *within* a subject tend to be similar). More precisely, there is a covariance between two observations on the same subject:

$$\begin{aligned} \text{cov}(\text{FECFAT}_{ij}, \text{FECFAT}_{ik}) &= \text{cov}(\text{SUBJECT}_i, \text{SUBJECT}_i) \\ &= \text{var}(\text{SUBJECT}_i) \qquad (8.3) \\ &= \sigma^2_{subj}. \end{aligned}$$

The first equality in (8.3) is because the μ and pill-type terms are assumed to be fixed constants and do not enter into the covariance calculation. The ε_{ij} terms drop out because they are assumed to be independent of the subject effects and of each other. The second equality is true because the covariance of any term with itself is a variance and the last equality is just the notation for the variance of the subject effects.

As we recall from Chapter 3, this is just one ingredient in the calculation of the correlation. We also need to know the standard deviations for the measurements. Model (8.2) also indicates how to calculate the variance and hence the standard deviation:

$$\begin{aligned} \text{var}(\text{FECFAT}_{ij}) &= \text{var}(\text{SUBJECT}_i) + \text{var}(\varepsilon_{ij}) \\ &= \sigma^2_{subj} + \sigma^2_\varepsilon \qquad (8.4) \end{aligned}$$

so that

$$\text{SD}(\text{FECFAT}_{ij}) = \sqrt{\sigma^2_{subj} + \sigma^2_\varepsilon},$$

which is assumed to be the same for all observations. The result, (8.4), is noteworthy by itself, since it indicates that the variability in the observations is being decomposed into two pieces, or components, the variability due to subjects and the residual, or error, variance.

We are now in a position to calculate the correlation as the covariance divided by the standard deviations:

$$
\text{corr}(\text{FECFAT}_{ij}, \text{FECFAT}_{ik}) = \frac{\text{cov}(\text{FECFAT}_{ij}, \text{FECFAT}_{ik})}{\text{SD}(\text{FECFAT}_{ij})\text{SD}(\text{FECFAT}_{ik})}
$$

$$
= \frac{\sigma^2_{subj}}{\sqrt{\sigma^2_{subj} + \sigma^2_\varepsilon}\sqrt{\sigma^2_{subj} + \sigma^2_\varepsilon}}
$$

$$
= \frac{\sigma^2_{subj}}{\sigma^2_{subj} + \sigma^2_\varepsilon}. \tag{8.5}
$$

While the methods of the calculations are not so important, the intuition and results are. Namely, that subject-to-subject variability simultaneously raises or lowers all the observations on a subject, thus inducing a correlation and that the variability of an individual measurement can be separated into that due to subjects and residual variance.

Looking at the ANOVA table in Table 8.3 we have an estimate of σ^2_ε, which is approximately 107.00. But what about an estimate for σ^2_{subj}? It would be almost correct to calculate the variance of the subject averages in the last column of Table 8.1, but this would be a bit too large since each subject average also has a small amount of residual variation as well. Taking this into account (see Problem 8.1), gives an estimate of 252.67.

Using this in (8.5) gives a correlation of $0.70 = 252.67/(252.67 + 107.00)$, not a particularly high value. So even a moderate value of the correlation can have a fairly dramatic effect on the analysis, which is why it is so important to recognize repeated measures or clustered data situations. In this instance the analysis ignoring the correlation led to results that were not statistically significant and inflated P-values. Unfortunately, the effect of ignoring the correlation can also make the P-values appear incorrectly small, as will be demonstrated below. So ignoring the correlation does not always produce a "conservative" result.

In this example, we are mainly interested in comparing the effect of the different pill types and the correlation within subjects must be accommodated in order to perform a proper analysis. The correlation is more of a nuisance. In other studies the correlation will be the primary focus of the analysis, such as repeatability or validation studies or in analysis of familial aggregation of a disease. In the knee osteoarthritis example, the same radiographs were sent to different reading centers to check consistency of results across the centers. One of the primary parameters of interest was the correlation of readings taken on the same image.

8.1.3 Estimates of the Effects of Pill Type

What about estimating the effects of the various pill types or differences between them? The simple averages across the bottom of Table 8.1 give the estimates of the mean fecal fat values for each pill type. There is nothing better we can do in this balanced-data experiment. The same is true for comparing different pill types. For example, the best estimate of the difference between a coated capsule and a regular capsule would be the simple difference in means: $31.07 - 17.42 = 13.65$. That is, we do nothing different than we would with a one-way ANOVA (in which all the observations are assumed independent). This is an important lesson that we extend in Sect. 8.4: the usual estimates based on the assumption of independent data are often quite good. It is the estimation of the standard errors and the tests (like the F-test) that go awry when failing to accommodate correlated data.

8.2 Hierarchical Data

Common methods for the assessment of individual physicians' performance at diabetes care were evaluated in Hofer *et al.* (1999). They studied 232 physicians from three sites caring for a total of 3,642 patients, and evaluated them with regard to their ability to control HbA_{1c} levels and with regard to resource utilization. Various methods for obtaining physician level predictions are compared including age- and sex-adjusted averages, the calculation of residuals after adjusting for the case-mix of the patients, and hierarchical modeling. They find that the first two methods overstate the degree to which physicians differ. This could have adverse consequences in falsely suggesting that some physicians (especially those with small numbers of patients) are over-using resources or ineffectively treating patients.

As we will see explicitly later in the chapter, hierarchical analysis is more effective in this situation because it "borrows strength" across physicians in order to improve the predicted values for each physician. Said another way, we can use knowledge of the variation between and within physicians in order to quantify the degree of unreliability of individual physician's averages and, especially for those with small numbers of patients, make significant adjustments.

8.2.1 Analysis Strategies for Hierarchical Data

As has been our philosophy elsewhere in this book, the idea is to use simpler statistical methods unless more complicated ones are necessary or much more advantageous. That raises the basic question: do we need hierarchical models and the attendant more complicated analyses? An important idea is the following. Observations taken within the same subgroup in a hierarchy are often more similar to one another than to observations in different subgroups, other things being equal. Equivalently, data which are clustered together in the same level of the hierarchy (data on the same physician, or on the same patient or in the same hospital) are likely to be correlated. The usual statistical methods (multiple regression, basic ANOVA, logistic regression, and many others) assume observations are independent. And we have seen in Sect. 8.2 the potential pitfalls of completely ignoring the correlation.

Are there simple methods we can use that accommodate the correlated data? Simpler approaches that get around the issue of correlation include separate analyses for each subgroup, analyses at the highest level in the hierarchy, and analyses on "derived" variables. Let us consider examples of each of these approaches using the back pain example introduced in Chapter 1 (Korff *et al.*, 1994).

Analyses for Each Subgroup

Analysis for each subgroup would correspond to doing an analysis for each of the 44 doctors separately. If there were sufficient data for each doctor, this might be effective for some questions, for example, the frequency with which patients for that physician understood how to care for their back. For other questions it would be less satisfactory, for example, how much more it cost to treat older patients. To answer this question we would need to know how to aggregate the data across doctors. For yet other questions it would be useless. For example, comparing practice styles is a between-physician comparison and any within-physician analysis is incapable of addressing it.

Analysis at the Highest Level in the Hierarchy

An analysis at the highest level of the hierarchy would proceed by first summarizing the data to that level. As an example, consider the effect of practice style on the cost of treatment. Cost data would be averaged across all times and patients within a physician, giving a single average value. A simple analysis could then be performed, comparing the average costs across the three types of physicians. And by entering into the analysis a single number for each physician, we avoid the complication of having correlated data points through time on the same patient or correlated data within a physician.

There are several obvious drawbacks to this method. First, there is no allowance for differences in patient mix between physicians. For example, if those in the aggressive treatment group also tended to have older, higher-cost patients we would want to adjust for that difference. We could consider having additional variables such as average age of the patients for each physician to try to accommodate this. Or a case mix difference of another type might arise: some physicians might have more complete follow-up data and have different proportions of data at the various times after the index visit. Adjusting for differences of these sorts is one of the key reasons for considering multipredictor models.

A second drawback of analysis at the highest level of the hierarchy is that some physicians will have large numbers of patients and others will have small numbers. Both will count equally in the analysis. This last point bears some elaboration. Some data analysts are tempted to deal with this point by performing a weighted analysis where the physician receives a weight proportional to the number of observations that went into their average values or the number of patients that contributed to the average. But this ignores the correlated nature of the data. If the data are highly correlated within a physician, then additional patients from each physician contribute little additional information and all physicians' averages should be weighted equally regardless of how many patients they have. At the other extreme, if each patient counts as an independent data point, then the averages *should* be weighted by the numbers of patients.

If the data are correlated but not perfectly correlated, the proper answer is somewhere in between these two extremes: a physician with twice as many patients as another should receive more weight, but not twice as much. To determine precisely how much more requires estimation of the degree of correlation within a physician, i.e., essentially performing a hierarchical analysis.

Analysis on "Derived Variables"

A slightly more sophisticated method than simple averaging is what is sometimes called the use of "derived variables." The basic idea is to calculate a simple, focused variable for each cluster or subgroup that can be used in a more straightforward analysis. A simple and often effective example of this method is calculation of a change score. Instead of analyzing jointly the before and after treatment values on a subject (with a predictor variable that distinguishes them) we instead calculate the change score.

Here are two other examples of this methodology. In a pharmacokinetic study we might sample a number of subjects over time after administration of a drug and be interested in the average value of the drug in the bloodstream and how it changes with different doses of the drug. One strategy would be to analyze the entire data set (all subjects and all times), but then we would need to accommodate the correlated nature of the data across time within a person. A common alternative is to calculate, for each person, the area under the curve (AUC) of the concentration of the drug in the bloodstream versus time. This AUC value would then be subjected to a simpler analysis comparing doses (e.g., a linear regression might be appropriate). In the fecal fat example, the derived variable approach is quite effective. Suppose we were interested in the effect of coating a capsule. We can calculate the six differences between the capsule and the coated capsule (one for each person) and do a one-sample or paired *t*-test on the six differences. (See Problem 8.5). For the back pain example, the derived variable approach is not as successful. The unbalanced nature of the data makes it difficult to calculate an effective derived variable.

In summary, the use of hierarchical analysis strategies is clearly indicated in any of three situations:

1. when the correlation structure is of primary interest;
2. when we wish to "borrow strength" across the levels of a hierarchy in order to improve estimates; and
3. when dealing with highly unbalanced correlated data.

8.3 Longitudinal Data

In *longitudinal* studies we are interested in the change in the value of a variable within a "subject" and we collect data repeatedly through time. For example, a study of the effects of alcohol might record a measure of sleepiness before and after administration of either alcohol or placebo. Interest is in quantifying the effect of alcohol on the *change* in sleepiness. This is often a good design strategy since each subject acts as his or her own control, allowing the elimination of variability in sleepiness measurements from person to person or even occasion to occasion within a person. For this strategy to be effective, the before and after measurements need to be at least moderately strongly positively correlated (otherwise, taking differences increases the variability rather than reducing it).

8.3.1 Analysis Strategies for Longitudinal Data

In simple situations there is a straightforward approach to analyzing such data – calculate the difference scores (subtract the before measurement from the after measurement) as a derived variable and perform an analysis on the differences. In the alcohol example, we could simply perform a two-sample *t*-test using the difference scores as data to compare the alcohol and placebo subjects.

We consider three approaches to analysis of before/after data that are commonly used: 1) analysis of difference scores, 2) repeated measures analysis, and 3) analysis using the after measurement as the outcome and using the baseline measurement as a covariate or predictor. The justification for this last strategy is to "adjust for" the baseline value before looking for differences between the groups. How do these approaches compare?

8.3.2 Example: Birthweight and Birth Order

We consider an analysis of birthweights of first-born and last-born infants from mothers (each of whom had five children) from vital statistics in Georgia. We are interested in whether birthweights of last-born babies are different from first-born and whether this difference depends on the age of the woman when she had her first-born.

For the first question we begin with the basic descriptive statistics given in Table 8.4, where `lastwght` in the variable containing the last-born birth-weights, `initwght` indicates the first-born, and `delwght` are the differences between last- and first-born within a woman. These show that last-born tend to be about 191 g heavier than first-born (the same answer is obtained whether you average the differences or take the difference between the averages). To

Table 8.4. Summary Statistics for First- and Last-Born Babies

```
summ initwght lastwght delwght
```

Variable	Obs	Mean	Std. Dev.	Min	Max
initwght	1000	3016.555	576.2185	815	4508
lastwght	1000	3208.195	578.3356	1210	5018
delwght	1000	191.64	642.3062	-1551	2700

accommodate the correlated data we either perform a one-sample t-test on the differences or, equivalently, a paired t-test of the first and last births. A paired t-test gives a t-statistic of 4.21, with 199 degrees of freedom (since there are 200 mothers) with a corresponding P-value that is approximately 0.

What about the relationship of the difference in birthweight to the mother's initial age? For this, we conduct a simple linear regression of the difference in birthweight regressed on initial age, where we have centered initial age (`cinitage`) by subtracting the mean initial age. The results are displayed in Table 8.5 with the interpretation that each increase of one year in initial

Table 8.5. Regression of Difference in Birthweight on Centered Initial Age

```
regress delwght cinitage
```

Source	SS	df	MS		Number of obs =	200
					F(1, 198) =	0.39
Model	163789.382	1	163789.382		Prob > F =	0.5308
Residual	82265156.7	198	415480.589		R-squared =	0.0020
					Adj R-squared =	-0.0031
Total	82428946.1	199	414215.809		Root MSE =	644.58

delwght	Coef.	Std. Err.	t	P>|t|	[95% Conf. Interval]	
cinitage	8.891816	14.16195	0.63	0.531	-19.03579	36.81942
_cons	191.64	45.57854	4.20	0.000	101.7583	281.5217

age is associated with an additional 8.9 g difference between the first and last birthweights. This is not statistically significant ($P = 0.53$). When centered age is used, the intercept term (`_cons`) is also the average difference.

To conduct a repeated measures analysis the data are first reordered to have a single column of data containing the birthweights and an additional column, birth order, to keep track of whether it is a first or last birth. The output for the repeated measures analysis using only the first and last births is displayed in Table 8.6, for which we leave the details to the next section. However many of the elements are similar to the regression analysis in Table 8.5. The term, _IbirXcini~5, is the interaction of birth order and centered initial age. It thus measures how the *difference* in birthweights between first- and last-born is related to centered initial age, that is, whether the difference score is related to initial age, the same question as the regression analysis. As is evident, the estimated coefficient is identical and the standard error is virtually the same. They are not exactly the same because slightly different modeling techniques are being used (regression versus GEE, short for generalized estimating equations). The overall difference between first and last born is also displayed in the repeated measures analysis (again with the same coefficient and a very similar standard error and P-value) and is associated with the birth order term in the model. Finally, the average for first births is displayed as the intercept (see Problem 8.8). So, at a cost of more complication, the repeated measures analysis answers both questions of interest.

Table 8.6. Repeated Measures Regression of Birthweight on Birth Order and Centered Initial Age

```
xi: xtgee bweight i.birthord cinitage i.birthord*cinitage, i(momid)
```

```
GEE population-averaged model          Number of obs      =      400
Group variable:              momid     Number of groups   =      200
Link:                     identity     Obs per group: min =        2
Family:                   Gaussian                   avg =      2.0
Correlation:           exchangeable                  max =        2
                                       Wald chi2(3)       =    26.47
Scale parameter:           323645.4    Prob > chi2        =   0.0000
```

bweight	Coef.	Std. Err.	z	P>|z|	[95% Conf. Interval]
_Ibirthord_5	191.64	45.35007	4.23	0.000	102.7555 280.5245
cinitage	25.13981	12.4992	2.01	0.044	.6418238 49.6378
_IbirXcini~5	8.891816	14.09096	0.63	0.528	-18.72596 36.50959
_cons	3016.555	40.22719	74.99	0.000	2937.711 3095.399

A different sort of analysis is to conduct a multiple regression with two predictor variables, initial age (centered) and first-born birthweight. The idea is to "adjust" the values of last-born weight by the first-born weight and then look for an effect due to initial age. Table 8.7 gives the results of that analysis, which are quite different than the previous analyses. Now, initial age has a much larger coefficient and is statistically significant ($P = 0.036$).

The intuitive explanation for why this analysis is so different starts with the observation that the coefficient for birthweight of the first-born is approximately .363. So, using BW_k to denote the birthweight of the kth born child,

Table 8.7. Regression of Final Birthweight on Centered Initial Age, Adjusting for First Birthweight

```
regress lastwght cinitage initwght if birthord==5

      Source |       SS       df       MS              Number of obs =     200
-------------+------------------------------           F(  2,    197) =   19.33
       Model | 10961363.1       2  5480681.54          Prob > F       =  0.0000
    Residual | 55866154.3     197   283584.54          R-squared      =  0.1640
-------------+------------------------------           Adj R-squared  =  0.1555
       Total | 66827517.4     199   335816.67          Root MSE       =  532.53


------------------------------------------------------------------------------
    lastwght |     Coef.   Std. Err.       t    P>|t|     [95% Conf. Interval]
-------------+----------------------------------------------------------------
    cinitage |  24.90948   11.81727     2.11    0.036     1.604886    48.21408
    initwght |  .3628564   .0660366     5.49    0.000      .232627    .4930858
       _cons |  2113.619   202.7309    10.43    0.000     1713.817     2513.42
------------------------------------------------------------------------------
```

we can think of the fitted model as

$$BW_5 = 2113.619 + .363BW_1 + 24.909 \text{ Centered initial age} \qquad (8.6)$$

or, taking BW_1 to the left side of the equation,

$$BW_5 - .363BW_1 = 2113.619 + 24.909 \text{ Centered initial age}. \qquad (8.7)$$

That is, this analysis is not purely looking at differences between last and first birthweight since we are only subtracting off a fraction of the initial birthweight. Since birthweights are more highly correlated with initial age than is the difference, this stronger relationship reflects that fact that the results are close to a regression of BW_5 on initial age.

In observational studies, such as this one, using baseline values of the outcome as a covariate is not a reliable way to check the dependence of the change in outcome on a covariate. In randomized studies, where there should be no dependence between treatment effects and the baseline values of the outcome, this may be a more reasonable strategy.

8.3.3 When To Use Repeated Measures Analyses

In the Georgia birthweight example, we see that analysis by difference scores or by a repeated measures analysis give virtually identical and reasonable results. The analysis using the baseline value as a covariate is more problematic to interpret.

If the analysis of difference scores is so straightforward, why consider the more complicated repeated measures analysis? For two time points and no (or little) missing data, there is little reason to use the repeated measures analysis. However, in the birthweight example there are three intermediate births we have ignored that should be included in the analysis. In the alcohol example it would be reasonable to measure the degree of sleepiness at numerous time points after administration of alcohol (or placebo) to track the speed of onset of sleepiness and when it wears off. When there are more than two repeated measures, when the measurements are recorded at different times and/or when there is missing data, repeated measures analysis can more easily accommodate the data structure than change score analyses. We now consider methods for multiple time points.

8.4 Generalized Estimating Equations

There are two main methods for accommodating correlated data. The first we will consider is a technique called *generalized estimating equations*, often abbreviated GEE. A key feature of this method is the option to estimate the correlation structure from the data without having to assume it follows a pre-specified structure.

Before embarking on an analysis we will need to consider five aspects of the data:

1. What is the distributional family (for fixed values of the covariates) that is appropriate to use for the outcome variable? Examples are the normal, binary, and binomial families.
2. Which predictors are we going to include in the model?
3. In what way are we going to link the predictors to the data? (Through the mean? Through the logit of the risk? Some other way?)
4. What correlation structure will be used or assumed temporarily in order to form the estimates?
5. Which variable indicates how the data are clustered?

The first three of these decisions we have been making for virtually every method described in this book. For example, the choice between a logistic and linear regression hinges on the distribution of the outcome variable, namely, logistic for binary outcome and linear for continuous, approximately normal outcomes. Chapter 5 discusses the choice of predictors to include in the model (and is a focus of much of this book) and the third has been addressed in specific contexts, e.g, the advantage of modeling the log odds in binary data. The new questions are really the fourth and fifth and have to do with how we will accommodate the correlations in the data. We start by considering an example.

8.4.1 Birthweight and Birth Order Revisited

We return to the Georgia birthweight example and now consider all five births. Recall that we are interested in whether birthweight increases with birth order and mothers' age. Fig. 8.1 shows a plot of birthweight versus birth order with

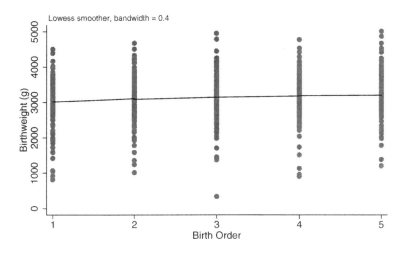

Fig. 8.1. Plot of Birthweight Versus Birth Order

both the average birthweights for a given birth order and a LOWESS smooth superimposed. Inspection of the plot suggests we can model the increase as a linear function. A simple linear regression analysis of birthweight versus birth order gives a t-statistic for the slope coefficient of 3.61, which is highly statistically significant. But this analysis would be wrong (why?).

Recall that the paired t-test using just the first and last births gave a t-statistic of 4.21, even more highly statistically significant. This is perhaps a bit surprising since it discards the data from the three intermediate births.

The explanation for this apparent paradox is that the paired t-test, while using less of the data, does take advantage of the fact that birth order is a within-mother comparison. It exploits the correlation of birthweights within a mom in order to make a more precise comparison. Of course, an even better analysis is to use all of the data and accommodate the correlated structure of the data, which we now proceed to do.

Analysis

To analyze the Georgia babies data set we need to make the decisions outlined above. The outcome variable is continuous, so a logical place to start is to assume it is approximately normally distributed. Fig. 8.2 shows boxplots of birthweight by birth order, suggesting that the normality and equal variance assumptions are reasonable. Fig. 8.1 has suggested entering birth order as a linear function, which leaves us with the accommodation of the correlation structure.

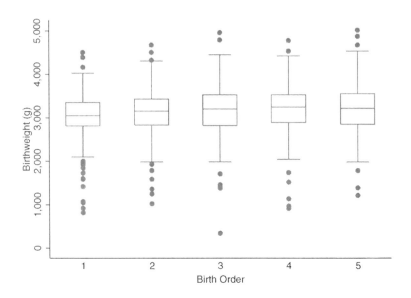

Fig. 8.2. Boxplots of Birthweight (g) Versus Birth Order

The data are correlated because five birthweights come from each mother and hence the clustering aspect is clear, leaving us with the decision as to how to model the correlation of measurements taken through time. Fig. 8.3 gives a matrix plot of each birthweight against each of the others, while Table 8.8 gives the values of the correlation coefficients. Correlations with the first birthweight might be a bit lower, but the graphs suggest that a tentative assumption of all the correlations being equal would not be far off.

Table 8.8. Correlation of Birthweights for Different Birth Orders

```
. corr bweight1 bweight2 bweight3 bweight4 bweight5 (obs=200)

             | bweight1 bweight2 bweight3 bweight4 bweight5
-------------+---------------------------------------------
    bweight1 | 1.0000
    bweight2 | 0.2282 1.0000
    bweight3 | 0.2950 0.4833 1.0000
    bweight4 | 0.2578 0.4676 0.6185 1.0000
    bweight5 | 0.3810 0.4261 0.4233 0.4642 1.0000
```

8.4.2 Correlation Structures

Dealing with correlated data typically means making some type of assumption about the form of the correlation among observations taken on the same

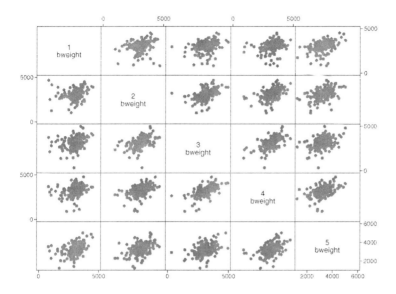

Fig. 8.3. Matrix Plot of Birthweights for Different Birth Orders

subject, in the same hospital, on the same mouse, etc. For the Georgia babies data set in the previous section, we noted that assuming all the correlations to be equal might be a reasonable assumption. This form of correlation is termed exchangeable and means that all correlations (except those variables with themselves) are a common value, which is typically estimated from the data. This type of structure is suitable when there is nothing to distinguish one member of a cluster from another (e.g., patients within a physician) and is the genesis for its name (patients within a doctor can be regarded as interchangeable or exchangeable). This sort of assumption is appropriate in the absence of other data structure, such as measurements taken through time or space.

If measurements are taken through time on the same person it may be that observations taken more closely in time are more highly correlated. Another common correlation structure is the autoregressive structure, which exhibits this feature. In the simplest form of an autoregressive process (first order or AR(1)) the correlation between observations one time unit apart is a given value ρ, that between observations two time units apart ρ^2, three time units apart ρ^3, etc. Simple arithmetic calculation shows this drops off rapidly to zero (e.g., $0.6^5 = 0.08$), so this assumption would only be appropriate if the correlation between observations taken far apart in time was small and would not be appropriate in cases where stable-over-time characteristics generated the association. For example, systolic blood pressure would be relatively stable over time for an individual. Even though observations taken more closely together in time would be slightly more highly correlated, an exchangeable correlation structure might come closer to the truth than an autoregressive one.

Other, less structured, assumptions can be made. In Stata, other options are *unstructured, nonstationary,* and *stationary.* All are related to the idea of observations within a cluster being ordered, such as by time. As its name suggests, the unstructured form estimates a separate correlation between observations taken on each pair of "times." The nonstationary form is similar, but assumes all correlations for pairs separated far enough in time are zero. The stationary form assumes equal correlation for all observations a fixed time apart and, like nonstationary, assumes correlations far enough apart in time have correlation zero. For example, stationary of order 2 would assume that observations taken at time points 1 and 3 would have the same correlation as time points 2 and 4, but this might be different from the correlation between observations taken at times 2 and 3. Also, correlations for observations 3 or more time periods apart would be assumed to be zero.

If the correlation structure is not the focus of the analysis, it might seem that the unstructured form is best, since it makes no assumptions about the form of the correlation. However, there is a cost: even with a small number of time points, we are forced to estimate quite a large number of correlations. For instance, with measurements on five time points for each subject, there are ten separate correlations to estimate. This can cause a decrease in the precision of the estimated parameters of interest, or, worse yet, a failure in being able to even fit the model.

This is especially true in situations where the data are not collected at rigid times. For example, in the Nutritional Prevention of Cancer trials (Clark *et al.*, 1996), long-term follow-up was attempted every six months. But the intervals varied widely in practice and quickly were out of synchronization. Estimation of the correlations between all pairs of distinct times would require literally hundreds of estimated correlations. Use of the unstructured, and, to some extent, the stationary and nonstationary correlation assumptions should be restricted to situations where there are large numbers of clusters, e.g., subjects, and not very many distinct pairs of observation times.

Diagnosis and specification of the "correct" correlation structure is very difficult in practice. One method of addressing these problems is via a *working* correlation assumption and the use of "robust" standard errors, which is the next topic.

8.4.3 Working Correlation and Robust Standard Errors

Given the difficulty of specifying the "correct" correlation structure, a compromise is possible using what are called *robust standard errors.* The idea is to make a temporary or working assumption as to the correlation structure in order to form the estimates but to adjust those estimates properly for the correlation in the data. For example, we might temporarily assume the data are independent and conduct a standard logistic regression. The estimates from the logistic regression will be fairly good, even when used with correlated data, but the standard errors will be incorrect, perhaps grossly so. The solution is to use the estimates but empirically estimate their proper standard errors. Another possibility is to make a more realistic assumption, such as an exchangeable working correlation structure; in some circumstances a gain in efficiency may result.

Then, after the model coefficients have been estimated using the working correlation structure, within-subject residuals are used to compute robust standard errors for the coefficient estimates. Because these standard errors are based on the data (the residuals) and not the assumed working correlation structure, they give valid (robust) inferences for large sized samples as long as the other portions of the model (distribution, link and form of predictors) are correctly specified, even if our working correlation assumption is incorrect. Use of robust standard errors is not quite the same as using an unstructured correlation since it bypasses the estimation of the correlation matrix to obtain the standard errors directly. Avoiding estimation of a large number of correlations is sometimes an advantage, though in cases where both approaches can be used they often give similar results.

The key to the use of this methodology is to have sufficient numbers of subjects or clusters so that the empirical estimate of the correlation is adequate. The GEE approach, which goes hand in hand with estimation using robust standard errors, will thus work best with relatively few time points and relatively more subjects. It is hard to give specific guidelines, but this technique could be expected to work well with 100 subjects, each measured at 5 time points but much less well with 20 subjects, each measured at 12 time points, especially if the times were not the same for each subject.

8.4.4 Hypothesis Tests and Confidence Intervals

Hypothesis testing with GEE uses Wald tests, in which the estimates divided by their robust standard errors are treated as approximately normal to form Z-statistics. Likewise, approximate confidence intervals are based on normality by calculating the estimate plus or minus 1.96 standard errors. Table 8.9 shows the analysis with an exchangeable working correlation structure and robust standard errors. Some comments are in order about the form of the command. xtgee is a regression type command with numerous capabilities. In its basic form, exhibited in Table 8.9, it performs a linear regression (link of identity) of birthweight (bweight) on birth order (birthord) and mother's age at first birth (initage) with an assumed exchangeable correlation structure (corr(exch)) within mother (i(momid)). The robust option requests the use of robust standard errors.

For the sake of comparison, Table 8.10 gives the analysis without robust standard errors. There is little difference, though this is to be expected since

Table 8.9. Generalized Estimating Equations Model With Robust Standard Errors

```
. xtgee bweight birthord initage, i(momid) corr(exch) robust

Iteration 1: tolerance = 7.180e-13

GEE population-averaged model          Number of obs      =    1000
Group variable:            momid       Number of groups   =     200
Link:                      identity    Obs per group: min =       5
Family:                    Gaussian                   avg =     5.0
Correlation:               exchangeable               max =       5
                                       Wald chi2(2)       =   27.95
Scale parameter: 324458.3              Prob > chi2        =   0.000

                       (standard errors adjusted for clustering on momid)
------------------------------------------------------------------------------
             |              Semi-robust
     bweight |    Coef.     Std. Err.      z     P>|z|    [95% Conf. Interval]
-------------+----------------------------------------------------------------
     birthord |   46.608    10.02134     4.65    0.000    26.96653    66.24947
      initage |  26.73226   10.1111      2.64    0.008    6.914877    46.54965
        _cons |  2526.622   177.2781    14.25    0.000    2179.164    2874.081
------------------------------------------------------------------------------
```

Table 8.10. Generalized Estimating Equations Model Without Robust Standard Errors

```
. xtgee bweight birthord initage, i(momid) corr(exch)

Iteration 1: tolerance = 7.180e-13

GEE population-averaged model          Number of obs      =    1000
Group variable:            momid       Number of groups   =     200
Link:                      identity    Obs per group: min =       5
Family:                    Gaussian                   avg =     5.0
Correlation:               exchangeable               max =       5
                                       Wald chi2(2)       =   30.87
Scale parameter: 324458.3              Prob > chi2        =   0.000

                       (standard errors adjusted for clustering on momid)
------------------------------------------------------------------------------
             |              Semi-robust
     bweight |    Coef.     Std. Err.      z     P>|z|    [95% Conf. Interval]
-------------+----------------------------------------------------------------
     birthord |   46.608    9.944792     4.69    0.000    27.11657    66.09943
      initage |  26.73226   8.957553     2.98    0.003    9.175783    44.28874
        _cons |  2526.622   162.544     15.54    0.000    2208.042    2845.203
------------------------------------------------------------------------------
```

the preliminary look at the data suggested that the exchangeable assumption would be a reasonable one.

Looking at the analysis with the robust standard errors, the interpretation of the coefficient is the same as for a linear regression. With each increase of initial age of one year, there is an associated increase in average birthweight of about 26.7 g. This result is highly statistically significant, with a P-value of 0.008.

Lest the reader think that the analysis is impervious to the correlational assumptions, Table 8.11 shows what happens to the estimates and standard

Table 8.11. Results for Initial Age by Type of Working Correlation and Standard Error

Working Correlation	Robust SE?	Coefficient estimate	Standard error	Z-statistic	P-value
Independence	No	26.73	5.60	4.78	0.000
Exchangeable	No	26.73	8.96	2.98	0.003
Autoregressive(1)	No	27.41	7.82	3.51	0.000
Independence	Yes	26.73	10.11	2.64	0.008
Exchangeable	Yes	26.73	10.11	2.64	0.008
Autoregressive(1)	Yes	27.41	9.69	2.83	0.005

errors under three different correlation structures both with and without the use of robust standard errors. As expected, the estimates are all similar (the independence and exchangeable are equal because of the balanced nature of the data – five observations per mom with the same values of birth order), though there are slight variations depending on the assumed working correlation. The estimates are unaffected by the use of robust standard errors.

However, the standard errors and hence Wald statistics and P-values are quite different. Those using the incorrect assumptions of independence or autoregressive structure (given in the rows without robust standard errors) are too small, yielding Wald statistics and P-values that are incorrect (P-values falsely small in this case, though they can, in general, be incorrect in either direction). Looking at the rows corresponding to the use of robust standard errors shows how the incorrect working assumptions of independence or autoregressive get adjusted and are now much more alike. As with any different methods of estimation slight differences do, however, remain.

8.4.5 Use of xtgee for Clustered Logistic Regression

As mentioned above, xtgee is a very flexible command. Another of its capabilities is to perform logistic regression for clustered data. We again analyze the Georgia birthweight data but instead use as our outcome the binary variable low birthweight (lowbrth), which has value one if the birthweight is less than 3,000 g and zero otherwise. Since the data are binary, we adapt xtgee for logistic regression by specifying family(binomial) and link(logit). As before, we specify i(momid) to indicate the clustering, corr(exch) for an exchangeable working correlation, and robust to calculate robust standard errors; also we add the option ef to get odds ratios instead of log odds. Table 8.12 displays the analysis. The estimated odds ratio for birth order is about 0.92, with the interpretation that the odds of a low-birthweight baby decrease by 8% with each increase in birth order. We see that initage is still statistically significant, but less so than in the analysis of actual birthweight. This

Table 8.12. Generalized Estimating Equation Logistic Model

```
. xtgee lowbrth birthord initage, i(momid) corr(exch) family(binomial) link(logit)
>  robust ef

Iteration 1: tolerance = .00603648
Iteration 2: tolerance = .00003423
Iteration 3: tolerance = 1.861e-07

GEE population-averaged model              Number of obs      =      1000
Group variable:                    momid   Number of groups   =       200
Link:                              logit   Obs per group: min =         5
Family:                         binomial                  avg =       5.0
Correlation:                 exchangeable                  max =         5
                                           Wald chi2(2)       =     10.64
Scale parameter:                       1   Prob > chi2        =    0.0049

                         (standard errors adjusted for clustering on momid)
------------------------------------------------------------------------------
             |               Semi-robust
   lowbrth   | Odds Ratio   Std. Err.      z    P>|z|     [95% Conf. Interval]
-------------+----------------------------------------------------------------
   birthord  |  .9204098     .03542     -2.16   0.031     .8535413    .9925168
    initage  |  .9148199    .0312663    -2.60   0.009     .8555464    .9781999
------------------------------------------------------------------------------
```

serves as a warning as to the loss of information possible by unnecessarily dicohotomizing a variable.

8.5 Random Effects Models

The previous section discussed the use of generalized estimating equations for the accommodation of correlated data. This approach is limited in that–

1. It is restricted to a single level of clustering.
2. It is not designed for inferences about the correlation structure.
3. It does not give predicted values for each cluster or level in the hierarchy.

A different approach to this same problem is the use of what are called *random effects* models.

First we need to consider two different modeling approaches that go by the names *marginal* and *conditional*. These are two common modeling strategies with which to incorporate correlation into a statistical model:

Marginal: Assume a model, e.g., logistic, that holds averaged over all the clusters (sometimes called population-averaged). Coefficients have the interpretation as the average change in the response (over the entire population) for a unit change in the predictor. Alternatively, we can think of the coefficient as the difference in the mean values of randomly selected subjects that differ by one unit in the predictor of interest (with all the others being the same).

Conditional: Assume a model specific to each cluster (sometimes called subject-specific). Coefficients have the interpretation as the change in the response for each cluster in the population for a unit change in the predictor. Alternatively, we can think of the coefficient as representing the change within a subject when the predictor of interest is increased by one (holding all the others constant).

In the conditional modeling approach marginal information can be obtained by averaging the relationship over all the clusters.

On the face of it these would seem to be the same. But they are not. Here is a hypothetical example. Suppose we are modeling the chance that a patient will be able to withstand a course of chemotherapy without serious adverse reactions. Patients have very different tolerances for chemotherapy, so the curves for individual subjects are quite different. Those patients with high tolerances are shifted to the right of those with low tolerances (see Fig. 8.4). The individual curves are subject-specific or conditional on each person. The population average or marginal curve is the the average of all the individual curves and is given by the solid line in Fig. 8.4 and has quite a different slope than any of the individual curves. This emphasizes that is important to keep

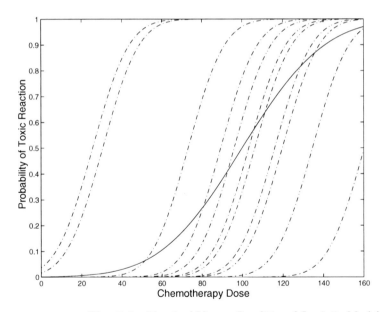

Fig. 8.4. Marginal Versus Conditional Logistic Models

straight which type of model is being used so as to be able to provide proper interpretations and comparisons.

The generalized estimating equations approach most always (always when using `xtgee`) fits a marginal model. Random effects models typically adopt the conditional approach.

Conditional models are usually specified by declaring one or more of the categorical predictors in the model to be *random factors*. (Otherwise they are called *fixed factors*.) Models with both fixed and random factors are called *mixed models*.

> Definition: If a distribution is assumed for the levels of a factor it is a *random factor*. If the values are fixed, unknown constants (to be estimated as model coefficients) it is a *fixed factor*.

The declaration of a factor to be random has several ramifications:

- Scope of inference: Inferences can be made on a statistical basis to the population from which the levels of the random factor have been selected.
- Incorporation of correlation in the model: Observations that share the same level of the random effect are being modeled as correlated.
- Accuracy of estimates: Using random factors involves making extra assumptions but gives more accurate estimates.
- Estimation method: Different estimation methods must be used.

How do we decide in practice as to which factors should be declared random versus fixed? The decision tree in Table 8.13 may be useful in deciding whether the factor is to be considered as fixed or random.

8.5.1 Re-Analysis of Birthweight and Birth Order

For the Georgia babies data set, a random effects assumption for the moms is quite reasonable. We want to regard these particular moms as a sample from a larger sample of moms. Correspondingly the moms' effects on birthweights are easily envisioned as being selected from a distribution of all possible moms.

Stata has a number of commands for conducting random effects analyses; we will focus on two of them: `xtreg` and `xtlogit`. The command structure is similar to that for `xtgee` except that we specify `mle` to do the maximum likelihood estimation for the random effects model.

The random effects model we fit is similar to that of (8.2):

$$\text{BWEIGHT}_{ij} = \text{birthweight of baby } j \text{ for mom } i$$
$$= \beta_0 + \text{MOM}_i + \beta_1 \text{BIRTHORD}_{ij} + \beta_2 \text{INITAGE}_i + \varepsilon_{ij},$$

with

$$\varepsilon_{ij} \sim \text{i.i.d } \mathcal{N}(0, \sigma_\varepsilon^2) \tag{8.8}$$
$$\text{MOM}_i \sim \text{i.i.d } \mathcal{N}(0, \sigma_u^2).$$

Table 8.13. Decision Tree for Deciding Between Fixed and Random

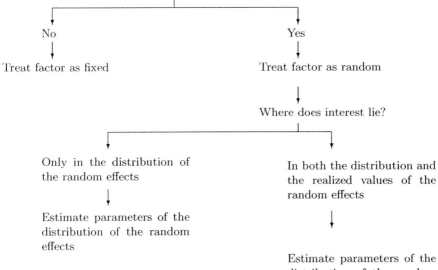

Is it reasonable to assume levels of the factor come from a probability distribution?

No → Treat factor as fixed

Yes → Treat factor as random → Where does interest lie?

Only in the distribution of the random effects ↓ Estimate parameters of the distribution of the random effects

In both the distribution and the realized values of the random effects ↓ Estimate parameters of the distribution of the random effects and calculate predictors of realized values of the random effects

Table 8.14. Random Effects Linear Regression Model for Birthweight

```
. xtreg bweight birthord initage, i(momid) mle

Random-effects ML regression          Number of obs    =    1000
Group variable (i):  momid            Number of groups =     200
Random effects u_i ~ Gaussian         Obs per group: min =      5
                                                     avg =    5.0
                                                     max =      5
                                      Wald chi2(2)     =   30.38
Log likelihood = -7659.9893           Prob > chi2      = 0.0000

------------------------------------------------------------------------------
   bweight |    Coef.   Std. Err.     z    P>|z|    [95% Conf. Interval]
-----------+------------------------------------------------------------------
  birthord |   46.608   9.944555    4.69   0.000   27.11703  66.09897
   initage | 26.73226   8.957566    2.98   0.003   9.175756  44.28877
     _cons | 2526.622   162.5441   15.54   0.000   2208.042  2845.203
-----------+------------------------------------------------------------------
  /sigma_u |  355.894    23.5171   15.13   0.000   309.8014  401.9867
  /sigma_e | 444.7446   11.11795   40.00   0.000   422.9538  466.5354
-----------+------------------------------------------------------------------
       rho | .3903754   .0349168                   .3239581  .4601633
------------------------------------------------------------------------------
Likelihood ratio test of sigma_u=0:chibar2(01)=207.81 Prob>=chibar2=0.000
```

Table 8.14 gives the analysis fitting this clustered data linear regression model. For a linear regression model, the random effects assumption is equivalent to an exchangeable correlation structure as demonstrated in (8.5). Furthermore, for linear models with identity link functions, the marginal and conditional models are equivalent. Hence the random effects analysis reproduces the analysis with an assumed exchangeable correlation structure as given in Table 8.10.

We do, however, have extra output in the random effects analysis. First, the standard deviation of the mom effects, sigma_u, is equal to 355.9, and the within-mom correlation, rho, is 0.39. The interpretation of the standard deviation of the mom effects is the standard deviation in the average birthweight across moms. Second is an estimate of the residual standard deviation of 444.7. And third, a test of whether the mom to mom variation can be considered to be zero, which can be easily rejected using a $\bar{\chi}^2$ test (given at the bottom of the Stata output and labeled chibar2, short for chi-bar-squared), which has a P-value of approximately 0.

8.5.2 Prediction

One of the advantages of the random effects approach is the ability to generate predicted values for each of the random effects, which we do not get to observe directly. For our example, this means predicted values for each of the mom effects, MOM_i.

First, let us consider how we might go about estimating the mom effect from first principles. The first mom in the data set had an initial age of 15 and hence, using the estimated coefficients from Table 8.14, has predicted values for the five births (in g) of 2,974, 3,021, 3,067, 3,114, and 3,161 – for example the first of these is $2.974 = 2,526.622 + 46.608(1) + 26.732(15)$ – and actual values of 3,720, 3,260, 3,910, 3,320, and 2,480, respectively. Her residuals, defined as actual minus predicted, were 746, 239, 843, 206, and −681, with an average of 241. So we might guess that this mom has babies that are, on average, about 241 g heavier than the "average" mom.

Using software to get the predicted effect (deviation from average) for the first mom gives 206, only about 76% of the raw data value. Calculation for the other moms shows that all the predicted values are closer to zero than the raw data predicts. Why?

Predicted values from random effects models are so-called *shrinkage estimators* because they are typically less extreme than estimates based on raw data. The shrinkage factor depends on the degree of similarity between moms and, for simple situations, is given by

$$\text{shrinkage factor} = \frac{\sigma_u^2}{\sigma_u^2 + \sigma_\varepsilon^2/n_i}, \tag{8.9}$$

where n_i is the sample size for the ith cluster. In our case this is approximately correct and that factor is equal to (taking the estimates from Table 8.14)

$$\text{shrinkage factor} = \frac{355.894^2}{355.894^2 + 444.7446^2/5}$$

$$= \frac{126660.5}{126660.5 + 39,559.6} = 0.76. \tag{8.10}$$

It is instructive to consider the form of (8.9). Since all the terms in the equation are positive, the shrinkage factor is greater than zero. Further, since the denominator is bigger than the numerator by the factor σ_ε^2/n_i, the shrinkage factor is less than 1. So it always operates to shrink the estimate from the raw data to some degree.

What is the magnitude of the shrinkage? If σ_u^2 is much larger than σ_ε^2/n_i then the shrinkage factor is close to 1, i.e., almost no shrinkage. This will occur when–

- subjects are quite different (i.e., σ_u^2 is large);
- results are very accurate and σ_ε^2 is small;
- the sample size per subject, n_i, is large.

So little shrinkage takes place when subjects are different or when answers are accurate or when there is much data.

On the other hand, in cases where subjects are similar (and hence σ_u^2 is small) there is little reason to believe that any individual person deviates from the overall. Or in cases of noisy data (σ_ε^2 large) or small sample sizes, random fluctuations can make up the majority of the raw data estimate of the effect and are naturally de-emphasized with this shrinkage approach.

The advantage of the shrinkage predictions are twofold. First, they can be shown theoretically to give more accurate predictions than those derived from the raw data. Second (which is related), they use the data to balance the subject-to-subject variability, the residual variance, and the sample size to come up with the best combination of the subject-specific information and the overall data.

Examples of uses of this prediction technology include prediction for prostate cancer screening (Brant et al., 2003) and the use of shrinkage estimators in the rating of individual physicians (Hofer et al., 1999) in treatment of diabetes.

8.5.3 Logistic Model for Low Birthweight

Turning to the binary outcome variable lowbrth we use the Stata command xtlogit. This model is similar to (8.8) with the needed changes for a logistic model for binary data. This model is:

$$\text{LOWBRTH}_{ij} = 1 \text{ if baby } j \text{ for mom } i \text{ is } < 3,000 \text{ g and } 0 \text{ otherwise}$$
$$\sim \text{Bernoulli}(p_{ij})$$

with

$$\text{logit}(p_{ij}) = \beta_0 + \text{MOM}_i + \beta_1\text{BIRTHORD}_{ij} + \beta_2\text{INITAGE}_i, \quad (8.11)$$

and

$$\text{MOM}_i \sim \text{i.i.d } \mathcal{N}(0, \sigma_u^2).$$

This analysis is given in Table 8.15 where we use the option **re** to invoke

Table 8.15. Random Effects Logistic Regression Model for Low Birthweight

```
    .        xtlogit lowbrth birthord initage, i(momid) re or nolog

Random-effects logistic regression          Number of obs      =      1000
Group variable (i): momid                    Number of groups   =       200

Random effects u_i ~ Gaussian                Obs per group: min =         5
                                                            avg =       5.0
                                                            max =         5

                                             Wald chi2(2)       =     11.96
Log likelihood  =  -588.0519                 Prob > chi2        =    0.0025

------------------------------------------------------------------------------
     lowbrth |        OR   Std. Err.      z    P>|z|     [95% Conf. Interval]
-------------+----------------------------------------------------------------
     birthord |  .8872496   .0500749    -2.12   0.034     .7943382    .9910286
      initage |  .8798436   .0406491    -2.77   0.006     .8036736    .9632328
-------------+----------------------------------------------------------------
    /lnsig2u |  .9532353   .2088377                       .543921     1.36255
-------------+----------------------------------------------------------------
     sigma_u |  1.610617   .1681788                      1.312535    1.976396
         rho |  .4408749   .0514794                       .3436825    .5428204
------------------------------------------------------------------------------
Likelihood-ratio test of rho=0: chibar2(01) =   123.25 Prob >= chibar2 = 0.000
```

the random effects analysis and the option **or** to get odds ratios. This gives somewhat different results than the GEE analysis, as expected, since it is fitting a conditional model. More specifically (as predicted from Figure 8.4) the coefficients in the conditional analysis are slightly farther from 1 than the marginal coefficients, for example the odds ratio for birth order is now 0.89 as compared to 0.92 in the marginal model. The tests are, however, virtually the same, which is not unusual.

The interpretation of the **birthord** coefficient in the conditional model is that the odds of a low birthweight baby decreases by about 11% for each increase of birth order of one for each woman.

This is opposed to the interpretation of the odds-ratio estimate from the marginal fit given in Table 8.12 of 0.92. The interpretation in the marginal model is the decrease in the odds (averaged across all women) is about 8% with an increase in birth order of one.

8.5.4 Marginal Versus Conditional Models

The previous section has demonstrated that, for nonlinear models like the logistic model, it is important to distinguish between marginal and conditional models since the model estimates are not expected to be equal. Conditional models have a more mechanistic interpretation, which can sometimes be useful (being careful, or course, to remember that many experiments do not strongly support mechanistic interpretations, no matter what model is fit). Marginal models have what is sometimes called a "public health" interpretation since the conclusions only hold averaged over the entire population of subjects.

8.6 Example: Cardiac Injury Following Brain Hemorrhage

Heart damage in patients experiencing brain hemorrhage has historically been attributed to pre-existing conditions. However, more recent evidence suggests that the hemorrhage itself can cause heart damage through the release of norepinephrine following the hemorrhage. To study this, Tung *et al.* (2004) measured cardiac troponin, an enzyme released following heart damage, at up to three occasions after patients were admitted to the hospital for a specific type of brain hemorrhage (subarachnoid hemorrhage or SAH).

The primary question was whether severity of injury from the hemorrhage was a predictor of troponin levels, as this would support the hypothesis that the SAH caused the cardiac injury. To make a more convincing argument in this observational study, we would like to show that severity of injury is an independent predictor, over and above other circulatory and clinical factors that would predispose the patient to higher troponin levels. Possible clinical predictors included age, gender, body surface area, history of coronary artery disease (CAD), and risk factors for CAD. Circulatory status was described using systolic blood pressure, history of hypertension (yes/no) and left ventricular ejection fraction (LVEF), a measure of heart function. The severity of neurological injury was graded using a subject's Hunt-Hess score on admission. This score is an ordered categorical variable ranging from 1 (little or no symptoms) to 5 (severe symptoms such as deep coma).

The study involved 175 subjects with at least one troponin measurement and between 1 and 3 visits per subject. Fig. 8.5 shows the histogram of troponin levels. They are *severely* right-skewed with over 75% of the values equal to 0.3, the smallest detectable value and many outlying values. For these reasons, the variable was dicohotomized as being above or below 1.0. Table 8.16 lists the proportion of values above 1.0 for each of the Hunt-Hess categories and Table 8.17 gives a more formal analysis using GEE methods, but including only the predictor Hunt-Hess score and not using data from visits four or greater (there were too few observations to use those for the later visits). The reference group for the Hunt-Hess variable in this analysis is a score of 1,

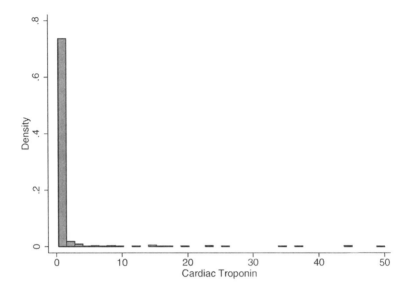

Fig. 8.5. Histogram of Cardiac Troponin Levels

Table 8.16. Proportion of Troponin Levels over 1.0 and Sample Size by Hunt-Hess Score

```
. table hunt, c(mean CTover1 n CTover1)

-------------------------------------------
Initial  |
Hunt-Hess | mean(CTover1)       N(CTover1)
----------+--------------------------------
       1 |      .0318471              157
       2 |      .0615385               65
       3 |      .1269841              126
       4 |      .1692308               65
       5 |      .6818182               22
-------------------------------------------
```

corresponding to the least injury. So the odds of heart damage, as evidenced by troponin values over 1, is over two times higher for a Hunt-Hess score of 2 as compared to 1 and the odds go up monotonically with the estimated odds of heart damage for a Hunt-Hess score of 5 being over 70 times those of a score of 1. Even though the odds ratio of a score of 5 is poorly determined, the lower limit of the 95% confidence interval is still over 16.

The primary goal is to assess the influence of a single predictor variable, Hunt-Hess score, which is measured only once per subject. Since it is only measured once, rather than repeatedly, a marginal model and the use of GEE methods is attractive. Since we are interested in a single predictor we will be more liberal in including predictors for adjustment. We certainly would like

Table 8.17. Effect of Hunt-Hess Score on Elevated Troponin Levels

```
. xi: xtgee CTo i.hunt if stday<4, i(stnum) family(binomial) ef

GEE population-averaged model              Number of obs       =        434
Group variable:                    stnum   Number of groups    =        168
Link:                              logit   Obs per group: min  =          1
Family:                         binomial                  avg  =        2.6
Correlation:                 exchangeable                  max  =          3
                                           Wald chi2(4)        =      39.03
Scale parameter:                       1   Prob > chi2         =     0.0000


------------------------------------------------------------------------------
    CTover1 | Odds Ratio   Std. Err.     z    P>|z|     [95% Conf. Interval]
------------+-----------------------------------------------------------------
   _Ihunt_2 |   2.036724    1.669731   0.87   0.386     .4084194     10.15682
   _Ihunt_3 |   4.493385    2.820396   2.39   0.017     1.313088     15.37636
   _Ihunt_4 |   6.542645    4.347658   2.83   0.005     1.778774       24.065
   _Ihunt_5 |   70.66887    52.16361   5.77   0.000     16.63111      300.286
------------------------------------------------------------------------------
```

to adjust for the amount of time after the SAH occurred, as captured by the visit number, `stday`, since troponin levels drop over time. We also want to adjust for fundamental differences that might be due to age, sex, and body surface area (`bsa`), which may be related to troponin levels.

In addition we choose to adjust for pre-existing conditions that might influence the troponin levels, including left ventricular ejection fraction (`lvef`), systolic blood pressure (`sbp`), heart rate (`hr`), and history of hypertension (`hxhtn`). Quadratic functions of left ventricular ejection fraction (`lvef2`) and systolic blood pressure (`sbp2`) are included to model nonlinear (on the logit scale) relationships.

Table 8.18 gives the output after dropping some non-statistically significant predictors from the model and using the `xtgee` command. It also gives an overall test of whether troponin levels vary with Hunt-Hess score. Even after adjustment for a multitude of characteristics, the probability of an elevated troponin level is associated with Hunt-Hess score. However, the picture is a bit different as compared to the unadjusted analysis. Each of the categories above 1 has an estimated elevated risk of troponin release, but it is not a monotonic relationship. Also, only category 5, the most severely damaged group, is statistically significantly different from category 1.

What is the effect of adjusting for the large number of predictors in this model? Table 8.19 gives the analysis after adjusting only for `stday`. We see that the pattern of estimated odds ratios as well as the standard errors are similar to the unadjusted analysis, though the unadjusted analysis might have overestimated the association with Hunt-Hess score slightly.

8.6.1 Bootstrap Confidence Intervals

We might also be concerned about the stability of the results reported in Table 8.18 given the modest-sized data set with a binary outcome and the

Table 8.18. Adjusted Effect of Hunt-Hess Score on Elevated Troponin Levels

```
. xi: xtgee CTo i.hunt i.stday sex lvef lvef2 hxhtn sbp sbp2 if stday<4, i(stnum)
> family(binomial) ef
```

```
GEE population-averaged model              Number of obs      =      408
Group variable:                     stnum  Number of groups   =      165
Link:                               logit  Obs per group: min =        1
Family:                          binomial                 avg =      2.5
Correlation:                 exchangeable                 max =        3
                                           Wald chi2(12)      =    44.06
Scale parameter:                        1  Prob > chi2        =   0.0000
```

CTover1	Odds Ratio	Std. Err.	z	P>|z|	[95% Conf. Interval]	
_Ihunt_2	1.663476	1.334533	0.63	0.526	.3452513	8.014895
_Ihunt_3	1.830886	1.211797	0.91	0.361	.5003595	6.699471
_Ihunt_4	1.560879	1.241708	0.56	0.576	.3282638	7.421908
_Ihunt_5	74.9901	69.48432	4.66	0.000	12.19826	461.0098
_Istday_2	.5258933	.2163491	-1.56	0.118	.2348112	1.177813
_Istday_3	.374303	.1753685	-2.10	0.036	.1494232	.9376232
sex	8.242845	6.418322	2.71	0.007	1.791785	37.92001
lvef	5.66e-14	5.59e-13	-3.09	0.002	2.28e-22	.000014
lvef2	1.68e+08	1.30e+09	2.45	0.014	43.94656	6.41e+14
hxhtn	3.11661	1.572135	2.25	0.024	1.15959	8.376457
sbp	1.143139	.0771871	1.98	0.048	1.001438	1.30489
sbp2	.9995246	.0002293	-2.07	0.038	.9990753	.9999742

```
. testparm _Ihunt*

 ( 1)  _Ihunt_2 = 0
 ( 2)  _Ihunt_3 = 0
 ( 3)  _Ihunt_4 = 0
 ( 4)  _Ihunt_5 = 0

           chi2(  4) =    23.87
         Prob > chi2 =    0.0001
```

Table 8.19. Effect of Hunt-Hess Score on Elevated Troponin Levels Adjusting Only for stday

```
. xi: xtgee CTo i.hunt i.stday if stday<4, i(stnum) family(binomial) ef
```

```
GEE population-averaged model              Number of obs      =      434
Group variable:                     stnum  Number of groups   =      168
Link:                               logit  Obs per group: min =        1
Family:                          binomial                 avg =      2.6
Correlation:                 exchangeable                 max =        3
                                           Wald chi2(6)       =    40.75
Scale parameter:                        1  Prob > chi2        =   0.0000
```

CTover1	Odds Ratio	Std. Err.	z	P>|z|	[95% Conf. Interval]	
_Ihunt_2	2.136339	1.711752	0.95	0.343	.4442634	10.27306
_Ihunt_3	4.312505	2.68268	2.35	0.019	1.274157	14.59609
_Ihunt_4	6.41448	4.228072	2.82	0.005	1.762367	23.34676
_Ihunt_5	60.09793	44.25148	5.56	0.000	14.19385	254.4595
_Istday_2	.5564922	.1968294	-1.66	0.098	.2782224	1.113079
_Istday_3	.5170812	.2016593	-1.69	0.091	.2407654	1.110512

large number of predictors. This is exactly a situation in which bootstrapping can help understand the reliability of standard errors and confidence intervals.

Correspondingly, we conducted a bootstrap analysis and we focus on the stability of the result for the comparison of Hunt-Hess score of 5 compared to a value of 1. Bootstrapping is conducted for the log odds (which can be transformed easily back to the odds scale) since that is the basis of the calculation of confidence intervals.

A complication with clustered data is what to resample. By default, bootstrapping will resample the individual observations. However, the basis of sampling in this example (which is common to clustered data situations) is subjects. We thus need to resample *subjects* not observations. Fortunately, this can be controlled within Stata by using a `cluster` option on the bootstrap command. Table 8.20 gives a portion of the output for two coefficients, namely, the comparison of Hunt-Hess score 5 with 1 and the comparison of

Table 8.20. Bootstrap Confidence Intervals for Adjusted Hunt-Hess Model

```
. bootstrap '"xi: xtgee CTo i.hunt i.stday sex lvef lvef2 hxhtn sbp sbp2 if stday
> <4, i(stnum) family(bin)"' _b, reps(1000) cluster(stnum)

Bootstrap statistics                      Number of obs   =     408
                                          N of clusters   =     165
                                          Replications    =    1000

-----------------------------------------------------------------------------
Variable    | Reps  Observed    Bias  Std. Err. [95% Conf. Interval]
------------+----------------------------------------------------------------

  ...

  b__Ihunt_5 |  743  4.317356  .6467686  1.36048   1.646507  6.988205   (N)
             |                                     3.108775  7.89118    (P)
             |                                     2.635067  6.01446    (BC)
  b__Istday_2 |  743 -.6426569 -.2134325 .4667512 -1.558967  .2736533   (N)
             |                                    -1.816595  .04441     (P)

  ...

-----------------------------------------------------------------------------
Note:  N  = normal
       P  = percentile
       BC = bias-corrected
```

study day 2 with 1. The bias-corrected bootstrap for the Hunt-Hess comparison gives a confidence interval ranging from 2.64 to 6.01 for the log odds, which corresponds to a confidence interval from 13.9 (which is the exponential of 2.635) to 409.3. This compares with the interval from 12.2 to 461.0 from Table 8.18 in the original analysis. The results are quite similar and give qualitatively the same results, giving us confidence in our original analysis.

8.7 Summary

The main message of this chapter has been the importance of incorporating correlation structures into the analysis of clustered, hierarchical, longitudinal and repeated measures data. Failure to do so can have serious consequences. Two main methods have been presented, generalized estimating equations and random effects models.

8.8 Further Notes and References

For those readers desiring more detailed information on longitudinal and repeated measures analyses, there are a number of book-length treatments, especially for continuous, approximately normally distributed data. Notable entries include Raudenbush and Bryk (2001), Goldstein (2003), Verbeke and Molenberghs (2000), McCulloch and Searle (2000), Diggle *et al.* (2002), and Fitzmaurice *et al.* (2004). Unfortunately many are more technical than this book.

Missing data

The techniques in this chapter handle unequal sample sizes and unequal spacing of observations in time with aplomb. However, sample sizes are often unequal and observation times unequal because of missing outcome data. And data are often missing for a reason related to the outcome under study. As examples, sicker patients may not show up for follow-up visits, leading to overly optimistic estimates based on the data present. Or those patients staying in the hospital longer may be the sicker ones (with the better-off patients having been discharged). This might lead us to the erroneous conclusion that longer stays in the hospital produce poorer outcomes, so why check in in the first place?

To a limited extent, the methods in this chapter cover the situation in which the missing data are systematically different from the data available. If the fact that data are missing is related to a factor in the model (i.e., more missing data for males, which is also a factor in the model) then there is little to worry about. However, the methods described here do *not* cover the situation where the missing data are related to predictors not in the model and can give especially misleading results if the fact the data are missing is related to the value of the outcome that would have been measured.

Not surprisingly, it is difficult to build a reliable model for observations that are missing in ways not predictable from the data on hand. At best, models which attempt to correct for such missing data (called *informative missing data* or data that are *not missing at random*) can be regarded as sensitivity analyses. There is an extensive literature on these models, codified in the book-length treatment by Little and Rubin (2002). Methods of analysis include various imputation (or filling in missing data) strategies, building subsidiary models for the reasons why the data are missing (e.g., Diggle and Kenward, 1994), and inverse weighting strategies (e.g., Preisser *et al.*, 2000).

Computing

Stata has a wide array of clustered data techniques, but they are mainly limited to one level of clustering. So, for example, they can handle repeated measures data on patients, but not repeated measures data on patients clustered within doctors. Other software packages and their extensions have additional capabilities. For continuous, approximately normally distributed data, SAS Proc MIXED can handle a multitude of models (Littell *et al.*, 1996) and SAS Proc GENMOD can fit models using generalized estimating equations and, for binary data, can fit two-level clustered binary data with a technique called alternating logistic regression (Carey *et al.*, 1993). An add-in package for Stata, called GLLAMM (Rabe-Hesketh *et al.*, 2004) extends Stata's capability to two levels and allows outcomes of disparate distributions. MLWin and HLM are two other clustered data packages with additional capabilities.

8.9 Problems

Problem 8.1. Using the fecal fat data in Table 8.1 calculate the sample variance of the subject averages. Subtract from this the residual variance estimate from Table 8.3 divided by four (why four?) to verify the estimate of σ^2_{subj} given in the text.

Problem 8.2. Using the fecal fat data in Table 8.1 verify the F-tests displayed in Tables 8.2 and 8.3.

Problem 8.3. From your own area of interest, describe a hierarchical data set including the outcome variable, predictors of interest, the hierarchical levels in the data set and the level at which each of the predictors is measured. Choose a data set for which not all of the predictors are measured at the same level of the hierarchy.

Problem 8.4. Could you successfully analyze the data from the fecal fat example using the idea of "analysis at the highest level of the hierarchy"? Briefly say why or why not.

Problem 8.5. For the fecal fat example of Table 8.1 analyze the difference between capsule and coated capsules in two ways. First use the "derived variable" approach to perform a paired t-test. Second, in the context of the two-way ANOVA of Table 8.3, test the contrast of coated vs. standard capsule. How do the two analyses compare? What differences do you note? Why do they come about? What are the advantages and disadvantages of each?

Problem 8.6. Use formula (8.5) to verify the calculation of the correlation (rho) displayed in Table 8.14.

Problem 8.7. Consider an example (like the Georgia birthweight example) with before and after measurements on a subject. If the variability of the before and after measurements each have variance σ^2 and correlation ρ, then it is a fact that the standard deviation of the difference is $\sigma\sqrt{2(1-\rho)}$.

1. The correlation of the first and last birthweights is about .381. Using Table 8.4, verify the above formula (approximately).
2. If we were to compare two groups, based on the difference scores or just the last birthweights (say, those with initial age greater than 17 versus those not), which analysis would have a larger variance and hence be less powerful? By how much?

Problem 8.8. The model corresponding to the analysis for Table 8.6 has an intercept, a dummy variable for the fifth birth, a continuous predictor of centered age (age minus the average age) and the product of the dummy variable and centered age.

1. Write down a model equation.
2. Verify that the intercept is the average for the first born, and that the coefficient for the dummy variable is the difference between the two groups, both of these when age is equal to its average.
3. Verify that the coefficient for the product measures how the change in birthweight from first to last birth depends on age.

Problem 8.9. Verify the calculation of the predicted values and residuals in Sect. 8.7.1.

Problem 8.10. Compare the bootstrap-based confidence interval for the comparison of study day 1 and study day 2 from Table 8.20 to the confidence interval from the original analysis reported in Table 8.18. Do the agree substantively? Do they lead to different conclusions?

8.10 Learning Objectives

1. Recognize a hierarchical data situation and explain the consequences of ignoring it.
2. Decide when hierarchical models are necessary versus when simpler analyses will suffice.
3. Define the terms hierarchical, repeated measures, clustered, longitudinal, robust variance estimator, working correlation structure, generalized estimating equations, fixed factor, and random factor.
4. Interpret Stata output for generalized estimating equation and random effects analyses in hierarchical analyses for linear regression or logistic regression problems.

5. Explain the difference between marginal and conditional models.
6. Decide if factors should be treated as fixed or random.
7. Explain the use of shrinkage estimators and best prediction for random factors.

9

Generalized Linear Models

A new program for depression is instituted in the hopes of reducing the number of visits to the emergency room in the year following treatment. Predictors include (among many others) treatment (yes/no), race, and drug, and alcohol usage indices. A common and minimally invasive treatment for jaundice in newborns is exposure to light. Yet the cost of this (mainly because of longer hospital stays) was estimated as long ago as 1984 at over $600 per infant. Predictors of the cost include race, gestational age, and birthweight.

These analyses require special attention both because of the nature of the outcome variable (counts in the depression example and costs, which are positive and right-skewed, for the jaundice example) and because the models we would typically employ are not the straightforward linear models of Chapter 4.

On the other hand, many features of constructing an analysis are the same as we have seen previously. We have a mixture of categorical (treatment, race) and continuous predictors (drug usage, alcohol usage, gestational age, birthweight). There are the same issues of determining the goals of inference (prediction, risk estimation, and testing of specific parameters) and winnowing down of predictors to arrive at a final model as we discussed in Chapter 5. And we can use tests and confidence intervals in ways that are quite similar to those for previously described analyses.

We begin this chapter by discussing the two examples in a bit more detail and conclude with a look at how those examples, as well as a number of earlier ones, can be subsumed under the broader rubric of *generalized linear models*.

9.1 Example: Treatment for Depression

A new case-management program for depression is instituted in a local hospital that often has to care for the poor and homeless. A characteristic of this population is that they often access the health care system by arriving in the emergency room – an expensive and overburdened avenue to receive treatment. Can the new treatment reduce the number of needed visits to the emergency room as compared to standard care? The recorded outcome variable is the number of emergency room visits in the year following treatment.

The primary goal of the analysis is to assess the treatment program, but emergency room usage varies greatly according to whether the subjects are drug or alcohol users. A secondary goal is to assess racial differences in usage of the emergency room and impact of the new treatment.

9.1.1 Statistical Issues

From a statistical perspective, we need to be concerned with the nature of the outcome variable: in the data set that motivated this example, about one-third of the observations are 0 (did not return to the emergency room within the year) and over half are either 0 or 1. This is highly non-normal and cannot be transformed to be approximately normal – any transformation by an increasing function will merely move the one-third of the observations that are exactly 0 to another numerical value, but there will still be a "lump" of observations at that point consisting of one-third of the data. For example, a commonly recommended transformation for count data with zeros is $\log(y+1)$. This transformation leaves the data equal to 0 unchanged since $\log(0 + 1) = 0$ and moves the observations at 1 to $\log(1 + 1) = \log(2)$, not appreciably reducing the non-normality of the data. Over half the data take on the two values 0 and $\log(2)$.

Even if we can handle the non-normal distribution, a typical linear model (as in Chap. 4) for the mean number of emergency room visits will be untenable. The mean number of visits must be a positive number and a linear model, especially with continuous predictors, may, for extreme values of the covariates, predict negative values. This is the same problem we encountered with models for the probability of an event in Sect. 6.3.

Another bothersome aspect of the analysis is that this is a hard-to-follow, transient population in generally poor health. It is not at all unusual to have subjects die or be unable to be contacted for obtaining follow-up information. So some subjects are only under observation (and hence eligible for showing up for emergency room visits) for part of the year.

Since not all the subjects are followed for the same periods of time, it is natural to think of a multiplicative model. In other words, if all else is equal, a subject that is followed for twice as long as another subject will have, on average, twice the emergency room utilization. This consideration, as well as the desire to keep the mean response positive, leads us to consider a model for the log of the mean response. Note that this is different from the mean of the log-transformed responses (See Problem 9.1, also Sects. 4.6.6 and 4.7.5).

9.1.2 Model for the Mean Response

To begin to write down the model more carefully, define Y_i as the number of emergency room visits for patient i and let $E[Y_i]$ represent the average number of visits for a year. For the moment we will ignore the fact that the observation periods are unequal. The model we are suggesting is

$$\log E[Y_i] = \beta_0 + \beta_1 \text{RACE}_i + \beta_2 \text{TRT}_i + \beta_3 \text{ALCH}_i + \beta_4 \text{DRUG}_i, \qquad (9.1)$$

or equivalently (using an exponential, i.e., anti-log)

$$E[Y_i] = \exp\{\beta_0 + \beta_1 \text{RACE}_i + \beta_2 \text{TRT}_i + \beta_3 \text{ALCH}_i + \beta_4 \text{DRUG}_i\}, \qquad (9.2)$$

Where β_0 is an intercept, RACE_i is 1 for non-whites and 0 for whites, TRT_i is 1 for those in the treatment group and 0 for usual care, ALCH_i is a numerical measure of alcohol usage and DRUG_i is a numerical measure of drug usage. We are primarily interested in β_2, the treatment effect.

Since the mean value is not likely to be exactly zero (otherwise there is nothing to model), using the log function is mathematically acceptable (as opposed to trying to log transform the original counts, many of which are zero). Also, we can now reasonably hypothesize models like (9.1) that are linear (for the log of the mean) in ALCH_i and DRUG_i since the exponential in (9.2) keeps the mean value positive.

This is a model for the number of emergency room visits per year. What if the subject is only followed for half a year? We would expect their counts to be, on average, only half as large. A simple way around this problem is to model the mean count per unit time instead of the mean count, irrespective of the observation time. Let t_i denote the observation time for the ith patient. Then the mean count per unit time is $E[Y_i]/t_i$ and (9.1) can be modified to be

$$\log(E[Y_i]/t_i) = \beta_0 + \beta_1 \text{RACE}_i + \beta_2 \text{TRT}_i + \beta_3 \text{ALCH}_i + \beta_4 \text{DRUG}_i, \qquad (9.3)$$

or equivalently (using the fact that $\log[Y/t] = \log Y - \log t$)

$$\log E[Y_i] = \beta_0 + \beta_1 \text{RACE}_i + \beta_2 \text{TRT}_i + \beta_3 \text{ALCH}_i + \beta_4 \text{DRUG}_i + \log t_i. \quad (9.4)$$

The term $\log t_i$ on the right-hand side of (9.4) looks like another covariate term, but with an important exception: there is no coefficient to estimate analogous to the β_3 or β_4 for the alcohol and drug covariates. Thinking computationally, if we used it as a predictor in a regression-type model, a statistical program like Stata would automatically estimate a coefficient for it. But, by construction, we know it must enter the equation for the mean with a coefficient of exactly 1. For this reason it is called an *offset* instead of a covariate and when we use a package like Stata, it is designated as an offset and not a predictor.

9.1.3 Choice of Distribution

Lastly, we turn to the non-normality of the distribution. Typically we describe count data using the Poisson distribution. Directly modeling the data with a distribution appropriate for counts recognizes the problems with discreteness of the outcomes (e.g, the "lump" of zeros). While the Poisson distribution is hardly ever ultimately the correct distribution to use in practice, it gives us a place to start.

We are now ready to specify a model for the data, accommodating the three issues: non-normality of the data, mean required to be positive, and unequal observation times. We start with the distribution of the data. Let λ_i denote the mean rate of emergency room visits per unit time, so that the mean number of visits for the ith patient is given by $\lambda_i t_i$. We then assume that Y_i has a Poisson distribution with log of the mean given by

$$
\begin{aligned}
\log E[Y_i] &= \log[\lambda_i t_i] \\
&= \log \lambda_i + \log t_i \\
&= \beta_0 + \beta_1 \mathrm{RACE}_i + \beta_2 \mathrm{TRT}_i + \beta_3 \mathrm{ALCH}_i + \beta_4 \mathrm{DRUG}_i + \log t_i.
\end{aligned} \tag{9.5}
$$

This shows us that the main part of the model (consisting of all the terms except for the offset $\log t_i$) is modeling the rate of emergency room visits per unit time:

$$
\log[\lambda_i] = \beta_0 + \beta_1 \mathrm{RACE}_i + \beta_2 \mathrm{TRT}_i + \beta_3 \mathrm{ALCH}_i + \beta_4 \mathrm{DRUG}_i, \tag{9.6}
$$

or, exponentiating both sides,

$$
\lambda_i = \exp\{\beta_0 + \beta_1 \mathrm{RACE}_i + \beta_2 \mathrm{TRT}_i + \beta_3 \mathrm{ALCH}_i + \beta_4 \mathrm{DRUG}_i\}. \tag{9.7}
$$

9.1.4 Interpreting the Parameters

The model in (9.7) is a multiplicative one, as we saw for the Cox model, and has a similar style of interpretation. Recall that RACE_i is 1 for non-whites and 0 for whites and suppose the race coefficient is estimated to be $\hat{\beta}_1 = -0.5$. The mean rate per unit time for a white person divided by that of a non-white (assuming treatment group, and alcohol and drug usage indices are all the same) would be

$$
\begin{aligned}
&\frac{\exp\{\beta_0 + 0 + \beta_2 \mathrm{TRT} + \beta_3 \mathrm{ALCH} + \beta_4 \mathrm{DRUG}\}}{\exp\{\beta_0 - 0.5 + \beta_2 \mathrm{TRT} + \beta_3 \mathrm{ALCH} + \beta_4 \mathrm{DRUG}\}} \\
&= \frac{e^{\beta_0} e^0 e^{\beta_2 \mathrm{TRT}} e^{\beta_3 \mathrm{ALCH}} e^{\beta_4 \mathrm{DRUG}}}{e^{\beta_0} e^{-0.5} e^{\beta_2 \mathrm{TRT}} e^{\beta_3 \mathrm{ALCH}} e^{\beta_4 \mathrm{DRUG}}} \\
&= \frac{e^0}{e^{-0.5}} \\
&= e^{0.5} \approx 1.65.
\end{aligned} \tag{9.8}
$$

So the interpretation is that, after adjustment for treatment group and alcohol and drug usage, whites tend to use the emergency room at a rate 1.65 that of the non-whites. Said another way, the average rate of usage for whites is 65%

higher than that for non-whites. Similar, multiplicative, interpretations apply to the other coefficients.

In summary, to interpret the coefficients when modeling the log of the mean, we need to exponentiate them and interpret them in a multiplicative or ratio fashion. In fact, it is often good to think ahead to the desired type of interpretation. Proportional increases in the mean response due to covariate effects are sometimes the most natural interpretation and are easily incorporated by planning to use such a model.

9.1.5 Further Notes

Models like the one developed in this section are often called Poisson regression models, named after the distribution assumed for the counts. A feature of the Poisson distribution is that the mean and variance are required to be the same. So, if the mean number of emergency room visits per year is 1.5, for subjects with a particular pattern of covariates, then the variance would also be 1.5 and the standard deviation would be the square root of that or about 1.23 visits per year. Ironically, the Poisson distribution often fails to hold in practice since the variability in the data often exceeds that of the mean. A common solution (where appropriate) is to assume that the variance is proportional to the mean, not exactly equal to it, and estimate the proportionality factor, which is called the *scale parameter*, from the data. For example, a scale parameter of 2.5 would mean that the variance was 2.5 times larger than the mean and this fact would be used in calculating standard errors, hypothesis tests, and confidence intervals. When the scale parameter is greater than 1, meaning that the variance is larger than that assumed by the named distribution, the data are termed *overdispersed*. Another solution is to choose a different distribution. For example, the Stata package has a negative binomial (a different count data distribution) regression routine, in which the variance is modeled as a quadratic function of the mean.

The use of log time as an offset in model (9.5) may seem awkward. Why not just divide each count by the observation period and analyze Y_i/t_i? The answer is that it makes it harder to think about and specify the proper distribution. Instead of having count data, for which there are a number of statistical distributions to choose from, we would have a strange, hybrid distribution, with "fractional" counts, e.g., with an observation period of 0.8 of a year, we would could obtain values of $0, 1.25$ (which is 1 divided by 0.8), $2.5, 3.75$, etc. With a different observation period, a different set of values would be possible.

9.2 Example: Costs of Phototherapy

About 60% of newborns become jaundiced, i.e., the skin and whites of the eyes turn yellow in the first few days after birth. Newborns become jaundiced because they have an increase in bilirubin production due to increased red blood cell turnover and because it takes a few days for their liver (which helps eliminate bilirubin) to mature. Newborns are treated for jaundice because of the possibility of bilirubin-induced neurological damage. What are the costs associated with this treatment and are costs also associated with race, the gestational age of the baby, and the birthweight of the baby?

Our outcome will be the total cost of health care for the baby during its first month of life. Cost is a positive variable and is almost invariably highly skewed to the right. A common remedy is to log transform the costs and then fit a multiple regression model. This is often highly successful as log costs are often well-behaved statistically, i.e., approximately normally distributed and homoscedastic. This is adequate if the main goal is to test whether one or more risk factors are related to cost.

However, if the goal is to understand the determinants of the actual *cost* of health care, then it is only the mean cost that is of interest (since mean cost times the number of newborns is the total cost to the health care system). One strategy is to perform the analysis on the log scale and then back transform (using an exponential) to get things back on the original cost scale.

However, since the log of the mean is not the same as the mean of the log, back-transforming an analysis on the log scale does not directly give results interpretable in terms of mean costs. Instead they are interpretable as models for median cost (Goldberger, 1968). The reasoning behind this is as follows. If the log costs are approximately normally distributed, then the mean and median are the same. Since monotonic transformations preserve medians (the log of the median value *is* the median of the log values) back-transforming using exponentials gives a model for median cost. There are methods for getting estimates of the mean via adjustments to the back transformation (Bradu and Mundlak, 1970) but there are also alternatives.

One alternative is to adopt the approach of the previous section: model the mean and assume a reasonable distribution for the data. What choices would we need to make for this situation?

A reasonable starting point is to observe that the mean cost must be positive. Additive and linear models for positive quantities can cause the problem of negative predicted values and hence multiplicative models incorporating proportional changes are commonly used. For cost, this is often a more natural characterization, i.e., "low birthweight babies cost 50% more than normal birthweight babies" and is likely to be more stable than modeling absolute changes in cost (locations with very different costs of care are unlikely to have the same differences in costs, but may have the same ratio of costs). As in the previous section, that would lead to a model for the log of the mean cost (similar to but not the same as log-transforming cost).

9.2.1 Model for the Mean Response

More precisely, let us define Y_i as the cost of health care for infant i during its first month and let $E[Y_i]$ represent the average cost. Our model would then be

$$\log E[Y_i] = \beta_0 + \beta_1 \text{RACE}_i + \beta_2 \text{TRT}_i + \beta_3 \text{GA}_i + \beta_4 \text{BW}_i, \qquad (9.9)$$

or equivalently (using an exponential)

$$E[Y_i] = \exp\{\beta_0 + \beta_1 \text{RACE}_i + \beta_2 \text{TRT}_i + \beta_3 \text{GA}_i + \beta_4 \text{BW}_i\}, \qquad (9.10)$$

where β_0 is an intercept, RACE_i is 0 for whites and 1 for non-whites, TRT_i is 1 for those receiving phototherapy and 0 for those who do not, GA_i is the gestational age of the baby, and BW_i is its birthweight. We are primarily interested in β_2, the phototherapy effect.

9.2.2 Choice of Distribution

The model for the mean for the jaundice example is virtually identical to that for the depression example in Sect. 9.2. But the distributions need to be different since cost is a continuous variable, while number of emergency room visits is discrete. There is no easy way to know what distribution might be a good approximation for such a situation, without having the data in hand. However, it is often the case that the standard deviation in the data increases proportionally with the mean. This situation can be diagnosed by looking at residual plots (as described in Chap. 4) or by plotting the standard deviations calculated within subgroups of the data versus the means for those subgroups. In such a case, a reasonable choice is the gamma distribution, which is a flexible distribution for positive, continuous variables that incorporates the assumption that the standard deviation is proportional to the mean.

When we are willing to use a gamma distribution as a good approximation to the distribution of the data, we can complete the specification of the model as follows. We assume that Y_i has a gamma distribution with mean, $E[y_i]$, given by

$$\log E[Y_i] = \beta_0 + \beta_1 \text{RACE}_i + \beta_2 \text{TRT}_i + \beta_3 \text{GA}_i + \beta_4 \text{BW}_i. \qquad (9.11)$$

9.2.3 Interpreting the Parameters

Since the model is a model for the log of the mean, the parameters have the same interpretation as in the previous section. For example if $\hat{\beta}_2 = 0.5$ (positive since phototherapy increases costs) then the interpretation would be that, adjusting for race, gestational age and birthweight, the cost associated with babies receiving phototherapy was $\exp(0.5) \approx 1.65$ as high as those not receiving it.

9.3 Generalized Linear Models

The examples in Sects. 9.2 and 9.3 have been constructed to emphasize the similarity of the models (compare subsections 9.1.4 and 9.2.3) for two very different situations. So even with very different distributions (Poisson versus gamma) and different statistical analyses, they have much in common.

A number of statistical packages, including Stata, have what are called *generalized linear model* commands that are capable of fitting linear, logistic, Poisson regression and other models. The basic idea is to let the data analyst tailor the analysis to the data rather than having to transform or otherwise manipulate the data to fit an analysis. This has significant advantages in situations like the phototherapy cost example where we want to model the outcome without transformation.

Fitting a generalized linear model involves making a number of decisions:

1. What is the distribution of the data (for a fixed pattern of covariates)?
2. What function will be used to *link* the mean of the data to the predictors?
3. Which predictors should be included in the model?

In the examples in the preceding sections we used Poisson and gamma distributions, we used a log function of the mean to give us a linear model in the predictors and our choice of predictors was motivated by the subject matter. Note that choices on the predictor side of the equation are largely independent of the first two choices.

In previous chapters, we have covered linear and logistic regression. In linear regression we modeled the mean directly and assumed a normal distribution. This is using an *identity link function*, i.e., we modeled the mean identically, without transforming it. In logistic regression, we modeled the log of the odds, i.e., $\log(p/[1-p])$, and assumed a binomial or binary outcome. If the outcome is coded as zero for failure and one for success, then the average of the zeros and ones is p, the probability of success. In that case we used a *logit link* to link the mean, p, to the predictors.

Generalized linear model commands give large degrees of flexibility in the choice of each of the features of the model. For example, current capabilities in Stata are to handle six distributions (normal, binomial, Poisson, gamma, negative binomial, and inverse gaussian), and ten link functions (including identity, log, logit, probit, power functions)

9.3.1 Example: Risky Drug Use Behavior

Here is an example of modeling risky drug use behavior (sharing syringes) among drug users. The outcome is the number of times the drug user shared a syringe (shsyr) in the past month (values ranged from 0 to 60!) and we will consider a single predictor, whether or not the drug user was homeless. Table 9.1 gives the results assuming a Poisson distribution. The Stata command, glm, specifies a Poisson distribution and a log link. The output contains a number of standard elements, including estimated coefficients, standard errors, Z-tests, P-values, and confidence intervals. The homeless coefficient is highly statistically significant, with a value of about 0.605, meaning that being homeless is associated with $\exp(0.605) \approx 1.83$ times more use of shared syringes.

Table 9.1. Count Regression Example Assuming a Poisson Distribution

```
. xi: glm shsyr i.homeless, family(poisson) link(log)
i.homeless        _Ihomeless_0-1      (naturally coded; _Ihomeless_0 omitted)

Iteration 0:   log likelihood = -305.54178
Iteration 1:   log likelihood = -297.95538
Iteration 2:   log likelihood =  -297.9521
Iteration 3:   log likelihood =  -297.9521

Generalized linear models                No. of obs      =        27
Optimization       : ML: Newton-Raphson  Residual df     =        25
                                         Scale param     =         1
Deviance        =  496.8993451           (1/df) Deviance = 19.87597
Pearson         =  599.1655782           (1/df) Pearson  = 23.96662

Variance function: V(u) = u              [Poisson]
Link function    : g(u) = ln(u)          [Log]
Standard errors  : OIM

Log likelihood   = -297.9520971          AIC             =  22.21867
BIC              =  414.5034234

------------------------------------------------------------------------
      shsyr |     Coef.   Std. Err.     z    P>|z|    [95% Conf. Interval]
------------+-----------------------------------------------------------
_Ihomeless_1 |   .6047529   .1218444   4.96   0.000    .3659422   .8435636
      _cons |   2.186051   .1059998  20.62   0.000    1.978296   2.393807
------------------------------------------------------------------------
```

However, these data are highly variable and the Poisson assumption of equal mean and variance is dubious. If we specify the scale(x2) option, which estimates the scale parameter using the Pearson residuals, then the standard errors are increased by the square root of 23.9662, or about 4.9 times. In the terminology of generalized linear models, these data are highly overdispersed, because the variance is much larger than that assumed for a Poisson distribution. Table 9.2 gives the results with the scaled standard errors, which are no longer statistically significant.

This example serves as a warning not to make strong assumptions, such as those embodied in using a Poisson distribution, blindly. It is wise at least to make a sensitivity check by estimating the scale parameter for count data as well as for binomial data with denominators other than 1 (with binary data, with a denominator of 1, no overdispersion is possible). Also, when there are just a few covariate patterns and subjects can be grouped according to their covariate values, it is wise to plot the variance within such groups versus the mean within the group to display the variance to mean relationship graphically.

Table 9.2. Count Regression Example With Scaled Standard Errors

```
. xi: glm shsyr i.homeless, family(poisson) link(log) scale(x2)
i.homeless          _Ihomeless_0-1      (naturally coded; _Ihomeless_0 omitted)

Iteration 0:   log likelihood = -305.54178
Iteration 1:   log likelihood = -297.95538
Iteration 2:   log likelihood =  -297.9521
Iteration 3:   log likelihood =  -297.9521

Generalized linear models                      No. of obs      =        27
Optimization       : ML: Newton-Raphson        Residual df     =        25
                                               Scale param     =         1
Deviance         =  496.8993451                (1/df) Deviance =  19.87597
Pearson          =  599.1655782                (1/df) Pearson  =  23.96662

Variance function: V(u) = u                    [Poisson]
Link function    : g(u) = ln(u)                [Log]
Standard errors  : OIM

Log likelihood   = -297.9520971                AIC             =  22.21867
BIC              =  414.5034234

------------------------------------------------------------------------------
      shsyr |      Coef.   Std. Err.      z    P>|z|     [95% Conf. Interval]
------------+-----------------------------------------------------------------
_Ihomeless_1 |   .6047529   .5964981     1.01   0.311    -.5643619    1.773868
      _cons |   2.186051   .5189296     4.21   0.000     1.168968    3.203135
------------------------------------------------------------------------------
```

(Standard errors scaled using square root of Pearson X2-based dispersion)

9.3.2 Relationship of Mean to Variance

The key to use of a generalized linear model program is the specification of the relationship of the mean to the variance. This is the main information used by the program to fit a model to data when a distribution is specified. As noted above, this relationship can often be assessed by residual plots or plots of subgroup standard deviations versus means. Table 9.3 gives the assumed variance to mean relationship, distributional name, and situations in which the common choices available in Stata would be used.

9.3.3 Nonlinear Models

Not every model fits under the generalized linear model umbrella. Use of the method depends on finding a transformation of the mean for which the predictors enter as a linear model, which may not always be possible. For example, a common model in drug pharmacokinetics is to model the mean concentration of the drug in blood, Y, as a function of time, t, using the following model:

$$E[Y] = \mu_1 \exp\{-\lambda_1 t\} + \mu_2 \exp\{-\lambda_2 t\}. \tag{9.12}$$

In addition to time, we might have other predictors such as drug dosage or gender of the subject. However, there is no transformation that will form a

Table 9.3. Common Distributional Choices for Generalized Linear Models in Stata

Distribution	Variance to Mean [a]	Sample situation
Normal	Constant σ^2	Linear regression
Binomial	$\sigma^2 = n\mu(1-\mu)$	Successes out of n trials
OD[b] Binomial	$\sigma^2 \propto n\mu(1-\mu)$	Clustered success data
Poisson	$\sigma^2 = \mu$	Count data, variance equals mean
OD Poisson	$\sigma^2 \propto \mu$	Count data, variance proportional to mean
Negative binomial	$\sigma^2 = \mu + \mu^2/k$	Count data, variance quadratic in the mean
Gamma	$\sigma \propto \mu$	Continuous data, standard deviation proportional to mean

[a]Mean is denoted by μ and the variance by σ^2
[b]Over-dispersed

linear predictor, even without the inclusion of dose and gender effects, and so a generalized linear model is not possible.

9.4 Summary

The purpose of this chapter has been to outline the topic of generalized linear models, a class of models capable of handling a wide variety of analysis situations. Specification of the generalized linear model involves making three choices:

1. What is the distribution of the data (for a fixed pattern of covariates)? This must be specified at least up to the the variance to mean relationship.
2. What function will be used to link the mean of the data to the predictors?
3. Which predictors should be included in the model?

Generalized linear models are similar to linear, logistic, and Cox models in that much of the work in specifying and assessing the predictor side of the equation is the same no matter what distribution or link function is chosen. This can be especially helpful when analyzing a study with a variety of different outcomes, but similar questions as to what determines those outcomes. For example, in the depression example we might also be interested in cost, with a virtually identical model and set of predictors.

9.5 Further Notes and References

There are a number of book-length treatments of generalized linear models, including Dobson (2001) and McCullagh and Nelder (1989). In Chapter 8 we extended the logistic model to accommodate correlated data by the use of generalized estimating equations and by including random effects. The generalized linear models described in this chapter can similarly be extended and the `xtgee` command in Stata and GENMOD procedure in SAS can be used with a variety of distributions for generalized estimating equations fits. Random effects models can be estimated for a number distributions using the cross-sectional time-series commands in Stata (these commands are prefixed by `xt`) and with the NLMIXED procedure in SAS.

9.6 Problems

Problem 9.1. We made the point in Sect. 9.2 that a log transformation would not alleviate non-normality. Yet we model the log of the mean response. Let's consider the differences.

1. First consider the small data set consisting of 0, 1, 0, 3, 1. What is the mean? What is the log of the mean? What is the mean of the logs of each data point?
2. Even if there are no zeros, these two operations are quite different. Consider the small data set consisting of 2, 3, 32, 7, 11. What is the log of the mean? What is the mean of the logs of the data? Why are they different?
3. Repeat the above calculation, but using medians.

Problem 9.2. What would you need to add to model (9.5) to assess whether the effect of the treatment was different in whites as compared to non-whites?

Problem 9.3. Suppose the coefficient for $\hat{\beta}_2$ in (9.6) was -0.2. Provide an interpretation of the treatment effect.

Problem 9.4. For each of the following scenarios, describe the distribution of the outcome variable (Is it discrete or approximately continuous? Is it symmetric or skewed? Is it count data?) and which distribution(s) might be a logical choice for a generalized linear model.

1. A treatment program is tested for reducing drug use among the homeless. The outcome is injection drug use frequency in the past 90 days. The values range from 0 to 900 with an average of 120, a median of 90, and a standard deviation of 120. Predictors include treatment program, race (white/non-white), and sex.
2. In a study of detection of abnormal heart sounds the values of brain natriurtic peptide (BNP) in the plasma are measured. The outcome, BNP, is sometimes used as a means of identifying patients who are likely to have signs and symptoms of heart failure. The BNP values ranged from from 5 to 4,000 with an average

of 450, a median of 150, and a standard deviation of 900. Predictors include whether an abnormal heart sound is heard, race (white/non-white), and sex.

3. A clinical trial was conducted at four clinical centers to see if alendronate (a bone-strengthening medication) could prevent vertebral fractures in elderly women. The outcome is total number of vertebral fractures over the follow-up period (intended to be 5 years for each woman). Predictors include drug versus placebo, clinical center, and whether the woman had a previous fracture when enrolled in the study.

Problem 9.5. For each of the scenarios outlined in Problem 9.4, write down a preliminary model by specifying the assumed distribution, the link function, and how the predictors are assumed to be related to the mean.

9.7 Learning Objectives

1. State the advantage of using a generalized linear models approach.
2. Given an example, make reasonable choices for distributions and link functions.
3. Given output from a generalized linear models routine, state whether predictors are statistically significant and provide an interpretation of their estimated coefficients.

10

Complex Surveys

Suppose we wanted to estimate the prevalence of diabetes among adults in the U.S., as well as the effects of diabetes risk factors in this broad target population, both with minimum bias – that is, in such a way that the estimates were truly representative of the target population. Observational cohorts that might be used for these purposes are usually convenience samples, and are often selected from subsets of the population at elevated risk. This would make it difficult to generalize sample diabetes prevalence to the broader target population. We might be more comfortable assuming that sample associations between risk factors and diabetes were valid for the broader population, but the assumption would be hard to check (Problem 10.1).

Observational studies as well as randomized trials use convenience samples for compelling reasons, among them reducing cost and optimizing internal validity. But when unbiased representation of a well-defined target population is of paramount importance, special methods for obtaining and analyzing the sample must be used. Crucial features of such a study are that *all members of the target population must have some chance of being selected for the sample*, and that *the probability of inclusion can be defined for each element of the sample.* Using data from a sample which meets these two criteria, we could in principle compute unbiased estimates of the number and percent prevalence of diabetes cases in the U.S. adult population, as well as of the effects of measured diabetes risk factors.

Studies implemented by the National Center for Health Statistics (NCHS), including the National Health and Nutrition Examination Survey (NHANES), the National Hospital Discharge Survey (NHDS), and the National Ambulatory Medical Care Survey (NAMCS), are prominent examples of surveys that meet these criteria.

However, obtaining representative samples, even from a local population of interest, as in the San Francisco Men's Health Study (Winkelstein *et al.*, 1987), is a difficult and expensive undertaking. To reduce costs, a *complex sampling design* is often used. Essentially this means initially sampling clusters, known as primary sampling units (PSUs), rather than individuals; only at some later stage are individual study participants selected. This is in contrast to a simple random sample (SRS), in which individuals are directly and independently sampled.

From Chapter 8, it should be clear that the initial sampling of clusters may affect precision, because outcomes for the observations within a cluster are positively correlated in most cases. The change in precision means that for many purposes a larger sample will be required to achieve a given level of statistical certainty. Nonetheless, the complex survey design is cost-effective, because cluster sampling can be implemented in concentrated geographic areas, rather than having to cover the entire area where the target population is found. Moreover, some of the information required to define probability of inclusion need only be obtained for the selected clusters. Especially for nationally representative samples, the savings can be considerable.

In *multi-stage* designs, there may be several levels of cluster sampling; for example, counties may initially be sampled, and then census tracts within counties, city blocks with census tracts, and households within blocks. Only at the final stage are individual study participants sampled within households. The rationale is again to reduce costs by making the survey easier to implement.

An additional feature of many complex surveys is that clusters may be selected from within mutually exclusive and exhaustive *strata*, usually geographic, which cover the entire target population. To the extent that subsets of the target populations are more similar within than across strata, the result is increased precision.

Another feature of many complex surveys is *unequal probability of inclusion*. In some cases, subgroups of special interest may be oversampled: that is, they are sampled at higher rates, so that they comprise a larger proportion of the sample than they do of the target population. The rationale is to ensure adequate precision of estimates both within the subgroup and in contrasting the subgroup to other parts of the larger population, by increasing their numbers in the sample.

As a result of their design, complex surveys can provide almost unbiased and often very precise estimates of the parameters of a target population. However, to obtain these estimates and compute valid standard errors, confidence intervals, and P-values, such surveys have to be analyzed using methods that take account of the special features of the design. In particular, the analysis must account for

- stratification
- cluster sampling
- probability of inclusion.

Fortunately a number of software packages make it straightforward to carry out descriptive as well as multipredictor regression analyses using complex survey data. These packages include

- Stata (Stata Corp., College Station, TX; www.stata.com),

- SUDAAN (Research Triangle Institute, Research Triangle Park, NC; www.rti.org),
- SAS (SAS Institute, Cary, NC; www.sas.com),
- WESVAR (Westat, Inc., Rockville MD; www.westat.com).

In the following sections we give an overview of how these packages account for the special features of a complex design.

10.1 Example: NHANES

The National Health and Nutrition Examination Survey (NHANES) is a series of complex, multi-stage probability samples representative of the civilian, non-institutionalized U.S. population. Interviews and physical exams are used to ascertain a wide range of demographic, risk-factor, laboratory, and disease outcome variables. In NHANES III, conducted between 1988 and 1994, the PSUs were primarily counties. Thirteen large PSUs were selected with certainty, and the remaining 68 were selected with probability proportional to PSU population size, two from each of 34 geographic strata. At the second stage of cluster sampling in NHANES III, area segments, often composed of city or suburban blocks, were selected. In the first half of the survey, special segments were defined for new housing built since the 1980 census, so that no portion of the target population would be systematically excluded; in the second half, more recent information from the 1990 census made this unnecessary. The third stage of sampling was households, which were carefully enumerated within the area segments. At the fourth and final stage, survey participants were selected from within households.

At each stage sampling rates were controlled so that the probability of inclusion for each participant could be precisely estimated. Children and people over 65 as well as African Americans and Mexican Americans were oversampled. Almost 34,000 people were interviewed and of these roughly 31,000 participated in the physical exam. NHANES data are available from the NCHS website http://www.cdc.gov/nchs and can be properly analyzed using any of the four major software packages with routines for complex surveys. Data from NHANES III have been used in many epidemiologic and clinical investigations.

10.2 Probability Weights

We pointed out that in selecting a representative sample, every member of the target population has to have some chance of being selected for the sample. To put it another way, no part of the target population can be systematically excluded. In addition, we said that for every element of the sample, the probability of having been selected must be known. Essentially this is what is meant by a "probability sample." Analysis of such samples makes use of information about probability of inclusion to produce unbiased estimates of the parameters of the target population.

To see how this works, consider a simple random sample of size 100, drawn at random from a target population of size 100,000. In this simple case, each member of the sample had a one-in-a-thousand chance of being included in the sample. We would say that the *sampling fraction*, another term for the probability of inclusion, was 0.001 for this sample, and constant across observations. Furthermore, we could think of each member of the sample as "representing" 1,000 members of the target population.

If we wanted to estimate the percent prevalence of diabetes in the target population, the proportion with diabetes in the sample would work fine in this case, for reasons that we explain below. Likewise the average age of the sample would be an unbiased estimate of mean age in the population. But consider the more interesting case of estimating the *number* of diabetics in the population. Suppose there were five diabetics in the sample. Since each represents 1,000 members of the target population, an unbiased (though obviously noisy) estimate of the population number of diabetics would be 5,000.

Essentially what we have done is to compute a "weighted" sum of the number of the diabetics in the sample, where each gets weight 1,000, or the number in the population that each sample participant represents. Formally, the weight is the reciprocal of the sampling fraction of 0.001. Note that the overall sum of these sample "probability" weights equals the population size.

> Definition: *Probability weights* are the reciprocal of the probability of inclusion, and are intepretable as the number of elements in the target population which each sampled observation "represents."

Now consider the more typical case where the probability of inclusion varies across participants. To make this concrete, suppose that women and men each number 100,000 in the target population, but that the sample includes 100 women and 200 men, for sampling fractions of 0.001 and 0.002, respectively. In this sample each woman represents 1,000 women in the population, but each man represents only 500 men.

In this case, to estimate means for the whole target population, we would need to use *weighted* sample averages. These would no longer equal their unweighted counterparts, in which women would be under-represented. The formula for the weighted average is

$$\hat{E}_w[Y] = \frac{\sum_i w_i y_i}{\sum_i w_i}, \tag{10.1}$$

where $\hat{E}_w[Y]$ denotes the weighted average of the outcome variable Y, y_i is the value of Y for participant i, and w_i is the corresponding probability weight. You can demonstrate for yourself that if all the weights are equal ($w_i \equiv w$), then the weighted average reduces to the usual sample average $\sum_i y_i/n$ (Problem 10.2).

Furthermore, if Y were a binary indicator variable coded 1 = diabetic and 0 = non-diabetic, then (10.1) also holds for estimating the population proportion with diabetes. As we pointed in Sect. 4.3, this equivalence between averages and proportions only holds with the 0-1 indicator coding of Y.

In addition, with this coding of Y, the weighted estimate of the total number in the population with diabetes is simply $\sum w_i y_i$ – the sum of the weights for the diabetics in the sample.

Analogous weighting in inverse proportion to sampling probabilities is easily extended to multipredictor linear, logistic, and Cox regression analyses. And all statistical packages suitable for analyses of complex survey data make it easy to account for the weights.

In every case, taking account of the weights, which are included in the NHANES, NHDS, NAMCS, and other NCHS data sets, is essential for obtaining unbiased estimates. The differences between the weighted and unweighted estimates can be considerable. For example, the unweighted proportion with diabetes among adult respondents in NHANES III is 7.4%, but the weighted proportion is 4.8%. While this is not an immediately striking difference in percentage point terms, the corresponding unweighted estimate of the number of adult diabetics at the time of NHANES III was 12.5 million, as compared to a weighted estimate of 8.1 million – obviously not a trivial difference.

In NHANES as in many complex surveys, the probability weights are adjusted to account for non-response in such a way as to minimize the potential for bias. The non-response rates in NHANES III were 17% for the interview and 21% for the physical exam – acceptably low for a contemporary survey, but substantial enough to introduce bias. The potential for bias arises because the non-responders usually differ systematically from responders; that is, the non-responders are not *missing completely at random.*

Specifically, the adjustment of the weights is carried out within relatively homogeneous demographic subgroups, within which it is reasonable to suppose that the non-responders more nearly resemble the responders. In formal terms, we assume that within subgroups, the data for the non-responders *are* missing at random. In practical terms, the weights for the responders are inflated by a fixed factor for each subgroup such that the adjusted weights for the responders sum to the subgroup total of the original probability weights for both responders and non-responders. In NHANES a second post-stratification adjustment is made to ensure that the weights sum appropriately to regional totals for the target population, which are known from the U.S. Census.

A final note on probability weights: these should be distinguished from variance weights, which are used when the variance of the outcome differs across observations. This happens when the outcome is an average of multiple measurements, as is commonly done with noisy variables like blood pressure. In a sample where the number of measurements contributing to the average varies across participants, outcomes based on larger numbers of measurements will be relatively precise. An efficient analysis will weight the more precise outcomes more heavily – in proportion to the number of measurements each represents. Variance weights (in Stata called *analytic* weights or `aweights`) do this. Use of variance weights has the same effect on point estimates as use of probability weights, but the resulting standard errors, confidence intervals, and P-values would not be correct for complex survey data.

Taking account of the probability weights in analyzing a complex survey is primarily required to ensure that the resulting estimates are unbiased (or nearly so) for the parameters of the target population. The survey regression routines in Stata, SUDAAN, and SAS accommodate probability weights. Closely related to the generalized estimating equation (GEE) methods with independence working correlation introduced in Chapter 8, these routines give estimates (but not standard errors) identical to the estimates that would be obtained from standard regression routines that accommodate weights. A secondary effect is that weighting may inflate the standard errors, but this is only substantial if the weights are highly variable across observations.

10.3 Variance Estimation

In contrast to accounting for the probability weights, which is required mainly to avoid bias, taking account of the stratification and clustering of observations due to the complex sampling design is required solely to get the standard errors, confidence intervals, and P-values right, and has no effect on the point estimates. Unlike the point estimates, standard errors accounting for the special characteristics of a complex survey do differ from what would be obtained in standard weighted regression routines, sometimes in ways that are crucial to the conclusions of the analysis. In fact, they are essentially the "robust" standard errors provided by GEE regression routines, and thus account, as with longitudinal and hierarchical data, for clustering. In Stata, the main difference is that for testing whether each estimated regression coefficient differs from zero, the survey routines use a t-test with degrees of freedom equal to the number of PSUs minus the number of strata, rather than the asymptotic Z-test used in GEE. In addition, stratification is taken into account, but the effect is usually slight. For reference, this method of obtaining standard errors, confidence intervals, and P-values is referred to as *Taylor series linearization*.

Table 10.1 shows three logistic models for prevalent diabetes estimated using data from NHANES III. The predictors are age (per 10 years), ethnicity, and sex. The reference group for ethnicity is whites. Note that the odds-ratio estimates given by unweighted logistic regression (Model 1) differ both quantitatively and qualitatively from the results of the weighted and survey analyses (Models 2 and 3), which are identical. In the unweighted model, women appear to be at about 20% higher risk, but this does not hold up after accounting for probability of inclusion; similarly, the increased risk among African Americans and Mexican Americans is less substantial after accounting for the weights. The standard errors differ across all three models, in part because the survey model takes proper account of clustering within PSUs. In

Table 10.1. Unweighted, Weighted, and Survey Logistic Models for Diabetes

```
* Model 1: Unweighted logistic model ignoring weights and clustering

Logit estimates                                 Number of obs   =      18140
                                                LR chi2(5)      =    1148.81
                                                Prob > chi2     =     0.0000
Log likelihood = -4206.1375                     Pseudo R2       =     0.1202
------------------------------------------------------------------------------
   diabetes | Odds Ratio   Std. Err.      z    P>|z|     [95% Conf. Interval]
------------+-----------------------------------------------------------------
      age10 |   1.679618   .0284107    30.66   0.000     1.624847    1.736235
     aframer |   2.160196   .1651838    10.07   0.000     1.859535     2.50947
     mexamer |   2.784521   .2125534    13.42   0.000      2.39759    3.233896
    othereth |    1.25516   .2297553     1.24   0.214     .8767739    1.796843
      female |   1.200066   .0713788     3.07   0.002     1.068013    1.348447
------------------------------------------------------------------------------


* Model 2: Weighted logistic model, still ignoring clustering

Logit estimates                                 Number of obs   =      18140
                                                LR chi2(5)      =     783.05
                                                Prob > chi2     =     0.0000
Log likelihood = -3092.1644                     Pseudo R2       =     0.1124
------------------------------------------------------------------------------
   diabetes | Odds Ratio   Std. Err.      z    P>|z|     [95% Conf. Interval]
------------+-----------------------------------------------------------------
      age10 |   1.704453   .0344345    26.39   0.000     1.638282    1.773297
     aframer |   1.823747   .1883457     5.82   0.000     1.489559    2.232912
     mexamer |   1.915197   .3011782     4.13   0.000     1.407201    2.606579
    othereth |   1.031416   .1599616     0.20   0.842     .7610644    1.397803
      female |   .9805769   .0706968    -0.27   0.786     .8513584    1.129408
------------------------------------------------------------------------------


* Model 3: survey model accounting for weights, stratification, and clustering.

pweight:  wtpfqx6                               Number of obs   =      18140
Strata:   sdpstra6                              Number of strata =        49
PSU:      sdppsu6                               Number of PSUs  =        98
                                                Population size = 1.685e+08
                                                F(   5,    45)  =      80.86
                                                Prob > F        =     0.0000
------------------------------------------------------------------------------
   diabetes | Odds Ratio   Std. Err.      t    P>|t|     [95% Conf. Interval]
------------+-----------------------------------------------------------------
      age10 |   1.704453   .0479718    18.95   0.000     1.610726    1.803634
     aframer |   1.823747   .1840178     5.96   0.000     1.489031    2.233704
     mexamer |   1.915197   .1934744     6.43   0.000     1.563321    2.346276
    othereth |   1.031416   .2259485     0.14   0.888     .6641163    1.601855
      female |   .9805769   .0921773    -0.21   0.836     .8117843    1.184466
------------------------------------------------------------------------------
```

summary, accounting for probability of inclusion affects the point estimates and secondarily the standard errors, while accounting for stratification and clustering only affects the latter.

Stata makes it easy to run a regression analysis taking account of the special features of a complex survey. Variables giving the stratum, PSU, and probability weight for each observation are first specified using the svyset command. Then logistic regression is run using the svylogit command, which is similar in almost every respect to the logit and logistic commands used

for ordinary logistic regression analysis of a binary outcome from a simple
random sample. Analogous svy regression commands are provided for linear,
Poisson, negative binomial, and other commonly used regression models.

10.3.1 Design Effects

Because of positive correlation with clusters, the standard errors of parameter estimates from a complex survey are often (but not always) inflated as
compared to estimates from a simple random sample of the same size. This
inflation can be summarized by a *design effect*:

> Definition: The *design effect* is the ratio of the true variance of a pa-
> rameter estimate from a complex survey to the variance of the estimate
> if it were based on data from a simple random sample.

Note that design effects can vary for different parameters estimated in the
same survey, because some predictors may be more highly concentrated and
outcomes more highly correlated within clusters than others. Furthermore,
design effects in regression may vary with the degree to which the regression
effect is estimated by contrasting observations within as opposed to between
clusters, as we show below.

Most of the survey routines in Stata optionally provide estimates of the
design effect for each parameter estimate. In the survey logistic model for
prevalent diabetes shown in Table 10.1, the design effects are 2.7 for age, 0.9
for African American, 0.4 for Mexican American, 2.0 for other ethnicity, and
1.7 for sex. The increase in precision for the coefficient for Mexican Americans
results from the strong concentration of this subgroup in a few PSUs, so that
the comparison with whites rests primarily on within-cluster contrasts. In
contrast, women are about half of respondents in all PSUs, so that more
of the information for the comparison with men comes from between-PSU
contrasts (Problems 10.3 and 10.4).

Design effects have a useful interpretation in sample size planning, specifying an inflation factor for a sample size estimate based on methods which
assume a simple random sample. For instance, standard methods show that
a sample of 626 would provide 80% power to detect the effect of an exposure
on a binary outcome if half the sample is exposed and the true population
prevalence of the outcome in the unexposed and exposed groups is 20% and
30%, respectively. In a complex survey with an estimated design effect of 1.5,
a typical value, a sample size of $626 \times 1.5 = 939$ would be required to provide
80% power. In this context, 626 is called the *effective sample size* for the survey. Note that estimation of the design effect in advance is difficult, requiring
hard-to-come-by prior estimates of within-cluster outcome correlations and
the distribution of predictors across clusters.

10.3.2 Simplification of Correlation Structure

We pointed out earlier that NHANES is a *multi-stage* complex survey, meaning that area segments are selected within PSUs, then blocks with segments and households within blocks, before individuals are finally selected. In effect clusters are nested within clusters. For the NCHS surveys, multi-stage design is typical. However, only the stratum and PSU identifiers are provided with the NHANES III data; in part to protect the confidentiality of survey respondents, no information is provided about area segment or block. Moreover, the survey routines in Stata and SAS, like their more general GEE routines, make no provision for using the extra information about the true correlation structure, if it were provided. SUDAAN is an exception in this regard, making it possible to account more completely for the effects of multi-stage cluster sampling.

The implicit assumption of the standard error estimates in the Stata and SAS survey routines is that observations within a PSU are exchangeable and thus equally correlated with all other observations in the same PSU. However, it is reasonable to expect that within-cluster homogeneity and thus correlation would increase at each stage of the cluster sampling; all observations within a PSU might be correlated, but observations from different area segments would not in general be as highly correlated as observations sampled from the same block. Under the simplified model, the correlation within PSUs can be thought of as an average over these different levels of correlation. While this approximation may be robust, its effects on the size of standard errors and resulting confidence intervals and P-values depends on the specifics of the case. In particular, it will depend on the degree to which information about the comparison being made comes from within or between the nested clusters.

10.3.3 Other Methods of Variance Estimation

NHANES 2000, next in the series after NHANES III, began collecting data in 1999 and will continue though 2005, using a similar complex multi-stage design. A nationally representative sample of approximately 5,000 participants is obtained each year, and data for the first two years were available in mid-2003. Because the sample was still relatively small, the stratum and PSU identifiers were not included in the public data set at that point, to protect the confidentiality of study participants. (Stratum and PSU were made available with the recent release of data from the first four years.) Other surveys that do not provide stratum and PSU identifiers include the National Hospital Discharge Survey (NHDS), and until recently, the National Ambulatory Medical Care Survey (NAMCS).

Effectively this means that Stata and SAS cannot be used to analyze the data from any of these surveys correctly. From the NHDS, constants for computing *relative standard errors* are provided with the documentation, so that approximate confidence intervals for means and proportions can be calculated, but regression analysis is not possible. In NAMCS, which systematically samples patient visits within medical practices sampled within strata and PSU, it is possible to treat the practice as the PSU, but borderline statistically significant inferences would need to be regarded with extra caution (Problem 10.5).

NHANES 2000 does provide variables required to use an alternative method of variance estimation that is implemented in the SUDAAN and WESVAR packages. Briefly, this *jackknife* method uses a re-sampling procedure to estimate variability. The complete sample is split into 52 groups in such a way as to reflect the complex sampling structure but obscure geographic location. A total of 52 sets of jackknife weights are provided. One of these 52 weights is set to zero for all the members of one of the 52 disjoint groups, and adjusted for the remaining 51 groups, using adjustment methods already described for dealing with non-response. The analysis is then carried out 53 times, once with the original weights and once with each of the 52 sets of jackknife weights. It should be clear that the group with jacknife weights equal to zero will be omitted from that analysis. Then the variance of the overall estimates is estimated by variability among the jacknife estimates, appropriately scaled (Rust, 1985; Rust and Rao, 1996). A related method for variance estimation called balanced repeated replication (BRR) is also implemented in SUDAAN and WESVAR, but is beyond the scope of this chapter.

10.4 Summary

Complex surveys, unlike many convenience samples, can provide representative estimates of the parameters of a target population. However, to obtain these estimates and compute valid standard errors, confidence intervals, and *P*-values, such surveys have to be analyzed using methods that take account of the special features of the design, including stratification, multi-stage cluster sampling, and varying probability of inclusion. A number of software packages make it straightforward to carry out multipredictor regression analyses using complex survey data.

10.5 Further Notes and References

Book-length introductions to complex survey sampling include Korn and Graubard (1999) and Scheaffer (1996). Standard references for survey data include Cochran (1977) and Kish (1995).

Missing Data

Missing data are an even more important problem in complex surveys than in other areas of statistics, and one that we have only touched on briefly in describing adjustment of probability weights for non-response. While we are often reasonably comfortable estimating the associations between variables in the subsets of convenience samples that provide complete data, unbiased estimation of population totals and proportions is much more vulnerable to missing data, especially when the response of interest is sensitive. For example, the Centers for Disease Control and Prevention abandoned the idea of using probability surveys to estimate prevalence of HIV infection in the face of preliminary evidence from a feasibility study that non-response bias would invalidate the resulting estimates (Horvitz *et al.*, 1990).

In addition to *unit* non-response – sampled people who are completely missing from the survey, but accounted for in the adjustment of the weights for non-response – there is also *item* non-response, or missing responses on particular questions by study participants. One of the most important approaches to item non-response has been *multiple imputation* (Rubin, 1987, 1996). In this approach, probability models are used to impute the values of missing items from the non-missing responses for participants with the missing item and parameter estimates based on other observations with complete data. These imputations are carried out multiple times, the analysis is carried out in each of the resulting five or ten "completed" data sets, and summary estimates are averaged over them. Furthermore, unlike single imputation methods which treat the imputations as if they were known, multiple imputation uses information from the variability of the estimates across the different completed data sets to obtain standard errors, confidence intervals, and *P*-values that accurately reflect the extra uncertainty introduced by imputation (as opposed to ascertainment) of the missing items. Schafer (1999) provides an introduction to multiple imputation as well as an excellent book on modern methods for missing data (Schafer, 1995).

Multipredictor Models Using Survey Data

In the current version of Stata, survey routines have been implemented for linear, logistic, and several other generalized linear models, but not for the Cox proportional hazards model. SAS is more restrictive in offering a survey routine for linear regression only. To our knowledge, only SUDAAN currently has a proportional hazards routine for complex survey data.

10.6 Problems

Problem 10.1. Taking HIV infection as an example, explain why it might be more problematic to generalize estimates of prevalence from a convenience sample than to generalize estimates of risk factor effects. For the latter, we essentially have to assume that there is little or no interaction between the risk factor and being represented in the sample. Does this make sense?

Problem 10.2. Show that (10.1) reduces to the unweighted average $\sum y_i/n$ when $w_i \equiv w$.

Problem 10.3. Judging from the logistic model shown in Table 10.1, which was used to assess risk factors for diabetes, design effects greater than 1.0 appear to be more common than design effects less than 1.0. Describe what would happen in these two cases to model standard errors, confidence intervals, and P-values, if we were to analyze the survey data incorrectly, ignoring the clustering. In which case would we be more likely to make a type-I error? In which case would we be likely to dismiss an important risk factor? Can we reliably predict whether the design effect will be greater or less than 1.0?

Problem 10.4. In contrast to the design effects in regression analyses, design effects for means, proportions, and totals are almost always greater than 1.0. Explain why this should be the case.

Problem 10.5. Suppose you attempt to analyze data from the NAMCS, treating the physician practice as the PSU, ignoring correlation between different practices in the same actual survey PSU (which until recently were not identified on the publicly available data set). Probably the correlation between observations from the same practice is much stronger than the correlation between observations from different practices within the same PSU. In view of the simplified treatment of correlation structures in Stata and SAS, how does this affect your thinking about the analysis of NAMCS?

10.7 Learning Objectives

1. Describe the rationale for and special features of a complex survey.
2. Identify what can go wrong if the analysis of a complex survey ignores probability weights, strata, and cluster sampling.
3. Know where to begin with data from NHANES III or a similar complex survey to estimate the parameters of multipredictor linear and logistic regression models validly, as well as standard errors, confidence intervals, and P-values.

11

Summary

11.1 Introduction

Our goal in writing this book was to provide investigators with a practical guide to the analysis of data from research studies focusing on the relationship between outcomes and multiple predictor variables. Through our experience as co-investigators and instructors at the University of California, San Francisco, we have observed that researchers from many fields can benefit greatly from being able to conduct their own data analyses. In addition to reducing dependence on professional statisticians, mastering these skills promotes better study designs as well as clearer and more informative papers and presentations. Admittedly, encouraging investigators to analyze their own data is also somewhat self-serving on our part, because collaborations with investigators who are experienced in analyzing their own data are often more focused and productive.

Despite the mathematical underpinnings of the subject of statistics, the prerequisites needed to acquire adequate data analysis skills are surprisingly nontechnical. Perhaps the most important one is critical thinking. As is true with many technical fields, the key ideas underlying the methods presented here are surprisingly simple, and become much clearer when applied in actual data analyses. All of them are characterized by a common structure that mirrors the majority of research questions arising in clinical research: the relationship between an outcome and measured explanatory variables.

In this chapter we provide a brief review of the general approach to data analysis developed in this book, and provide guidance on how to use it as a resource to address particular analytical issues. We also briefly discuss a number of topics relevant to investigators undertaking their own data analyses, including development of analysis plans and finding help with technical questions.

11.2 Selecting Appropriate Statistical Methods

Selection of the right statistical tool to apply in addressing a research question is not always easy. Despite a number of unsuccessful attempts to use concepts from artificial intelligence in the development of algorithms to automate this process, common sense and experience remain most important for choosing an appropriate analysis method. In this section we provide some general guidelines on selecting statistical methods, with references to appropriate chapters and sections in the book. In keeping with our overall theme, we assume that the research question and available data involve investigating the relationship between a specified outcome and one or multiple measured predictor variables.

The first step in most data analyses is to define clearly the candidate outcome and predictor variable(s) and choose an appropriate analytic approach. As described in Sect. 1.1, outcomes can generally be classified as being either numeric (e.g., measured characteristics such as cholesterol level or body weight) or categorical (e.g., disease status indicators). Table 11.1 uses this classification to distinguish the main types of outcomes considered in the book (that subsume the majority considered in health research applications), along with the standard regression approaches for each, and the chapters in which they are discussed. Clearly many outcomes do not fit cleanly into the

Table 11.1. Outcome, Regression Model, and Chapter Reference

Outcome classification	Outcome type	Regression model	Chapter reference
Numerical	Continuous	Linear	4
	Count	Poisson model	9
	Time-to-event	Proportional hazards	7
Categorical	Binary	Logistic	6
	Ordinal	Proportional odds	6
	Nominal	Polytomous logistic	6

categories provided in the table. For example, the severity score in the back pain example introduced in Chapter 1 could be considered as either continuous or as a categorical variable with ordinal categories. In many such cases, the decision of how to consider such variables for the purpose of analysis will be driven by practicality (e.g., available software) and/or convention. In cases where multiple approaches are available, it is often a good idea to try more than one to insure that results are not sensitive to the choice.

Although the type of outcome usually dictates the choice of which regression model to consider, further consideration of how the outcome is observed and measured is necessary before settling on an analysis approach. A fundamental consideration is whether individual outcomes can be viewed as independent or not. Examples of studies with independent outcomes include diagnosis of coronary heart disease in participants in the WCGS study (used for examples in Chaps. 2–4 and 6) and baseline glucose levels in women participating in the HERS study (Sect. 4.2). Dependence between outcomes can arise in a number of ways detailed in Chapter 8. These include repeated measures of outcomes measured in the same individuals, or outcomes on different individuals that are associated via a shared environment or genetic relationship (e.g., disease outcomes among members of the same family). Examples include repeated measures of fat content of feces (8.1) and birthweights of first- and last-born infants from the same mothers (Sect. 8.3). As described in Chapter 8, most of the regression approaches for independent outcomes have direct analogs applicable in the dependent outcome setting.

In addition to dependence between individual outcomes, it is also important to consider how individuals were selected for inclusion in the study being analyzed. Although for many studies it is reasonable to assume that study participants had equal chances of being selected, in some cases these chances are controlled by the investigator to obtain a sample with desired properties. Examples include case-control studies for binary outcomes and complex sample surveys. As illustrated in Sect 6.3 and Chapter 10, regression methods for such studies generally mirror those used for independent samples.

Finally, we want to stress that despite the large number of outcome types and corresponding approaches to regression modeling covered here, the tools used for model fitting and evaluation are quite similar in most cases. Key concepts and techniques in model construction and interpretation such as accounting for confounding, mediation, and interaction are shared across approaches as well. Experience with regression modeling for different types of outcomes and study designs will surely reinforce these points.

11.3 Planning and Executing a Data Analysis

Data analyses are usually complex and benefit from careful planning in order to proceed in a timely and organized fashion. In our experience, few analyses are limited to straightforward application of textbook procedures. Invariably, technical questions arise related to data structure and/or quality, application of particular techniques, use of software programs, and interpretation of results. In this section, we provide some advice on several topics related to conducting an efficient analysis.

11.3.1 Analysis Plans

Before beginning a data analysis, it is useful to formulate a plan for how the work will proceed. For randomized controlled trials, analysis plans are generally specified in advance by the study protocol. For observational and clinical studies, preliminary plans are often formulated at the proposal stage. However, even when existing plans are not available to guide analyses, a clear outline of the important issues and tasks can aid in organizing the process.

A detailed plan should include a summary of the study design, statements of the research hypotheses, descriptions of each stage of analysis, and clear procedures for record-keeping, data distribution, and security.

11.3.2 Choice of Software

Fortunately, there are a number of excellent software packages available that implement the majority of techniques discussed here. Although we have used Stata in our examples, SAS, S-PLUS, and SPSS all provide commercial alternatives that offer many of the same facilities and run on a variety of computer platforms and operating systems. Also, the R language for statistical computing and graphics (R Development Core Team, 2004) is freely available and includes most of the procedures presented here. Finally, there are a number of special-purpose programs providing methods not well-represented in the major packages, including StatXact and LogXact (exact inference for contingency tables and logistic regression), and SUDAAN (analysis of data from complex surveys).

11.3.3 Record Keeping and Organization

An important part of a complete data analysis includes keeping files of relevant commands and procedures used in each of the stages above. Because a typical data analysis involves a large number of steps, having all files necessary to recreate results can save work for revision of research publications. Adding comments and explanatory text to programs and keeping text files outlining the analysis procedures and cataloging the important files are very useful in this regard. This information should be kept in an identifiable place in your file system (preferably organized with other project-specific materials) and backed up in a secure location for disaster recovery.

11.3.4 Data Security

Records from research studies often contain sensitive patient information and must be protected from unauthorized access. Although studies generally have data security measures in place to protect primary data sources, data analyses often involve creation of multiple data sets that may be distributed between investigators. As a general rule, it is a good practice to keep analysis data sets physically separate from source data, with any variables that can be linked to participant identities removed. Make sure that all analysis and data distribution procedures conform to current government, institutional, and study-specific guidelines on data security.

11.3.5 Consulting a Statistician

As we have noted frequently in the text, there are many instances where analysis issues arise that do not fall in the neat categories typical of many of the examples. Complex sampling schemes, extensive missing data, unusual patterns of censoring, misclassification in measured outcomes and predictors – all are examples of situations where standard methods and attendant assumptions may not apply without modification. Being able to recognize these circumstances is an important step in addressing these issues. When faced with an analysis problem that appears to fall outside of the range of techniques covered here, access to a professional statistician is a valuable resource. For investigators at research institutions, the best way to insure the availability of sound statistical support is to include a statistician as a consultant or co-investigator in proposals. Participating in courses or workshops on specialized statistical methods is another way to gain access to expert advice on advanced topics.

11.3.6 Use of Internet Resources

The Internet provides a vast and very valuable resource to assist in selection of statistical methods and planning data analyses. Frequently, answers to questions about particular applications and methods can quickly be found via a search using one of the available Web search engines. Unfortunately, even judicious searches often yield too many results to review completely. Also, the relevance of returned results is influenced by factors completely unrelated to their scientific value. For these reasons, beginning with searches of established research resources such as the PubMed interface to the MEDLINE index and the Current Index to Statistics will often yield more focused searches. Many educational institutions and private companies provide online access to electronic scientific journals and technical reports, including search capabilities. Also, statistical software sites frequently have online documentation and message lists that can provide useful information on the use of particular methods. Finally, message boards related to particular software programs and academic interests can frequently be a good way to get answers to analysis questions. Of course, unless the qualifications of individuals posting are known, blindly following advice can be dangerous.

11.4 Further Notes and References

Considering the broad and rapidly evolving nature of medical research and the increasing power of modern computers and computational algorithms, the coverage of statistical methods in this book is necessarily incomplete. A review of topics represented in current statistical journals reveals that new methods and modifications of existing methods are constantly being developed. Although many of the questions arising in clinical and epidemiological research studies can be adequately addressed (at least in a preliminary fashion) with careful application of the techniques covered here, studies often raise analysis issues novel and/or complex enough to require alternative approaches. We conclude the book by providing some references to new developments in the field that are likely to influence the practice of regression analysis in the future.

Genomics is an example of a field of research that is influencing the development of new biostatistical tools and forcing modifications of existing approaches. Although many data analyses in this field can be viewed in the general outcome-predictor framework developed here, frequently a very large number of potential predictors may be involved. An example is provided by study of the use of gene expression data in the classification of two types of acute leukemia (myeloid and lymphoblastic) (Golub *et al.*, 1999). RNA from bone marrow samples from 38 patients (27 lymphoblastic and 11 myeloid) was hybridized to oligonucleotide microarrays, each containing probes for 6,817 genes. The research questions centered on the use of genes as predictors for leukemia type. Although some form of binary regression model relating the disease outcome to predictors is clearly appropriate in this example, the fact that the number of candidate predictors greatly outnumber the observations, and that the correlation between predictors may be quite complex (reflecting functional relationships between genes) raises a number of difficult computational and inferential issues. We refer readers to Hastie *et al.* (2001) for a book-length overview of some modern statistical approaches being applied in this area.

Another area of biostatistics that is experiencing rapid growth is in the field of causal inference for observational studies. Much of this work has been prompted by the observation that confounding in longitudinal studies may be a time-dependent phenomenon, and classical methods that attempt to control this via simple inclusion of potential confounders in a given model may may be ineffective. An example of this was raised in Sect. 7.3: assessment of the effectiveness of HAART treatment to delay progression to AIDS, based data from observational studies is complicated by the fact that the effect of treatment may be confounded by disease stage (e.g., patients that have been infected longer tend to be sicker and are therefore more likely to receive treatment). Attempts to control for this by adjusting for time-varying measures of immune status (e.g., CD4 count) may not be effective (i.e., may not yield a valid measure of the causal effect of treatment on development of the outcome) because these are also affected by prior treatment. Although there are a number of modified regression techniques available to apply in these situations, most require specialized software or additional programming to implement. See Robins *et al.* (2000) and Cole *et al.* (2003) for recent examples of work in this area.

References

Ades, A., Parker, S., Walker, J., Edginton, M., Taylor, G. P. and Weber, J. N. (2000). Human t cell leukaemia/lymphoma virus infection in pregnant women in the United Kingdom: population study. *British Medical Journal*, **320**, 1497–1501.

Allen, D. M. and Cady, F. B. (1982). *Analyzing Experimental Data by Regression*. Wadsworth, Belmont, CA.

Altman, D. G. and Andersen, P. K. (1989). Bootstrap investigation of the stability of the Cox regression model. *Statistics in Medicine*, **8**, 771–783.

Ananth, C. V. and Kleinbaum, D. G. (1997). Regression models for ordinal responses: a review of methods and applications. *International Journal of Epidemiology*, **26**, 1323–1333.

Aurora, P., Whitehead, B. and Wade, A. (1999). Lung transplantation and life extension in children with cystic fibrosis. *Lancet*, **354**, 1591–1593.

Baron, R. M. and Kenny, D. A. (1986). The moderator-mediator variable distinction in social psychological research: conceptual, strategic, and statistical considerations. *Journal of Personality and Social Psychology*, **51**(6), 1173–1182.

Beach, M. L. and Meier, P. (1989). Choosing covariates in the analysis of clinical trials. *Controlled Clinical Trials*, **10**, 161S–175S.

Begg, C. B., Cramer, L. D., Venkatraman, E. S. and Rosai, J. (2000). Comparing tumour staging and grading systems: a case study and a review of the issues, using thymoma as a model. *Statistics in Medicine*, **19**, 1997–2014.

Begg, M. D. and Lagakos, S. (1993). Loss in efficiency caused by omitted covariates and misspecifying exposure in logistic regression models. *Journal of the American Statistical Association*, **88**(421), 166–170.

Belsey, D. A., Kuh, E. and Welsch, R. E. (1980). *Regression Diagnostics*. John Wiley & Sons, New York, Chichester.

Bradu, D. and Mundlak, Y. (1970). Estimation in lognormal linear models. *Journal of the American Statistical Association*, **65**, 198–211.

Brant, L. J., Sheng, S. L., Morrell, C. H., Verbeke, G. N., Lesaffre, E. and Carter, H. B. (2003). Screening for prostate cancer by using random-effects models. *Journal of the Royal Statistical Society: Series A*, **166**, 51–62.

Breiman, L. (2001). Statistical modeling: the two cultures. *Statistical Science*, **16**(3), 199–231.

Breiman, L., Friedman, J. H., Olshen, R. A. and Stone, C. J. (1984). *Classification and Regression Trees*. Wadsworth Publishing Co., Inc, Belmont, CA.

Breslow, N. E. and Day, N. E. (1984). *Statistical Methods in Cancer Research Volume I: The Analysis of Case-Control Studies*. Oxford University Press, Lyon.

Brookes, S. T., Whitley, E., Peters, T. J., Mulheran, P. A., Egger, M. and Smith, G. D. (2001). *Subgroup analyses in randomised controlled trials: quantifying the risks of false-positives and false-negatives*. The National Coordinating Centre for Health Technology Assessment, University of Southampton, Southampton, UK.

Brookmeyer, R. and Gail, M. H. (1987). Biases in prevalent cohorts. *Biometrics*, **43**(4), 739–749.

Brookmeyer, R., Gail, M. H. and Polk, B. F. (1987). The prevalent cohort study and the acquired immunodeficiency syndrome. *American Journal of Epidemiology*, **26**(1), 14–24.

Brown, J., Vittinghoff, E., Wyman, J. F., Stone, K. L., Nevitt, M. C., Ensrud, K. E. and Grady, D. (2000). Urinary incontinence: does it increase risk for falls and fractures? Study of Osteoporotic Fractures Research Group. *Journal of the American Geriatric Society*, **B48**, 721–725.

Buchbinder, S. P., Douglas, J. M., McKirnan, D. J., Judson, F. N., Katz, M. H. and MacQueen, K. M. (1996). Feasibility of human immunodeficiency virus vaccine trials in homosexual men in the United States: risk behavior, seroincidence, and willingness to participate. *Journal of Infectious Diseases*, **174**(5), 954–961.

Buckland, S. T., Burnham, K. P. and Augustin, N. H. (1997). Model selection: an integral part of inference. *Biometrics*, **53**, 603–618.

Carey, V., Zeger, S. L. and Diggle, P. (1993). Modelling multivariate binary data with alternating logistic regressions. *Biometrika*, **80**, 517–526.

Carroll, R. J., Ruppert, D. and Stefanski, L. A. (1995). *Measurement Error in Nonlinear Models*. Chapman & Hall/CRC, London, New York.

Chatfield, C. (1995). Model uncertainty, data mining and statistical inference. *Journal of the Royal Statistical Society, Series A*, **158**, 419–466.

Clark, L., Jr., G. C., Turnbull, B., Slate, E., Chalker, D., Chow, J., Davis, L., Glover, R., Graharn, G., Gross, E., Krongrad, A., Lesher, J., Park, H., Jr., B. S., Smith, C. and Taylor, J. (1996). Effects of selenium supplementation for cancer prevention in patients with carcinoma of the skin: a randomized controlled trial. *Journal of the American Medical Association*, **276**(24), 1957–1963.

Clayton, D. and Hills, M. (1993). *Statistical Models in Epidemiology*. Oxford University Press, Oxford.

Cleveland, W. S. (1985). *The Elements of Graphing Data*. Wadsworth & Brooks/Cole, Pacific Grove, CA.

Cochran, W. G. (1977). *Sampling Techniques*. John Wiley & Sons, New York, Chichester, 3rd ed.

Cole, S. R. and Hernan, M. A. (2002). Fallibility in estimating direct effects. *International Journal of Epidemiology*, **31**, 163–165.

Cole, S. R., Hernan, M. A., Robins, J. M., Anastos, K., Chmiel, J., Detels, R., Ervin, C., Feldman, J., Greenblatt, R., Kingsley, L., Lai, S., Young, M., Cohen, M. and Munoz, A. (2003). Effect of highly active antiretroviral therapy on time to acquired immunodeficiency syndrome or death using marginal structural models. *American Journal of Epidemiology*, **158**, 687–694.

Collett, D. (2003). *Modelling Binary Data*. Chapman & Hall/CRC, London, New York.

Concato, J., Peduzzi, P. and Holfold, T. R. (1995). Importance of events per independent variable in proportional hazards analysis i. background, goals, and general strategy. *Journal of Clinical Epidemiology*, **48**, 1495–1501.

DeGruttola, V. and Tu, X. M. (1994). Modelling progression of cd4-lymphocyte count and its relationship to survival time. *Biometrics*, **50**, 1003–1014.

Devore, J. and Peck, R. (1986). *Statistics, the Exploration and Analysis of Data*. West Publishing Co., St. Paul, MN.

Dickson, E. R., Grambsch, P. M. and Fleming, T. R. (1989). Prognosis in primary biliary-cirrhosis - model for decision-making. *Hepatology*, **10**, 1–7.

Diggle, P., Heagerty, P., Liang, K.-Y. and Zeger, S. (2002). *Analysis of Longitudinal Data*. Oxford University Press, Oxford, 2nd ed.

Diggle, P. and Kenward, M. (1994). Informative drop-out in longitudinal data analysis (Disc: p73-93). *Applied Statistics*, **43**, 49–73.

Dobson, A. J. (2001). *An Introduction to Generalized Linear Models*. Chapman & Hall Ltd, London, 2nd ed.

Draper, N. R. and Smith, H. (1981). *Applied Regression Analysis*. John Wiley & Sons, New York, Chichester.

Efron, B. and Tibshirani, R. (1986). Bootstrap measures for standard errors, confidence intervals, and other measures of statistical accuracy. *Statistical Science*, **1**, 54–77.

Efron, B. and Tibshirani, R. (1993). *An Introduction to the Bootstrap*. Chapman & Hall Ltd, London, New York.

Ehrenberg, A. S. C. (1981). The problem of numeracy. *The American Statistician*, **35**, 67–71.

Fitzmaurice, G. M., Laird, N. M. and Ware, J. H. (2004). *Applied Longitudinal Data Analysis*. John Wiley & Sons, New York.

Fleiss, J. L. (1988). One-tailed versus two-tailed tests: rebuttal. *Controlled Clinical Trials*, **10**, 227–228.

Fleiss, J. L., Levin, B. and Paik, M. C. (2003). *Statistical Methods for Rates and Proportions, 3rd Edition*. John Wiley & Sons, New York, Chichester, 4th ed.

Freedman, D., Pisani, R., Purves, R. and Adhikari, A. (1991). *Statistics*. W. W. Norton & Co, Inc., New York.

Freedman, L. S., Graubard, B. I. and Schatzkin, A. (1992). Statistical validation of intermediate endpoints for chronic diseases. *Statistics in Medicine*, **11**, 167–178.

Freireich, E. J., Gehan, E., Frei, E. I., Schroeder, L. R., Wolman, I. J., Anbari, R., Burgert, E. O., Mills, S. D., Pinkel, D., Selawry, O. S., Moon, J. H., Gendel, B. R., Spurr, C. L. and Storrs, R. (1963). The effect of 6-mercaptopurine on the duration of steroid-induced remissions in acute leukemia: a model for the evaluation of other potentially useful therapy. *Blood*, **21**, 699–716.

Friedman, L. M., Furberg, C. D. and Demets, D. L. (1998). *Fundamentals of Clinical Trials*. Springer, New York, 3rd ed.

Frost, C. and Thompson, S. G. (2000). Correcting for regression dilution bias: comparison of methods for a single predictor variable. *Journal of the Royal Statistical Society, Series A, General*, **163**(2), 173–89.

Gail, M. and Simon, R. (1985). Testing for qualitative interactions between treatment effects and patient subsets. *Biometrics*, **41**, 361–372.

Gail, M. H., Tan, W. Y. and Piantodosi, S. (1988). Tests for no treatment effect in randomized clinical trials. *Biometrika*, **75**, 57–64.

Gail, M. H., Wieand, S. and Piantodosi, S. (1984). Biased estimates of treatment effect in randomized experiments with nonlinear regressions and omitted covariates. *Biometrika*, **71**, 431–444.

Glidden, D. V. and Vittinghoff, E. (2004). Modelling clustered survival data from multicentre clinical trials. *Statistics in Medicine*, **23**, 369–388.

Goldberger, A. S. (1968). The interpretation and estimation of Cobb-Douglas functions. *Econometrica*, **36**, 464–472.

Goldman, L., Cook, E. F., Johnson, P. A., Brand, D. A., Ronan, G. W. and Lee, T. H. (1996). Prediction of the need for intensive care in patients who come to the emergency departments with acute chest pain. *New England Journal of Medicine*, **334**(23), 1498–1504.

Goldstein, H. (2003). *Multilevel Statistical Models*. Hodder Arnold, London, 3rd ed.

Golub, T. R., Slonim, D. K., Tamayo, P., Huard, C., Gaasenbeek, M., Mesirov, J. P., Coller, H., Loh, M. L., Downing, J. R., Caligiuri, M. A., Bloomfield, C. D. and Lander, E. S. (1999). Molecular classification of cancer: class discovery and class prediction by gene expression monitoring. *Science*, **286**, 531–537.

Grady, D., Wenger, N. K., Herrington, D., Khan, S., Furberg, C., Hunninghake, D., Vittinghoff, E. and Hulley, S. (2000). Postmenopausal hormone therapy increases risk of venous thromboembolic disease. The Heart and Estrogen/progestin Replacement Study. *Annals of Internal Medicine*, **132**(9), 689–696.

Graham, D. Y. (1977). Enzyme replacement therapy of exocrine pancreatic insufficiency in man. Relations between in vitro enzyme activities and in vivo potency in commercial pancreatic extracts. *New England Journal of Medicine*, **296**, 1314–1317.

Greenland, S. (1989). Modeling and variable selection in epidemiologic analysis. *American Journal of Public Health*, **79**(3), 340–349.

Greenland, S. (1994). Alternative models for ordinal logistic regression. *Statistics in Medicine*, **13**, 1665–1677.

Greenland, S. and Brumback, B. (2002). An overview of relations among causal modeling methods. *International Journal of Epidemiology*, **31**(5), 1030–1037.

Greenland, S., Pearl, J. and Robins, J. M. (1999). Causal diagrams for epidemiologic research. *Epidemiology*, **10**, 37–48.

Grodstein, F., Manson, J. E. and Stampfer, M. J. (2001). Postmenopausal hormone use and secondary prevention of coronary events in the Nurses' Health Study. *Annals of Internal Medicine*, **135**, 1–8.

Harrell, F., Lee, K. and Mark, D. (1996). Multivariable prognostic models: issues in developing models, evaluation assumptions, and adequacy, and measuring and reducing errors. *Statistics in Medicine*, **15**, 361–387.

Hastie, T. and Tibshirani, R. (1990). *Generalized additive models*. Chapman & Hall/CRC, London, New York.

Hastie, T. and Tibshirani, R. (1999). *Generalized Additive Models*. Chapman & Hall Ltd, London, New York.

Hastie, T., Tibshirani, R. and Friedman, J. H. (2001). *The Elements of Statistical Learning: Data Mining, Inference, and Prediction*. Springer-Verlag, New York.

Hastie, T. J. and Tibshirani, R. J. (1986). Generalized additive models (with discussion). *Statistical Science*, **1**, 297–318.

Hauck, W. W., Anderson, S. and Marcus, S. M. (1998). Should we adjust for covariates in nonlinear regression analyses of randomized trials? *Controlled Clinical Trials*, **19**, 249–256.

Henderson, R. and Oman, P. (1999). Effect of frailty on marginal regression estimates in survival analysis. *Journal of the Royal Statistical Society, Series B, Methodological*, **61**, 367–379.

Hernan, M. A., Brumback, B. and Robins, J. M. (2001). Marginal structural models to estimate the joint causal effect of nonrandomized treatments. *Journal of the American Statistical Association*, **96**(454), 440–448.

Hoenig, J. M. and Heisey, D. M. (2001). The abuse of power: the pervasive fallacy of power calculations for data analysis. *The American Statistician*, **55**(1), 19–24.

Hoerl, A. E. and Kennard, R. W. (1970). Ridge regression: biased estimates for nonorthogonal problems. *Technometrics*, **12**, 55–67.

Hofer, T., Hayward, R., Greenfield, S., Wagner, E., Kaplan, S. and Manning, W. (1999). The unreliability of individual physician "report cards" for assessing the costs and quality of care of a chronic disease. *Journal of the American Medical Association*, **281**(22), 2098–2105.

Holcomb, W. L. J., Chaiworapongsa, T., Luke, D. A. and Burgdorf, K. D. (2001). An odd measure of risk: use and misuse of the odds ratio. *Obstetrics and Gynecology*, **98**, 685688.

Horvitz, D. G., Weeks, M. F., Visscher, W., Folsom, R. E., Hurley, P. L., Wright, R. A., Massey, J. T. and Ezzati, T. M. (1990). A report of the findings of the national household seroprevalence survey feasibility study. In *Proceedings of the Survey Research Methods Section*. Survey Methods Section, American Statistical Association.

Hosmer, D. W. and Lemeshow, S. (2000). *Applied Logistic Regression*. John Wiley & Sons, New York, Chichester.

Hulley, S., Grady, D., Bush, T., Furberg, C., Herrington, D., Riggs, B. and Vittinghoff, E. (1998). Randomized trial of estrogen plus progestin for secondary prevention of heart disease in postmenopausal women. The Heart and Estrogen/progestin Replacement Study. *Journal of the American Medical Association*, **280**(7), 605–613.

Jewell, N. P. (2004). *Statistics for Epidemiology*. Chapman & Hall/CRC, Boca Raton, FL.

Kalbfleisch, J. D. and Prentice, R. L. (1980). *The Statistical Analysis of Failure Time Data*. John Wiley & Sons, New York.

Kanaya, A., Vittinghoff, E., Shlipak, M. G., Resnick, H. E., Visser, M., Grady, D. and Barrett-Connor, E. (2004). Association of total and central obesity with mortality in postmenopausal women with coronary heart disease. *American Journal of Epidemiology*, **158**(12), 1161–1170.

Kish, L. (1995). *Survey Sampling*. John Wiley & Sons, New York, Chichester.

Klein, J. P. and Moeschberger, M. L. (1997). *Survival Analysis: Techniques for Censored and Truncated Data*. Springer.

Kleinbaum, D. G. (2002). *Logistic Regression: a Self-Learning Text*. Springer-Verlag Inc.

Korff, M., Barlow, W., Cherkin, D. and Deyo, R. (1994). Effects of practice style in managing back pain. *Annals of Internal Medicine*, **121**, 187–195.

Korn, E. L. and Graubard, B. I. (1999). *Analysis of Health Surveys*. John Wiley & Sons, New York, Chichester.

Lagakos, S. W. and Schoenfeld, D. A. (1984). Properties of proportional-hazards score tests under misspecified regression models. *Biometrics*, **40**, 1037–1048.

Le Cessie, S. and Van Houwelingen, J. C. (1992). Ridge estimators in logistic regression. *Applied Statistics*, **41**, 191–201.

Li, Z., Meredith, M. P. and Hoseyni, M. S. (2001). A method to assess the proportion of treatment effect explained by a surrogate endpoint. *Statistics in Medicine*, **20**, 3175–3188.

Lin, D. Y., Fleming, T. R. and De Gruttola, V. (1997). Estimating the proportion of treatment effect explained by a surrogate marker. *Statistics in Medicine*, **16**, 1515–1527.

Linhart, H. and Zucchini, W. (1986). *Model Selection*. John Wiley & Sons, New York, Chichester.

Littell, R. C., Milliken, G. A., Stroup, W. W. and Wolfinger, R. (1996). *SAS System for Mixed Models*. SAS Publishing, Cary, NC.

Little, R. J. A. and Rubin, D. B. (2002). *Statistical Analysis With Missing Data*. John Wiley & Sons, New York, Chichester.

Magder, L. S. and Hughes, J. P. (1997). Logistic regression when the outcome is measured with uncertainty. *American Journal of Epidemiology*, **146**, 195–203.

Maldonado, G. and Greenland, S. (1993). Simulation study of confounder-selection strategies. *American Journal of Epidemiology*, **138**, 923–936.

Marubini, E. and Valsecchi, M. G. (1995). *Analysing Survival Data from Clinical Trials and Observational Studies*. John Wiley & Sons, New York, Chichester.

McCullagh, P. and Nelder, J. A. (1989). *Generalized linear models*. Chapman & Hall Ltd, 2nd ed.

McCulloch, C. E. and Searle, S. R. (2000). *Generalized, Linear, and Mixed Models*. John Wiley & Sons, New York, Chichester.

McNutt, L., Wu, C., Xue, X. and P., H. J. (2003). Estimating the relative risk in cohort studies and clinical trials of common outcomes. *American Journal of Epidemiology*, **157**, 940–943.

Meier, P., Ferguson, D. J. and Karrison, T. (1985). A controlled trial of extended radical mastectomy. *Cancer*, **55**, 880–891.

Mickey, R. M. and Greenland, S. (1989). The impact of confounder selection on effect estimation. *American Journal of Epidemiology*, **129**(1), 125–137.

Miller, A. J. (1990). *Subset Selection in Regression*. Chapman & Hall Ltd, London, New York.

Miller, R. G., Gong, G. and Munoz, A. (1981). *Survival Analysis*. John Wiley & Sons, New York, Chichester.

Neuhaus, J. (1998). Estimation efficiency with omitted covariates in generalized linear models. *Journal of the American Statistical Association*, **93**, 1124–1129.

Neuhaus, J. and Jewell, N. P. (1993). A geometric approach to assess bias due to omitted covariates in generalized linear models. *Biometrika*, **80**, 807–815.

O'Brien, T. R., Busch, M. P., Donegan, E., Ward, J. W., Wong, L., Samson, S. M., Perkins, H. A., Altman, R., Stoneburner, R. L. and Holmberg, S. D. (1994). Heterosexual transmission of human immunodeficiency virus type i from transfusion recipients to their sexual partners. *Journal of AIDS*, **7**, 705–710.

Orwoll, E., Bauer, D. C., Vogt, T. M. and Fox, K. M. (1996). Axial bone mass in older women. *Annals of Internal Medicine*, **124**(2), 185–197.

Pagano, M. and Gavreau, K. (1993). *Principles of Biostatistics*. Wadsworth Publishing Co., Belmont, CA.

Pearl, J. (1995). Causal diagrams for empirical research. *Biometrika*, **82**, 669–688.

Peduzzi, P., Concato, J. and Feinstein, A. R. (1995). Importance of events per independent variable in proportional hazards regression analysis ii. accuracy and precision of regression estimates. *Journal of Clinical Epidemiology*, **48**, 1503–1510.

Peduzzi, P., Concato, J., Kemper, E., Holford, T. R. and Feinstein, A. R. (1996). A simulation study of the number of events per variable in logistic regression analysis. *Journal of Clinical Epidemiology*, **49**, 1373–1379.

Preisser, J. S., Lohman, K. K., Craven, T. E. and Wagenknecht, L. E. (2000). Analysis of smoking trends with incomplete longitudinal binary responses. *Journal of the American Statistical Association*, **95**, 1021–1031.

R Development Core Team (2004). *R: A language and environment for statistical computing*. R Foundation for Statistical Computing, Vienna, Austria. ISBN 3-900051-00-3.

Rabe-Hesketh, S., Pickles, A. and Skrondal, S. (2004). *Multilevel and Structural Equation Modeling for Continuous, Categorical, and Event Data*. Stata Press, College Station, TX.

Raudenbush, S. W. and Bryk, A. S. (2001). *Hierarchical Linear Models: Applications and Data Analysis Methods (Advanced Quantitative Techniques in the Social Sciences)*. Sage, Newbury Park, CA.

Robins, J. M. and Greenland, S. (1992). Identifiability and exchangeability for direct and indirect effects. *Epidemiology*, **3**, 143–155.

Robins, J. M., Hernan, M. and Brumback, B. (2000). Marginal structural models and causal inference in epidemiology. *Epidemiology*, **11**, 550–560.

Rosenman, R. H., Friedman, M., Straus, R., Wurm, M., Kositchek, R., Hahn, W. and Werthessen, N. T. (1964). A predictive study of coronary heart disease: the western collaborative group study. *Journal of the American Medical Association*, **189**, 113–120.

Rothman, K. J. and Greenland, S. (1998). *Modern Epidemiology*. Lippincott Williams & Wilkins Publishers, Philadelphi, PA, 2nd ed.

Rubin, D. B. (1987). *Multiple Imputation for Nonresponse in Surveys*. John Wiley & Sons, New York, Chichester.

Rubin, D. B. (1996). Multiple imputation after 18+ years. *Journal of the American Statistical Association*, **91**, 473–489.

Rust, K. (1985). Variance estimation for complex estimators in sample surveys. *Journal of Official Statistics*, **1**(4), 381–397.

Rust, K. and Rao, J. N. K. (1996). Variance estimation for complex surveys using replication techniques. *Statistical Methods in Medical Research*, **5**, 283–31–.

Schafer, J. L. (1995). *Analysis of Incomplete Multivariate Data by Simulation*. Chapman & Hall Ltd, London, New York.

Schafer, J. L. (1999). Multiple imputation: a primer. *Statistical Methods in Medical Research*, **8**, 3–15.

Scheaffer, R. L. (1996). *Elementary Survey Sampling*. Duxbury, Boston, 5th ed.

Schmoor, C. and Schumacher, M. (1997). Effects of covariate omission and categorization when analysing randomized trials with the Cox model. *Statistics in Medicine*, **16**, 225–237.

Schoenfeld, D. (1980). Chi-squared goodness-of-fit tests for the proportional hazards regression model. *Biometrika*, **67**, 145–153.

Scott, A. J. and Wild, C. J. (1997). Fitting regression models to case-control data by maximum likelihood. *Biometrika*, **84**, 57–71.

Self, S. and Pawitan, Y. (1992). Modeling a marker of disease progression and onset of disease. In *AIDS Epidemiology: Methodlogical Issues* (edited by N. Jewell, K. Dietz and V. Farewell). Birkhauser, Boston.

Steyerberg, E. W., Eijkemans, M. J. C. and Habbema, J. D. F. (1999). Stepwise selection in small datasets: a simulation study of bias in logistic regression analysis. *Journal of Clinical Epidemiology*, **52**, 935–942.

Steyerberg, E. W., Eijkemans, M. J. C., Harrell, F. E. and Habbema, J. D. F. (2000). Prognostic modelling with logistic regression analysis: a comparison of selection and estimation methods in small datasets. *Statistics in Medicine*, **19**, 1059–1079.

Sturges, H. A. (1926). The choice of a class interval. *Journal of the American Statistical Association*, **21**(153), 65–66.

Sun, G. W., Shook, T. L. and Kay, G. L. (1999). Inappropriate use of bivariable analysis to screen risk factors for use in multivariable analysis. *Journal of Clinical Epidemiology*, **49**, 907–916.

Therneau, T. M. and Grambsch, P. M. (2000). *Modeling Survival Data: Extending the Cox Model*. Springer, New York.

Tibshirani, R. (1997). The lasso method for variable selection in the Cox model. *Statistics in Medicine*, **16**, 385–395.

Tung, P., Kopelnik, A., Banki, N., Ong, K., Ko, N., Lawton, M. T., Gress, D., Drew, B. J., Foster, E., Parmley, W. W. and Zaroff, J. G. (2004). Predictors of neurocardiogenic injury after subarachnoid hemorrhage. *Stroke*, **35**(2), 548–551.

van der Laan, M. J. and Robins, J. M. (2003). *Unified Methods for Censored Longitudinal Data and Causality*. Springer, New York.

Vanderpump, M. P., Tunbridge, W. M., French, J. M., Appleton, D., Bates, D., Clark, F., Grimley Evans, J., Rodgers, H., Tunbridge, F. and Young, E. T. (1996). The development of ischemic heart disease in relation to autoimmune thyroid disease in a 20-year follow-up study of an english community. *Thyroid*, **6**, 155–160.

Verbeke, G. and Molenberghs, G. (2000). *Linear Mixed Models for Longitudinal Data*. Springer, New York.

Verweij, P. J. M. and Van Houwelingen, H. C. (1994). Penalized likelihood in Cox regression. *Statistics in Medicine*, **13**, 2427–2436.

Vittinghoff, E., Hessol, N. A., Bacchetti, P., Fusaro, R. E., Holmberg, S. D. and Buchbinder, S. P. (2001). Cofactors for hiv disease progression in a cohort of homosexual and bisexual men. *Journal of the Acquired Immunodeficiency Syndromes*, **27**(3), 308–314.

Vittinghoff, E., Shlipak, M. G., Varosy, P. D., Furberg, C. D., Ireland, C. C., Khan, S. S., Blumenthal, R., Barrett-Connor, E. and Hulley, S. (2003). Risk factors and secondary prevention in women with heart disease: The Heart and Estrogen/progestin Replacement Study. *Annals of Internal Medicine*, **138**(2), 81–89.

Volberding, P. A., Lagakos, S. W. and Koch, M. A. (1990). Zidovudine in asymptomatic human-immunodeficiency-virus infection - a controlled trial in persons with fewer than 500 cd4-positive cells per cubic millimeter. *The New England Journal of Medicine*, **322**(14), 941–949.

Vuong, Q. H. (1989). Likelihood ratio tests for model selection and non-nested hypotheses. *Econometrica*, **57**(2), 307–333.

Walter, L. C., Brand, R. J., Counsell, S. R., Palmer, R. M., Landefeld, C. S., Fortin-
    sky, R. H. and Covinsky, K. E. (2001). Development and validation of a prog-
    nostic index for 1-year mortality in older adults after hospitalization. *Journal of
    the American Medical Association*, **285**(23), 2987–2994.
Wei, L. J. and Glidden, D. V. (1997). An overview of statistical methods for multiple
    failure time data in clinical trials (with discussion). *Statistics in Medicine*, **16**(8),
    833–839.
Weisberg, S. (1985). *Applied Linear Regression*. John Wiley & Sons, New York,
    Chichester.
Winkelstein, W., Lyman, D. M., Padian, N., Grant, R., Samuel, M., Wiley, J. A.,
    Andersen, R. E., Lang, W., Riggs, J. and Levy, J. A. (1987). Sexual practices and
    risk of infection by the human immunodeficiency virus. The San Francisco Men's
    Health Study. *Journal of the American Medical Association*, **257**(3), 321–325.
Wulfsohn, M. S. and Tsiatis, A. A. (1997). A joint model for survival and longitudinal
    data measured with error. *Biometrics*, **53**, 330–339.
Zhang, J. and Yu, K. F. (1998). What's the relative risk? a method for correcting
    the odds ratio in cohort studies of common outcomes. *Journal of the American
    Medical Association*, **280**, 1690–1691.

Index

LaVergne, TN USA
20 November 2009

164692LV00004B/23/P